ERRATUM

Security of Natural Gas Supply through Transit Countries, edited by Jens Hetland and Teimuraz Gochitashvili

Nato Science Series II, Mathematics, Physics and Chemistry,
ISBN 1-4020-2076-7 Hardbound
ISBN 1-4020-2077-5 Paperback
ISBN 1-4020-2078-3 E-Book
Volume nr. 149

Following pages should replace the pages in the book of article: PROSPECTS FOR GAS SUPPLY AND DEMAND AND THEIR IMPLICATION WITH REFERENCE TO TRANSIT COUNTRIES AND THEIR POLICY ***– drawing upon recent experiences from European countries***

Author: Boyko Nitzov

Beginning with Equation 9 on page 281, the text until the end of Section 2.1 on page 283 should read:

"Similarly, for moving gas through a pipeline compressor horsepower *HP* is:

$$HP = MZ_g RT_f \left[\left(\frac{P_2}{P_1} \right)^a - 1 \right] \frac{1}{a}$$

Equation 9, adiabatic compressor station horsepower

Where M is the molar flow rate, Z_g is average gas compressibility factor, R = 1,544 (universal gas constant), T_f is gas flowing temperature (R), P_1 is the compressor inlet pressure and P_2 is outlet pressure (psia), and $a=(k-1)/k$ (k is adiabatic exponent). HP can also be written (with some assumptions) as:

$$HP = 5.552 * 10^{-11} \frac{HQM_g}{E_p}$$

Equation 10, compressor station horsepower

Where H is compressor head, Q is the volumetric gas flowrate in cubic feet per day, M_g is the molecular weight of the gas mixture and E_p is adiabatic efficiency. Head is given by:

$$H = \frac{Z_g RT_f kE_p}{M_g (k-1))} \left[\left(\frac{P_2}{P_1} \right)^{\frac{k-1}{kE_p}} - 1 \right]$$

Equation 11, compressor head

To simplify, assume $k=1.35$, $E_p=0.75$ and $M_g=18.465$[15], and $R=1,544$ (universal gas constant lb/mol Rankine), hence:

$$HP = 3.3067 * 10^{-7} QZ_g T_f \left[\left(\frac{P_2}{P_1} \right)^{0.34568} - 1 \right]$$

Equation 12, simplified compressor station horsepower as a function of volumetric flow

Volumetric flow, on the other hand, is given by:

$$Q = 737 \left(\frac{T_0}{P_0} \right)^{1.02} D^{2.53} \left(\frac{P_2^2 - P_1^2}{G^{0.961} T_f L Z_g} \right) E$$

Equation 13, modified Panhandle volumetric flowrate of gas in pipe (production function of moving gas through pipeline)

Where T_0 is standard temperature and P_0 is standard pressure, G is the relative density of gas to air, E is pipe efficiency ("roughness", <1) and the other parameters are as described above. Other formulas[16] may also be used to assess the production function, for example:

---

[15] Corresponding to gas composition by weight (percent): methane 94%, ethane 4.5%, propane 1.27%, i-butane 0.09% and n-butane and i-pentane 0.07% each. A PC-based model developed by using these and the subsequent equations actually calculates *HP*, *TC*, "transportation tariff" and other parameters for various values of *k*, E_p, gas composition and other variables; here the idea is to illustrate the derivation of the production, *TC* and *MC* functions of a gas pipeline.

[16] If the pressure drop in a pipeline is less than 40% of P_1, then the Darcy-Weisbach incompressible flow calculation may be more accurate than the Weymouth or Panhandles A and B for a short pipe or low flow. In main pipelines, compressible flow calculations are generally used.

$$Q = \left(0.03393 \frac{T_0}{P_0}\right)\left[\frac{(P_2^2 - P_1^2)D^5 R}{M_g T_s Z_g L f}\right]^{0.5}$$

Equation 14, Weymouth volumetric flow of gas in pipe (production function of moving gas through pipeline)

Where f is a friction factor[17] and the other parameters are as described above[18].

Rewrite *Equation 12* for horsepower by using Weymouth (*Equation 14*)[19]:

$$HP = 1.122 * 10^{-8} Z_g T_f \frac{T_0}{P_0}\left[\left(\frac{P_2}{P_1}\right)^{0.34568} - 1\right]\left[\frac{(P_2^2 - P_1^2)D^5 R}{M_g T_f Z_g L f}\right]^{0.5}$$

Equation 15

Construct the cost function:

$$TC = c_1 HP + \rho(r,t)(c_2 D + c_3 HP)$$

Equation 16

Where c_1 is the cost of operation per HP/day, c_2 is the construction cost per inch/mile, c_3 is the construction cost per installed horsepower and $\rho(r,t)$ is the imputed daily cost of the pipeline as a function of the interest rate r and project life t.

Marginal cost is given by

$$MC = f'(TC)$$

Equation 17

More precisely, MC is controlled by changes in many factors affecting pipeline productivity (throughput), so, for example:

$$MC = \frac{\partial TC}{\partial P_1} \qquad MC = \frac{\partial TC}{\partial P_2}$$

or , etc.

Equations 17a and 17b

Some of these factors are clearly beyond the control of pipeline operators (e.g. ambient temperature and pressure), and in this sense gas pipelines are more "sensitive" and "risky" to variations in the operational framework. A factor that is relatively easy to control is compressor outlet pressure P_2, which also entails control over the pressure at the end of the line section and the inlet of the next compressor unit[20], which entails varying compressor horsepower. The next sections of this paper use this model of gas pipelines to assess scale effects, cost of transportation, pipeline tariffs, netbacks and other parameters."

Please note the numbering of equations and footnotes on subsequent pages is affected (for example, Equation 15 on page 287 becomes Equation 18 and so on, Footnote 18 on page 283 becomes Footnote 21, and so on).

---

[17] Equation 14 parameters in SI dimensions where applicable. *Cf. Dahl et al.* in this volume.

[18] Note $P_2 > P_1$ (compressor outlet pressure exceeds inlet pressure). For a line section, the usual notation is the opposite (inlet pressure to the pipe section exceeds outlet pressure at the end of the pipe section). To avoid confusion, compressor-style notation is used in Equations 13, 14, etc.

[19] Conversion from SI to British units needed.

[20] Sample calculations of *TC* and *MC* with the help of a PC-based utility using the model are available on request.

Security of Natural Gas Supply through Transit Countries

NATO Science Series

A Series presenting the results of scientific meetings supported under the NATO Science Programme.

The Series is published by IOS Press, Amsterdam, and Kluwer Academic Publishers in conjunction with the NATO Scientific Affairs Division

Sub-Series

I. Life and Behavioural Sciences	IOS Press
II. Mathematics, Physics and Chemistry	Kluwer Academic Publishers
III. Computer and Systems Science	IOS Press
IV. Earth and Environmental Sciences	Kluwer Academic Publishers
V. Science and Technology Policy	IOS Press

The NATO Science Series continues the series of books published formerly as the NATO ASI Series.

The NATO Science Programme offers support for collaboration in civil science between scientists of countries of the Euro-Atlantic Partnership Council. The types of scientific meeting generally supported are "Advanced Study Institutes" and "Advanced Research Workshops", although other types of meeting are supported from time to time. The NATO Science Series collects together the results of these meetings. The meetings are co-organized bij scientists from NATO countries and scientists from NATO's Partner countries – countries of the CIS and Central and Eastern Europe.

Advanced Study Institutes are high-level tutorial courses offering in-depth study of latest advances in a field.
Advanced Research Workshops are expert meetings aimed at critical assessment of a field, and identification of directions for future action.

As a consequence of the restructuring of the NATO Science Programme in 1999, the NATO Science Series has been re-organised and there are currently Five Sub-series as noted above. Please consult the following web sites for information on previous volumes published in the Series, as well as details of earlier Sub-series.

http://www.nato.int/science
http://www.wkap.nl
http://www.iospress.nl
http://www.wtv-books.de/nato-pco.htm

Series II: Mathematics, Physics and Chemistry – Vol. 149

Security of Natural Gas Supply through Transit Countries

edited by

Jens Hetland
SINTEF Energy Research,
Trondheim, Norway

and

Teimuraz Gochitashvili
Georgian Technical University,
Tbilisi, Georgia

Kluwer Academic Publishers

Dordrecht / Boston / London

Published in cooperation with NATO Scientific Affairs Division

Proceedings of the NATO Advanced Research Workshop on
Security of Natural Gas Supply through Transit Countries
Tbilisi, Georgia
20–22 May 2003

A C.I.P. Catalogue record for this book is available from the Library of Congress.

ISBN 1-4020-2076-7 (HB)
ISBN 1-4020-2077-5 (PB)
ISBN 1-4020-2078-3 (e-book)

Published by Kluwer Academic Publishers,
P.O. Box 17, 3300 AA Dordrecht, The Netherlands.

Sold and distributed in North, Central and South America
by Kluwer Academic Publishers,
101 Philip Drive, Norwell, MA 02061, U.S.A.

In all other countries, sold and distributed
by Kluwer Academic Publishers,
P.O. Box 322, 3300 AH Dordrecht, The Netherlands.

Printed on acid-free paper

TABLE OF CONTENTS

PREFACE

In contrast to oil,

- natural gas is usually routed through pipeline systems stretched from the wellhead to the end-user – although liquefied natural gas (LNG) is gaining increased interest;
- the commercialisation of natural gas fields is inherently linked to rigid transportation systems that require huge investments in tangible assets fixed to specific locations;
- the supply of natural gas is constrained by the transportation system, and requires access to appropriate infrastructure for transport and distribution;
- the trading of natural gas is traditionally associated with long-term contracts, albeit the duration per se of gas transport contracts appears to be less important after the deregulation of the energy markets.

As diversification is strategically important to modern societies, the security of energy supplies becomes an inherent issue. In order to avoid situations of shortage, and to keep the price level stable, industrial nations are paying attention to the security of energy supplies. In brief terms this means that having more than one supplier of natural gas and more than one transport route would be strategically important. This also affects political issues and international law and regulations, economics, science and technology.

The purpose of this book is to address opportunities extended from science and research pertaining to the exploitation and international trading of natural gas that involves transit countries. This especially relates to the transport and handling of gas from remote regions and pipelines that are routed through third-party countries. It also pertains to strategic reserves and energy management on a broad basis.

As pipelines per se are fairly vulnerable to sabotage and terrorist actions, the social and political stability in the involved countries are of greatest importance for the security of natural gas supply. This especially applies to transit countries that – per definition - are not part of the gas supply contracts. It is therefore mandatory that transit countries are enabled to make use of their compensation in a prosperous manner for the sake of improving the social stability. It is assumed that this in the next run will increase the safety of energy supply that is crucial to all parties.

Although the subject of *security of natural gas supply through transit countries* may be conceived as well defined, the content of this book will constitute a powerful information source of disparate interests. The book aims at presenting the state of the art. In addition to providing new concepts and scientific results it also deals with strategic and geopolitical issues, and technological opportunities pertaining to reservoir geology and cryogenic research for the storage and distribution of natural gas on a local and regional basis.

ACKNOWLEDGEMENTS

Special acknowledgements are given to the NATO Science Programme and the Georgian International Oil Company, GIOC. NATO on the one hand funded the workshop under the Cooperative Science & Technology Sub-Programme Advanced Research Workshop, and made it possible to publish the scientific material in the NATO Science Series. GIOC on the other hand sponsored the workshop by making its impressive conference hall in Tbilisi available for the workshop.

INTRODUCTION

Some important topics have been carefully selected according to a logical approach in consideration of the key issues of implementing new transit lines and deploying new gas technologies in the Caucasus region. The overlying objective is the security of gas supply through the transit countries in reflection of economy and more generalised geopolitical aspects. It is believed that joint efforts and pronounced readiness of developing new opportunities in the transit countries will contribute to assuring the situation, and also contribute to improve the general social situation in the region.

Like most western countries, **there is an increased focus on reliable supplies of natural gas** in the Caucasus region. The reason is much due to lacking security of the existent infrastructure that is deemed quite detrimental to the industries in the region. Therefore gas pipelines have become political and strategic issues. Especially the new export pipeline from Baku to Turkey and later to Europe is vitally important to the transit countries. The compensation to these countries is mostly remunerated in kind reckoned by a certain percentage of the gas flow. Georgia will for instance receive 5% of the gas that passes its territory. This energy represents an important stimulus to societies that for historical and political reasons are in a deep recession. In that respect gas supplies may be important for the restructuring of the societies, and may, thereby, have a geopolitical impact.

In historical perspective it could be noted that the rapid growth of the world industry in the 19th century drew increased attention to energy carriers in the Caspian region. In 1904 the famous Nobel brothers and their partners constructed the first Caspian oil pipeline from Baku (Azerbaijan) to Batumi (Georgia), at that time the biggest pipeline of the world.

Exploitation of remote gas fields requires infrastructure and access to land that may affect third parties. In those cases where land-lease is compensated in kind some percentage of the gas is handed over to the transit country (herein referred to as *option gas*). Concerning security, two immediate issues appear: a) how the *option gas* fits to local energy plans and b) the economic and social stability of the transit country. To some extent b) relates to a) provided the implementation proves successful.

The markets for natural gas are gradually expanding and natural gas accounts for about 25% of the consumption of primary energy sources (2000). Because the reserves of natural gas are unevenly distributed, some regions are more dependent on imports than others. In Europe the demand by the year 2000 was 137 bcm, and is foreseen to grow to 475/515 in 2005 and 515/586 in 2010 and 650/720 in 2020 following a low/high hypothesis. By year 2010 the estimated European import dependency on natural gas may reach 62/70%.

As the main supply lines like the existent pipelines form Algeria through Morocco and Tunisia have rather limited capacities, Russia faces significant difficulties with its infrastructure for the development of new reserves mainly for financial reasons. At the same time Caspian reservoirs represent a huge potential for European supply. The export potential from the Caspian fields may range from 75/90 bcm in 2010 to 120/145 bcm in 2020 according to various development scenarios based on a low/high hypothesis.

A similar situation may apply to the transit countries like Turkey and Georgia and also to the producer countries like Azerbaijan and Turkmenistan. Whereas Turkey may present a growth of almost 400% since 1990, the South-Caucasian and Central-Asian countries must first recover their setback in capacity after the collapse of the Soviet Union. Hence, the prerequisites for safe supply to the end-user countries do not only mean differentiation of pipelines; Social and economic restructuring of the transit countries are deemed necessary and important as well.

At present Russia and Iran are the only suppliers of natural gas to the South-Eastern European region including Turkey and the South Caucasian countries. This 'one-sided' dependency is a de-facto-monopoly situation that may be used to impose political pressure on the countries involved - as experienced by Ukraine and Georgia over the last ten years. In order to improve this situation, natural gas supply from the Caspian region to Europe should be secured by diversification of source, supplier and transit country.

The South-Caucasian Gas Pipeline system stretching from Azerbaijan to Turkey is now being designed, and will according to plan go on stream in year 2006. Its capacity will gradually increase as the Caspian gas fields develop. Hence, the amount of the *option gas* to the transit countries may vary a) over the seasons, and b) from year to year.

Pipeline systems connecting Caspian gas to Europe

In Georgia, for instance, the energy dependency is twice as high as the average international practice. Experience shows that due to the lack of strategic reserves interaction or supply from monopolists like Russia was highly required. The prevailing practice, however, among European importers of fuel is to maintain a certain reserve requirement. This is seen as necessary in order to sustain critical situations. Provided additional volume is available, a commercial option also exists for buying the gas cheaper over the summer season in order to reduce the peak price during the winter.

International experience shows that underground storage facilities have a history of more than 50 years. The typical storage capacity in Europe varies from 16.5 – 38% of the annual demands in countries like Denmark, Austria, France, Germany, Italy, Slovakia, Hungary and the Czech Republic. Likewise, the reserves in the USA and Canada are 17% and 19.1% respectively.

New development in LNG technology may offer competitive solutions for decentralized LNG production. This, combined with atmospheric, insulated LNG storage tanks suggests a rational solution for a cheaper and simpler storage facility that includes distribution that is independent of pipeline systems. This is an interesting option because many countries are lacking distribution systems for some regions. LNG could be an alternative to developing a full coverage pipeline system, the latter being far more expensive. For remote mountainous areas, especially those that are lacking biomass sources, LNG would be an adequate solution. It should further be noted that LNG also is a fuel with high energy density that may be used for mobile applications.

The concept map shown on the figure below indicates the prospects of transporting gas and alternative energy carriers derived from natural gas - such as CNG, GTL/methanol, LNG and even electricity generated at the gas field and transmitted by high voltage DC. The map suggests a feasible range of such transport alternatives in terms of distance from plant to end-user versus quantities. As it appears LNG is the only option for gas transport over very long distances. It is also shown that new small-scale LNG facilities appear to become economically feasible in the lower left-hand area of the map. This means that small-scale LNG facilities represent a most promising solution for countries that lack the full coverage of their infrastructure for energy supply.

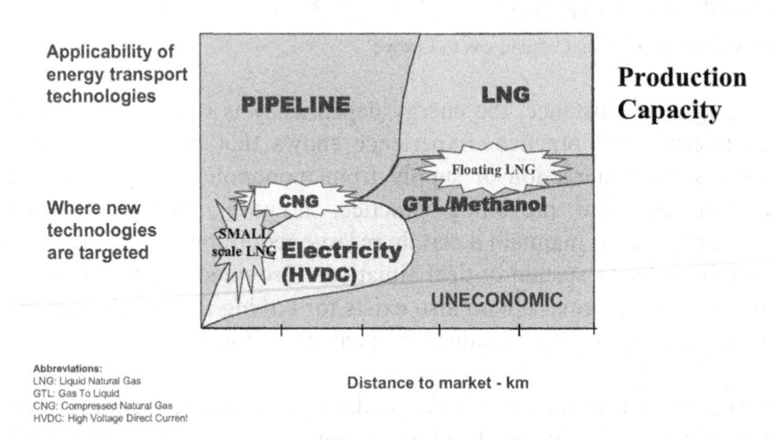

Concept map for the transport of gas and energy carries derived from natural gas, showing distance versus flow.

PART I: CURRENT STATUS OF PRODUCER COUNTRIES AND CONDITIONS FOR THE TRANSIT OF NATURAL GAS

PROSPECTS OF THE CASPIAN NATURAL GAS SUPPLY TO EUROPE AND PROBABLE IMPACT ON THE ENERGY SECURITY OF TRANSIT COUNTRIES

Teimuraz GOCHITASHVILI
Georgian Technical University, GEORGIA

Abstract: The market for natural gas is gradually expanding worldwide, including Europe. The supply of European market from the remote Caspian gas fields requires infrastructure development and introduction of modern technologies that may affect the transit countries' energy security and more general development.

The energy security of transit countries requires some significant issues that need immediate response: how does the transit fee for the received gas fit to the local energy market, how may the gas through transit countries' territory impact on the social and economic development of the local population, and what are the probable geopolitical consequences of the construction of new transit pipelines on local, regional and global levels.

This article addresses the above listed problems and related issues, offers recommendations for their rational solution. Namely, it has been defined that the main priorities of transit countries in the South Caucasus region in a foreseeable future are: diversification of imported fuel supply, enhanced use of local energy sources, improved self-sufficiency by rehabilitation/renovation of existing electricity and gas supply chains, and the planning of strategic reserves combining water, liquid fuel and gas.

Key words: Natural gas, Diversification of supply, Energy security, Transit country

J. Hetland and T. Gochitashvili (eds.),
Security of Natural Gas Supply through Transit Countries, 3–36.

1. INTRODUCTION

The global gas demand is expected to increase considerably during the next 25-50 years. Gas is a cheaper, environmentally-friendly fuel with a high calorific value. While the increase of gas consumption by the main industrial countries, their production will significantly decrease. To fill the gap between local production and demand new sources for additional gas imports are required, leading to a bilateral increasing dependence of producer and consumer countries. Providing reliable supplies from remote gas fields, transit countries, are also involved in this process. Therefore the political and economic instability of transit countries may play a quite detrimental role in the sustainable development processes totally. In order to improve the security of supply new export pipelines and other gas infrastructure for the diversification and strategic reserves may have a significant geopolitical impact on local, regional and global issues.

2. THE WORLD ENERGY DEVELOPMENT TENDENCIES

2.1 Fossil Fuel Resources and Consumption

The world energy development processes have a significant influence upon the global economic and political development tendencies and vice versa. The atrocious terrorist act referred to as 'September 11', and the anti-terrorist operations have significantly heightened the importance of energy security issues. Therefore analysis of the prospective development of the global energy resources acquires a strong economic and geostrategic significance for the entire world. This includes determination of alternative transportation routes and additional supply sources from remote areas, with an additional abundant hydrocarbons potential.

The global energy intensity was steadily increased during the last quarter of the twentieth century. However, from year 2000 to 2030 a doubling of the demand of primary resources is expected - from 9 to 18 btoe - as predicted by various development scenarios [1]. Furthermore, the forecast for 2050 shows a demand of roughly 25-30 btoe.

There is a considerable change in the structure of the utilization of the global primary energy resources. As coal prevailed until the 1970-ies, oil products were prone to gradually replace solid fuels. And, since the 1980-ies natural gas and renewable energy sources are likely to substitute oil.

It is still expected that fossil fuels will play a significant role in the future. By the year 2050 it is assumed that 2/3 of the total energy use will be needed to balance the demand – as compared to the present level of 85% [1].

Coal forms about 60% of the global fossil fuel resources that are relatively evenly distributed around the world in more than 90 countries. The relative share of the coal consumption had a peak of 70% around 1900, but since then the relative importance has been on decline. Nowadays approximately ¼ of the global fossil fuel consumption is by solids, from which 60 % is used for power generation.

Difficulties of transportation and adverse emissions, determines low quantity of coal imports on the international market (about 10 % of the total consumption). The future prospects for coal will largely depend on the impacts of the deregulation of the energy markets, and new policies to reduce greenhouse gas emissions, and also of technological innovations [2].

According to the latest assessments [1] the proven reserves of oil, mainly concentrated in the middle East, increased less intensively (from 86 btoe in 1973 to 140 btoe in 2000), than those of gas (52 Tcm in 1973, that was roughly tripled at beginning of the 21st century). The total proven reserves of crude oil and gas liquids at the end of 1999 aggregates roughly to 140 btoe that allows continuing current rate of production for 40 more years.

A major advantage of oil is its high energy density (calorific value – around 42-43 MJ/kg), easier transportation and consumption. The ratio of hydrogen atoms to carbon atom is typically around 2:1. That's why a complete combustion of oil emits less CO_2 than a complete combustion of coal per unit of energy (73 –75 gCO_2/MJ at stoichiometric combustion).

Oil still is a more important energy resource and meets almost 47 % of the global energy demand. The share of oil consumption had a peak of 50% around 1990, but is now declining.

The total proven reserves of natural gas amount approximately to 164 Tcm, that is sufficient for over 65 years by current rate of world production (2.5 Tcm at 2000) [3]. Natural gas has the highest energy density – calorific value of around 50-55 MJ/kg (35-40 MJ/m^3). The ratio of hydrogen atoms to carbon atom are typically around 4:1 that results in the lowest carbon dioxide emission – 56 gCO_2/MJ. A significant emission reduction can be achieved by switching from solid and liquid fuels to gas. This is much owing to the ability of natural gas to combine high firing temperature in modern gas turbines in combined cycles that lead to an increased interest in using gas in power generation. Hence, state-of-the art gas power plants offer efficiencies up to 58% in contrast to coal that offers maximum 47-48 %, and oil around 50 %.

The trend is that new technologies such as combined cycle gas fired power generation combined with CO_2 capture and sequestration, liquefaction of natural gas (LNG), hydrogen fuel cells, and market incentives promoted by the Kyoto Clean Development Mechanisms, shows that the demand for natural gas will continue to grow on a long term prospective. This especially applies to those countries that avail of large quantities – domestically or from neighbouring regions.

At the same time the relatively high cost of gas transportation requires a reconsideration of the gas field development perspectives. This will affect investment opportunities in transportation pertaining to pipeline systems and the production and distribution of LNG and related markets [4]. Thus, natural gas significantly differs from oil that is a universal commodity with international prices, whereas gas prices and contracts are often specific to local or regional markets. This would give the transit countries an advantage in buying natural gas at favourable conditions. This may affect the security of supply in a positive manner.

2.2 Export/Import Balances

The energy demand of the different regions of the world and the supply potential are rather unbalanced. As the importer countries lack local resources, the producer countries are able to produce more than their own demands. This has created an international market for natural gas that became mutually beneficial to both parties.

In 2001 the global export of gas amounted to 554.4 bcm, including natural gas exports (411.3 bcm) through pipelines. In 2001 Russia was the largest exporter with 126.9 bcm NG exported to Europe. Export from Canada amounted to 109.1 bcm, from Algeria – 57.7 bcm, from Norway – 50.5 bcm, from Netherlands – 45.2 bcm. 23 % of the global and 33 % of the European demand for gas was secured by import [3]. This process would become more intensive in the future. For example in Europe in the year 2010 about 70% of consumed solid fuels, about 65 % of natural gas and more than 90% of crude oil would be secured by import.

In order to avoid a deficiency in energy supply alternative ways of delivering the primary energies to the market are needed such as pipelines versus LNG, that results in a several supply conditions that, eventually, may have a bearing on the economics.

Depending of type and application the transport expenses of hydrocarbons varies within wide frames. Usually, overseas transportation of oil by tankers is more convenient than through pipelines. Therefore, the oil pipelines are used only from the place of extraction to the seashore or to the oil processing plant. For natural gas supply up to three thousand kilometers onshore pipelines tend to be preferable, whereas LNG carriers probably would be the only option for the transportation above ten thousand kilometers. And, for shorter distances sub-sea pipelines for gas transportation could be suitable.

The aggregated length of constructed pipelines around the world exceeds some tens of thousand kilometers per annum. In the year 2000 projects concerning roughly 225 significant main pipelines of crude oil, petroleum products and natural gas were planned, designed and constructed. The work on new pipelines pertained to a total length of 52600 km whereof 70% were gas pipelines. The pipelines were intensively developed in the USA and the Caspian region that accounted for about 18% and 17% respectively of the world activities [5].

2.3 Special situation

Some European countries strongly depend on imported gas and require special measures for the energy security. Most importantly this includes the creation of strategic reserves in underground gas storage facilities (UGS), which may be used not only in regular situations for peak shaving, but also in events of force-majeure. In order to sustain critical situations the practice among European importers of natural gas is that reserves should correspond to some 90 days of normal supply. According to the IEA Statistics the seasonal variation of the price from 1999 to 2000 varied within 10% - 13% in the USA and Germany; cycling of indicative cost for natural gas over 60 days is more significant according to the DRI-WEFA's analysis [6].

Underground storage facilities have a history of more than 50 years. Gas companies have made large investments to avoid the impact of the monopolistic supply, and to meet seasonal fluctuations of the demand of natural gas with minimum additional expenses. The typical storage capacity amounts to 16.5 – 38% of annual demands in countries like Denmark, Austria, France, Germany, Italy, Slovakia, the Czech Republic and Hungary. Likewise, the reserves in the USA and Canada are 17% and 19.1% respectively. Nevertheless, statistical analysis shows that European UGS have been under-utilized during the latest years. That's why it is recommended for operators to justify planning margins through cost-benefit analysis and other affecting factors, in the liberalized conditions of new gas markets.

The need for additional UGS is likely to be limited over the next two decades. The DRI-DEFA's estimate shows no need for additional storage capacity in Europe during the next few years. Additional capacity will not be needed in Europe until after 2010.

The progress in natural gas liquefaction technologies may offer competitive solutions for decentralized liquid gas production that may contribute to secure the energy supply [7]. Gas-to-liquid technologies used to convert natural gas to liquid fuels are under consideration as a competitive option to export gas from remote areas to the international market.

In general, some conclusions can be drawn:

- For the nearest foreseeable future, the international market will involve new players to secure the supply of natural gas. This means a globalisation and, hence, a liberalization of the energy markets;
- A growing inter-dependence will be developed between main exporters and importers. This will improve the security level and the economic stability of the producing and transit countries;
- Diversification of energy supply and actions to be taken towards the implementation of modern gas technologies. This may include strategic gas storage and energy management within the framework of technical and economic viability.
- Financial and political support to the alternative transport routes should be broadly applied. This applies to the implementation of new gas technologies. This may contribute to maintain stability of the international energy market, and, thus, to keep the comparative prices low.

3. CASPIAN GAS RESOURCES AND EXPORT POTENTIAL

For securing of future sustainable development, European net importers should avoid international gas market substantial dependence on large suppliers. At the same time, Russia, being the largest and one of the most influential exporter of natural gas, is trying to enhance its influence. Particularly within the former Soviet Republics and Eastern Europe the Russian business policy is de facto to operate as a monopolist in the natural gas market. Examples are the construction of a natural gas transportation system designated “Blue Stream” that has been politically substantiated, and the long term contracts recently made with Central Asian States on natural gas deliveries. Under these conditions, the Caspian region – being owner of abundant energy resources and South Caucasus being a prospective provider of transportation routes may play a significant strategic role in the future.

Estimated proven reserves of Caspian reservoirs total approximately 10 Tcm (whereof 8 Tcm belong to Azerbaijan, Kazakhstan, Turkmenistan and Uzbekistan). Additional proposed resources are 32 Tcm, which roughly correspond to 75-80 years of current European import level. This is somewhat lower than that of the major world exporters as only proven Russian reserves aggregate to some 47.7 Tcm, Iran 24.3 Tcm, and Qatar 10.9 Tcm. Likewise the Caspian reserves exceed the reserves of such big producers as the United Arab Emirates (6.0 Tcm), Saudi Arabia (5.8 Tcm), USA (4.7 Tcm), and Algeria (4.5 Tcm) [2].

Azerbaijan is one of the early developers of gas industry, using natural gas for industrial applications from the 1860-ies. Roughly a total of 470 bcm of natural gas has been produced in Azerbaijan over these years. Proven reserves of Azerbaijan amount to 1.37 Tcm (see Figure 1). Estimated resources that are of deemed economic aggregate to approximately to 1.9 Tcm. Other sources like Cedigas have stipulated the total gas resources of Azerbaijan to some 3-5 Tcm. Owing to the uncertainty due to a low survey rate applied to a widespread area of assumed gas reservoirs [8] such an optimistic quotation may be justified.

Azerbaijan is attracting interests of western companies not least for its rich hydrocarbon resources, but also because of its key location for the Caspian oil and gas export to international markets.

The local annual production of Azerbaijan amounted to 4.5-6 bcm over the recent years. Because of limitations of the existing offshore pipelines capacity, more than half of the current gas production is reportedly flared or vented. The country's total demand is secured by imports of Russian natural gas (4-6 bcm p.a.).

The production of the Azerbaijani State Oil Company will remain rather high until 2010. From 2005-2006 the country will receive additional amounts from the Shah Deniz and Azeri-Chirag-Guneshli fields (phase I).

Furthermore, Azerbaijan is planning an annual production of up to 24-30 bcm. The potential of already exploited oil and gas reservoirs for associated gas and other gas reservoirs that are under exploration, provides an export potential of approximately 16-20 bcm p.a. in foreseeable future.

According to the Oil and Gas Journal and Cedigaz the proven reserves of Kazakhstan amounts to 1.84 Tcm, including 1.3 Tcm of the giant Karachaganak field. Kazakhstan produces about 11 bcm per annum, and consumes approximately 9 bcm (2000) [3]. It further exports gas to Russia from its western production fields, at the same time gas is imported to the south-eastern regions from neighbouring Turkmenistan and Uzbekistan. Kazakhstan also possesses significant amounts of other resources with development prospective in the near future. This involves plans for increasing the gas production up to 60 bcm by the year 2010, whereof roughly 40 bcm is for export. The same situation is anticipated in Uzbekistan having a production rate of some 54 bcm of natural gas per year [9].

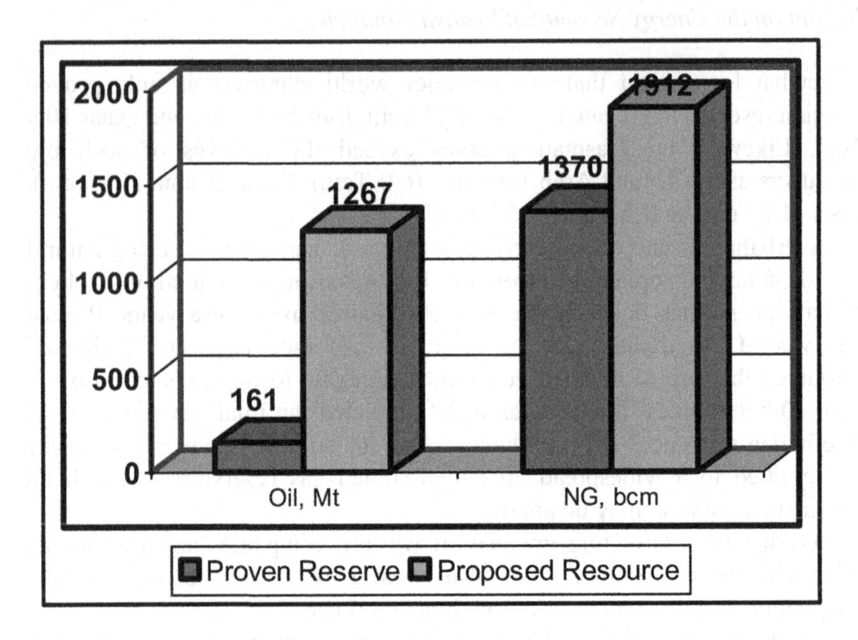

Figure 1. Azerbaijani Oil and Gas Reserves and Resources

During the 1980-ies and early 1990-ies Turkmenistan used to be the fourth largest natural gas producer next to the Russian Federation, the USA and Canada. The country possesses large gas fields such as the super giant Dauletabad field that offers roughly 40 bcm per year. Also some oil fields contain large volumes of associated gas [10]. After the fall of the Soviet Union, there was a recession in the Turkmen production of around 75 percent. The main reason for this dramatic decrease can be explained to be the denial of access to Russian pipelines, and due to the absence of alternative export pipelines to the major customers. Nowadays a main part of the Turkmen gas is exported through Russian pipelines, whereas a smaller part goes to Iran.

In 2001 the Turkmen production amounted to 48 bcm [11]. By the year 2010 the country plans to produce up to 100 bcm, and further increase its production up to 120-130 bcm by 2020. Most of these volumes are targeted for export, mainly to Europe that will receive up to 50 bcm in 2010 and 100 bcm in 2020 [12,13].

An estimated cost of the gas at the wellhead in Turkmenistan is less than US$20/1000m^3 (by some estimates up to 1 $/bbl of oil equivalent). In 1999 the Russian-based Gazprom solicited a multi year gas contract from Turkmenistan that was locked for some period, due to price [14]. The final agreement of 20 bcm of gas was entered into in the year 2000 with a contracted tariff of US$36/1000m^3. In 2001 Turkmenistan requested to raise the price to US$38/1000m^3 reckoned at the border with Kazakhstan. The US-based - but Russian controlled - company Itera was the main trader to deal with the Turkmen gas. Itera signed a contract at a higher price for an additional amount of 10 bcm gas. Recently a 25-year contract was signed by Gazprom and Turkmenistan for delivery of natural gas to Russia at US$44/1000m^3. As agreed upon 50% of the contract is to be paid in cash and 50% in kind.

The transportation tariff of Caspian gas through Russian pipelines, following the Resolution of the Russian State Federal Commission has been limited at US$16/1000m^3 per 1000 km. The added value to Central Asian gas on the Russian territory is quite impressive, as it increases by US$20-40/1000m^3 (depending on route). An average of US$25-35/1000m^3 will be paid for the Caspian gas to Western Europe or Turkey through territories of countries like Ukraine, the Balkan States, Slovakia, and the Czech Republic. Significant increase of added value for the Turkmen gas through Russian pipelines may weaken its competitiveness on international markets (see Study of Observatoire Mediterraneen de L'energy – OME Institution).

Perhaps Turkmenistan has assumed a strategic importance of a long-term contract with Russia on natural gas delivery. However, when Russia and Iran in the future may develop their production and pipeline capacity for their export of natural gas, no firm alternative exists to Turkmenistan but to sell its gas to said re-exporters at a price dictated by them. In a longer run Turkmenistan appears to be the looser of this deal as Turkmenistan is dependent on these two gas-rich countries [15].

Forecasts for the Caspian gas production made by BP and the Center for Global Energy Studies based on Wood and Mackenzie, show that there could be produced 320 bcm of gas per annum by year 2020. However, taking more recent development of the international energy market into consideration, this forecast is conceived to be rather exaggerated. Figure 2 presents another forecast for Caspian gas, based on a more conservative development scenario (including the higher and lower production rate of Caspian reservoirs) [16].

While estimating the Caspian gas prospects, the probable impact of the Russian gas sector should be taken into consideration. Also the Russian influence of the markets in the East and Central European Countries, Balkan, Turkey and South Caucasus should be taken into consideration. This includes long-term supply contracts and sometimes equity and management stakes in major supply entities. Currently Russia exports roughly 130 bcm/y of natural gas to Europe and 60-70 bcm to Ukraine, Belarus, Moldova, the Baltic and South Caucasian States.

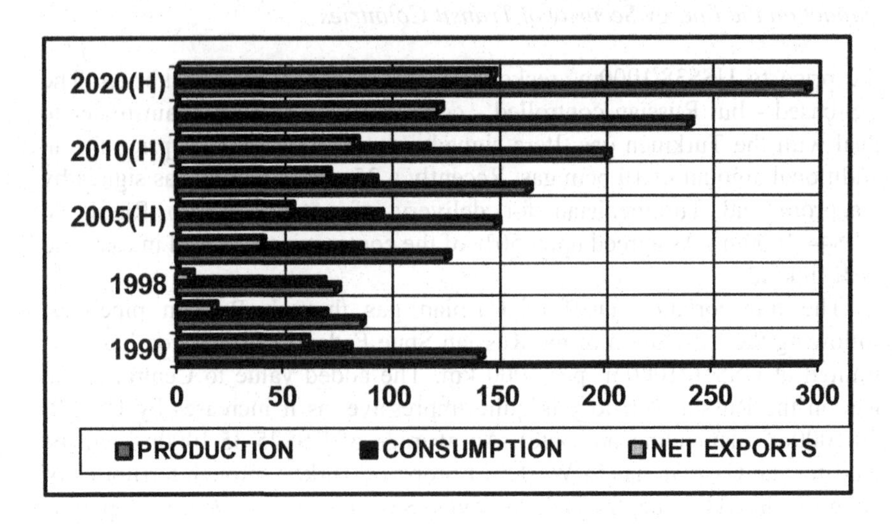

Figure 2. Caspian gas prospects

By the year 2010 Russia plans to produce totally 635-665 bcm of natural gas, including an export potential of 140-150 bcm for supplying the European market (excluding republics of the Former Soviet Union). By 2020 the production of Russian gas will reach roughly 680-730 bcm with about 160–165 bcm export to Europe per annum [17]. The planned level of exports to Europe seems realistic provided Central Asian gas is available at a price lower than the alternative price of gas from new Russian fields (for instance in Siberia, the Barents Sea, the Yamal peninsula and Sakhalin).

Some European countries have supported the most significant Russian project for the near future. This was announced during president Putin's visit to the United Kingdom (2003) as was stated that the UK is ready to invest some US$6 billion for the development of the North-European pipeline system to take gas from Russian Barents Sea to Europe via the Baltic Sea.

Gazprom was experiencing certain difficulties with its transit pipelines through Ukraine, and is now seeking to avoid this route by constructing a bypass through Belarus.

The "Blue Stream" sub-sea pipeline (BS) project was constructed to deliver gas from the Russian Black Sea region to Turkey without involving any transit country. Gas might be Russian, Turkmen or Kazakhstani. However, in contrast to Gazprom's expectation the delivery through the Blue Stream system has been delayed for a long time because of disputes with Turkey. To resolve this Blue Stream problem, the Russian presidential envoy to the Caspian Region, Mr. Viktor Kalyuzhnyi, offered to build a gas pipeline at Gazprom expenses to link the Azeri Shah Deniz field to the Blue Stream at Astrakhan compressor station on the Russian side. From this place the Azeri gas may be routed to Turkey – and eventually delivered at discounted prices. The reaction from Azeri state oil company, SOCAR, and the Shah Deniz project partners was quite negative as they claimed this proposal to kill the project of the South Caucasus Pipeline System [18].

In parallel Gazprom initiated negotiations with Georgian officials to squeeze out the Russian controlled company Itera from the local market. Moreover the state-owned Georgian Gas International Corporation, GGIC, planned to create a Joint Venture with Gazprom in order to share ownership of the Georgian main gas pipelines. This included a new pipeline segment through the Georgian territory to deliver Russian gas into Turkey. The immediate public reaction was quite negative, and the proposal was withdrawn - at least for a while. Eventually, this reaction was shared by US Shah Deniz project supporters. Meanwhile Georgia – owing to shortage of energy – has to rely upon Russian supplies until the Shah Deniz gas becomes available by late 2006.

Consequently Russia seems to continue harbouring its ambition to control natural gas exports from Central Asia and Azerbaijan. This pursuit is partially supported by the lack of competition and the absence of appropriate activities by western companies [19]. President Putin's policy to establish closer relations with Central Asia and South Caucasian neighbours may play an important role as well in the future.

4. MARKET FOR CASPIAN GAS

Turkey, the South Caucasus Countries and South-Eastern Europe seem to be the most promising markets for the Caspian gas. Recent statistics show that Armenia imported 100 %, whereas Georgia and Turkey roughly 95 %, and the South-Eastern European countries around 70% of its gas consumption (2000, 2001). All these countries currently import natural gas through existing pipelines mainly controlled by the Russian Gazprom or its subsidiaries.

4.1 Turkey

The consumption of natural gas in Turkey is growing rapidly. Owing to its proximity to the Caspian resources Turkey is seen as an attractive market. Traditionally Turkey has been supplied by gas from Russia via Ukraine and Balkans. In 1987 the Turkey gas consumption was 0.5 bcm, whereas the imports of the year 2002 amounted to reached approximately 19 bcm. The length of the existing high pressure pipelines total some 4700 km. At the completion ongoing projects the length will become some 8000 km [20]. The basic factors that support this significant growth are the dynamical economy of Turkey along with programs for industrialization, urbanization and environmental issues. Also the policy of diversification of energy sources is of some relevance.

Long term contracts have already been signed by Russia and the Turkish pipeline company BOTAS concerning the supply of 6 bcm from the western part of the Russian Federation and 8 bcm from the Russian company Turgusgaz through the existing pipelines. Additionally 16 bcm will be supplied through the Blue Stream pipeline during the next 23-25 years. In addition Turkey will receive 4 bcm as LNG from Algeria (till 2014) and another 1.2 bcm as LNG from Nigeria (till 2021). Turkey has also secured supplies of 55.4 bcm for 2010 and 61.3 for 2020 by additional export agreements with Iran (10 bcm till 2026), Turkmenistan (16 bcm, 2006-2036) and Azerbaijan (6.6 bcm, 2006-2020). Furthermore, considerations are made to take 10 bcm gas from the Iraqi reserves (3.7 Tcm). According to the recent estimates made by BOTAS the forecasted demand of Turkey is roughly 55 bcm in 2010 and over the 80 bcm in 2020, which exceed the contracted supply.

It appears that BOTAS is rather optimistic. The contracted amount of gas seems to exceed the level that the country can actually absorb. The point is that lots of forecasts on natural gas apply to power generation projects, whereas the power generation projects are progressing slowly as compared to previous plans. Moreover, instead of adding new power capacity in Turkey considerations should be given to possible supply of cheap electricity from Russia and South Caucasus. As shown in Table 1 the most reasonable development of the Turkish gas market by 2020 requires less gas than the contracts that BOTAS has entered into. At the same time significant changes in the proposed supply sources are anticipated due to obstacles like:

- The neighbouring coastal states cannot decided over the division of the Caspian Sea reservoirs;
- Problems relating to the distribution of Turkmen gas through the Russian system, and competition with the Azeri, Iraqi and Iranian gas in the framework of a future deregulated energy market;
- The Turkmen-Russian contract for long-term gas delivery to Russia is prone to weaken the Turkmen participation on the Turkish gas market.

Also under these conditions the gas delivery from ongoing Russian development projects through the Blue Stream pipeline seems problematic.

And, finally, the assumed cheaper Azeri gas from the Shah Deniz reservoir, and the prospects of deliveries from Iran and Iraq provide a guarantee for increased export into Turkey.

4.2 Europe

The European gas demand is fast growing. According to various estimations the import dependency of the EU countries will rise to some 65-70 % in 2020.

According to special studies conducted under the EU ETAP Programme the gas demand in Europe will rich 642 bcm by 2010 and 777 bcm by 2020 [21]. The plans of the European Union include expected consequences of the liberalised gas market and the strategy of diversified supplies, and aim at:

- Connecting the gas networks of the UK, the Netherlands, Germany and Russia;
- Construct new pipeline and increase the Algerian export potential to France and Spain;
- Provide import of the Caspian gas and gas from other remote regions into EU.;
- Create a new LNG chains to France, Italy, Portugal and Spain;
- Establish underground storage capacity for natural gas in Spain, Portugal and Greece.

Forecasts for the natural gas demand of Europe for the period 2005-2020 are shown on Table 1 [22].

The European countries have guaranteed supply contracts until 2010. Beyond this period Europe may expect a shortage of natural gas amounting to 100-200 bcm per annum (until 2020). The liberalised European energy market will become instrumental to the growing gas demand [23].

In the foreseeable future a strong requirement for gas supply to Europe from new sources, including those of the Caspian region, is likely to occur. In Europe the Caspian gas will have to compete primarily with the established suppliers. At the same time the North Sea reserves will be on decline. The pipelines for supply of the Algerian gas through Morocco and Tunisia have limited capacity. LNG imports from the Middle East and Africa will reach their limits. And Russia will face significant difficulties in developing new reserves in Siberia and the Barents Sea [24,25].

In co-operation with Turkey and the Balkan states the South Caucasus region can secure the supply of deficient natural gas from the Caspian region to Europe on favourable terms and conditions.

Table 1. The market for Caspian gas in bcm per annum [22]

	Europe	Turkey	Georgia	Armenia	Total
1990	319	3.5	5.3	4.5	323.3
2000	456	14.8	1.0	1.5	473
2005	475-515	23-30	1.5-2.0	1.7	501-548
2010	515-586	32-45	2.5-3.5	2.5	552-637
2015	550-655	40-57	4.0-5.0	4.2	698-721
2020	650-720	50-65	4.5-5.5	5.5	710-798

4.3 South Caucasus countries

The South Caucasus countries have well-developed distribution networks and related infrastructures.

By early 1990-ies the gas consumption in Azerbaijan totalled about 17 bcm. The Azeri gas distribution network covers around 85 % of the territory. The local gas production and the imports from Russia and Iran had significantly declined after the collapse of Soviet empire, and the supply to significant part of countries territory discontinued.

In Azerbaijan current gas supplies amount some 9-10 bcm p.a., including import from Russia. The planned consumption of natural gas in Azerbaijan will reach some 12-14 bcm throughout 2006-2010.

Armenia owns some 1800 km high-pressure gas transmission and approximately 9000 km gas distribution network. In 1990 Armenia consumed some 4.5 bcm of gas, which declined to 1.5 bcm p.a. (2000), of which more than 75% is used for power generation.

Following the energy demand forecasts [26], the gas consumption in Armenia will increase from about 2.5 bcm to 5.5 bcm in 2010 and 2020 respectively. The supply is planned to be secured from Russia using the existing North-South Caucasus Gas main transmission system via Georgia, and also from such new supply sources as Iran or others.

In the late 1980-ies, Georgia was one of the most intensive users of gas per capita in the world. The peak consumption in 1990 amounted to 6 bcm that was roughly 60 percent of the total energy demand in the country (See Figure 3).

The total gas transportation expenses from Russia, Central Asia and Iran to South Caucasus are significantly less than those to Turkey or to Europe. Accordingly, the price of the gas to the South Caucasian market will be favourable and, thus, a potential exists for a growing consumption. One scenario shows that within the next 20 years this market is expected to return to a consumption level that corresponds to the early 1990-ies. Another scenario of the Georgian gas demand was prepared within the framework of a TACIS project carried out by the Danish-based Ramboll. Here the gas consumption in Georgia is assumed at 1.5 bcm in 2010 and 3.2 bcm in 2020 [27]. This forecast is rather conservative as it underestimates the potential of alternative sources for natural gas supply.

The consumption is expected to increase significantly in the Georgian residential, industrial and power generation sectors. Moreover in order to eliminate its energy deficit Georgia will increase the reliability of its power system. This would include a substantial refurbishment of thermal power plants and enhanced utilisation of existing generation capacity. Eventually, Armenia has a nuclear power station that will have to be closed because of inherent safety issues. Owing to the low price of natural gas combined with the environmental advantages most future thermal plants in the region are expected to be gas-fired.

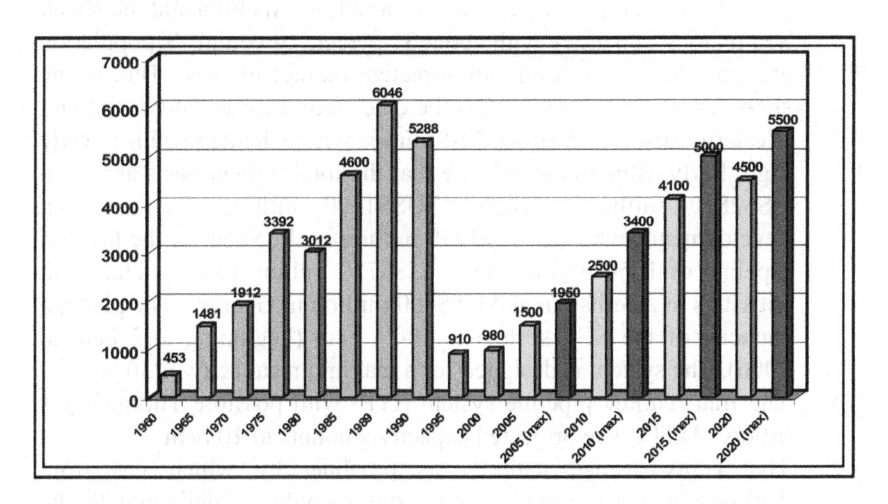

Figure 3. Gas demand of Georgia in Mcm (historical and planned)

5. TRANSIT PIPELINES AND DEVELOPMENT PROSPECTS

5.1 Export Pipelines

For the additional natural gas supply to Europe and Turkey from the Caspian region the following routes are considered:

- Russian owned, trans-border pipelines trough the Eastern European or Balkan states with planned capacity 14 bcm p.a.;
- The Blue Stream Black Sea sub-sea pipeline with planned capacity 16 bcm p.a.;
- The North–South Caucasus natural gas main system from Russia to Turkey through Georgia. This will require rehabilitation, upgrading and extension of the existing pipeline from Georgia to the Turkish towns Hopa or Erzerum, and further to Europe. The projected maximum export capacity is 14 bcm p.a., and the estimated investment for rehabilitation/refurbishment of the Georgian section amounts roughly to US$450 million;
- The South Caucasian gas pipeline system from Baku to Tbilisi and Erzerum. A bp-led consortium plans to construct the SCP system to deliver gas from Shah Deniz in the Caspian Sea to Georgia, Turkey and later to Europe. The South-East and East Anatolia and the Black Sea regions of Turkey with about 31 percent of country's population are expected to constitute an attractive market for gas. But, in the short-run, the pipeline needs to be extended towards the Central and Western Turkey and South-East Europe where the demand is already high. Early estimations indicate that the total project costs are up to US$2850 million, including US$1600 million for upstream developments, whereas US$1250 million is allocated for the 685 km pipeline of 1050 mm diameter. US$750 million is associated with activities in Azerbaijan and US$500 million in Georgia. The planned capacity of the SCP system is 18-24 bcm (beginning at 2 bcm in 2006). The system is designed for a maximum capacity of 30 bcm.
- The Iran–Turkey pipeline system (ITP) with possible Turkmen gas inflow (TkIT). The projected capacity amounts to 10 bcm;
- The Trans-Caspian natural gas pipeline, by which gas from Turkmenistan and other Caspian states could be delivered to the European market through Azerbaijan, Georgia and Turkey with a maximum projected capacity 30 bcm p.a. The estimated investment is up to US$2400 million.

The competitiveness of the discussed pipeline routes will largely depend on the transportation costs and the transit fees. Experience from North America and Great Britain shows that owing to the liberal market conditions economic competitiveness becomes a main criteria. This means that long-term contracts will be less attractive.

5.2 Economics of Export Pipelines

In order to stipulate the cost detailed calculations have been carried out for each route. The analyses include: production or purchase costs, specifications of pipelines (existing, planned and under-construction), transportation expenses and transit fees. The data used for the calculations were either obtained from different information sources or calculated based on international experience of pipeline exploitation.

Inputs to the calculations are as follows for the:

- Turkmen gas: US$42-44/1000m^3 (i.e. requested price from Turkmenistan);
- Azeri gas: US$55/1000m^3 (i.e. proposed price at Georgian border for 2006);
- Russian gas: For comparison different prices have been applied to the Russian gas – one for the gas received from the Volga-Ural region US$52/1000m^3, one from Western Siberia US$75/1000m^3 (including transportation cost on Russian territory), and one for the South Caucasus States supply US$56/1000m^3 priced at Georgian border in 2001;
- Iranian gas (Southern fields): US$70/1000m^3.

The actual gross calorific values of gas from each source were taken into consideration.

The transportation costs vary between 10-20 US$/1000m^3 for 1000 km, in accordance with pipelines capacity, their location (onshore/offshore) and total distance. The transit fee equal 5-10 % of transported gas price, and varies in accordance with international and currently existing practice for various countries. The tax exemptions for the Blue Stream pipeline on the Russian and Turkish territories, which allow substantial improvement of the economy, are also taken into consideration.

Figure 4 presents the results of the economic appraisal on relative terms. On this basis the competitiveness of each pipeline route can be estimated.

As presented in Figure 4 it becomes obvious that the most competitive project would be the South Caucasian Pipeline system, which would provide the cheapest transportation of the Caspian natural gas to South Caucasus, Turkey and Europe. However, the Blue Stream pipeline and the North-South Caucasus Pipeline appear to be the least beneficial.

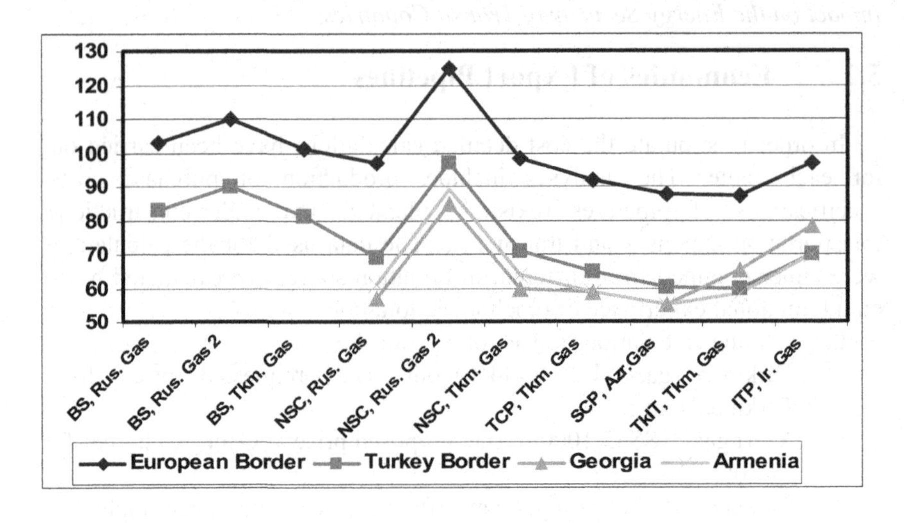

Figure 4. Marginal prices for Caspian gas by transportation through different Pipelines

Data obtained by BOTAS for the supply costs to Europe show quite acceptable similarity [20,21].

Furthermore, a comparative economic analysis of the same projects has been carried out, and the possible annual savings for the consumers in Europe, Turkey and the South Caucasus countries have been suggested [22].

As appears Turkey may obtain the highest incomes by realization of the South Caucasian Pipeline system, as well as Trans-Caspian Pipeline and the Iranian pipeline projects. Another income source for Turkey might be the transit fee for the Azeri and the Turkmen gas to Europe.

When comparing the South Caucasian Pipeline system for Azeri gas with the Blue Stream and the North-South Caucasus Pipelines for delivery of Turkmen gas calculations show that the best economic outcome for the Turkish consumers will be achieved by the former pipeline (SCP). The cheapest at wellhead Turkmen gas supply to Turkey and Europe using of Trans-Caspian Pipeline or the Turkmenistan-Iran-Turkey Pipelines is rather comparable, but the political isolation imposed upon Iran by the USA, and the latest agreement on delivery of the Turkmen gas to Russia make the further development prospect of these projects very doubtful. The proposed co-operation between two gas export pipeline projects – the South Caucasian Pipeline system and the Trans-Caspian Pipelines - for supply to Turkey and Europe via Azerbaijan and Georgia seem to be more attractive. In that case instead of construction of new pipeline sections on the Azeri and Georgian territories, a free capacity of South Caucasian Pipeline system can be used. This may result in a reduced investment cost up to US$1000 million.

The analysis shows that the maximum profit for the Georgian economy may be reached via the development of the South Caucasian Pipeline system. The total direct revenue in this case is calculated as the sum of capitalization of transit fees (corresponding to 5% of the transported gas) and incomes received because of difference between agreed special price for Georgia (55 $/1000 cub. m) and the forecasted price on local market as shown in Figure 5 for the examined period (2020).

Besides the direct revenue, Georgia will seek to improve its energy security. In addition guaranteed supply of gas will give boost to industrial activity. As analyses show, additional income in the Georgian economy by substitution of other energy sources with natural gas can be achieved, and as a result some sectors are expected to grow by 250% by 2010, and 360% by 2020 as compared with the year 2000 (refer Figure 6).

If all proposed pipelines were constructed, there is an apprehension of oversupply of regional markets, especially during the period 2006-2010. However, by consolidated development the South Caucasian Pipeline system appears to be the winner in a strong competition.

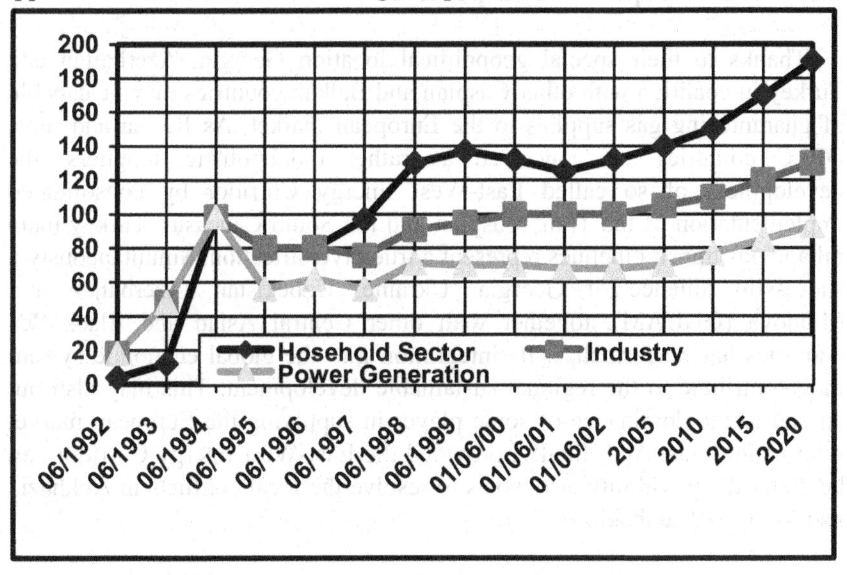

Figure 5. The price of natural gas in Georgia

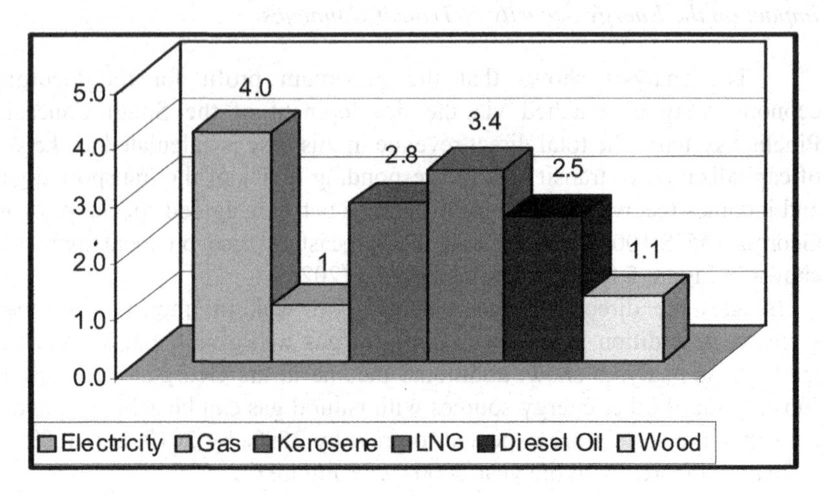

Figure 6. Comparative prices of energy resources in Georgia

5.3 Development Perspective

Thanks to their special geopolitical location Georgia, Azerbaijan and Turkey in coalition with other Caspian and Balkan countries may be capable of guaranteeing gas supplies to the European market. As Russia and most OPEC countries are conceived as rather monopolistic suppliers, the development of so called East-West Energy Corridor by consolidated implementation of the Trans-Caspian and the South Caucasus–Turkey main oil and gas transit pipelines represent a true diversification. Simultaneously a successful alliance of Georgia, Ukraine, Uzbekistan, Azerbaijan and Moldova (GUUAM), together with other Central Asian and Black Sea countries has be formalise. Its integration into the global economic system may contribute to the regional sustainable development. This may also put an end to the dominance of some player in supplying the European market with natural gas. And, finally, provided the East-West Energy Corridor can be realised a possibility also exists to resolve the local conflicts in Abkhazia and Nagorny Karabakh.

6. GEORGIAN ENERGY SECTOR

6.1 Georgian Energy Resources

6.1.1 Fossil Fuels

The total estimated oil resources, according to the official information of the Georgian National Oil Company (GNOC), are approximately 580 Million tons of oil. This corresponds to one third of which is proposed to be in Georgian Black Sea offshore reservoirs [28]. Currently, sixteen oil and gas fields are being developed in Georgia. The proven reserves of Georgian onshore reservoirs amount to approximately 32 Mt of oil (see Table 2). Nine oil companies (among them eight foreign) are now licensed to explore and produce oil and gas from these structures. Five of the companies are already producing oil and gas with a planned annual production of 1.2 Mt of oil in the year 2005 [29].

The Georgian onshore gas resources of Eocene and Cretaceous structures are assumed at 269 bcm. The proven "Near-Tbilisi-region" reserves in the Eocene structures of the Samgori, Rustavi and Ninotsminda fields approximately equal to 3.5 bcm. The data on gas prospective resources obtained by modern three-dimensional seismic survey over the Black Sea offshore structures are not available yet.

Table 2 **Georgian Energy resources**

Energy Recources	Reserve	Resource	Production Rate	
			2002	2005
Coal, M toe	185	301	0.003	0.020
Lignite, M toe	20	-	-	-
Oil. M toe	32	580	0.075	0.30
Gas, M toe	3	85-215	0.015	0.15
Hydro, TWh	32 (Econ. Pot.)	81 (Tech. Pot.)	6.7	8.5-9.0
Wind, GWh	500 (Econ. Pot.)	1500 (Tech. Pot.)	negl.	500
Solar, GWh	3 (Econ. Pot.)	18 (Tech. Pot.)	negl.	4.5
Bio (Wood), k toe	400	-	500	400
Bio(Wastes), k toe	150	450	negl.	150
Geothermal, k toe	40	113	negl.	40

The estimated daily rates for some wells in the East Georgian gas structures are up to 0.5-1.5 million m^3, and the planned annual output for 2005 may increase to 0.5 bcm.

The proven reserves of the Georgian sub-bituminous coal approximately amount to 354-404 Mt, and the lignite reserves are 71-76 Mt. Owing to its high ash content - up to 35-40 % - and its volatile content the Georgian coal is hampered with a relatively low calorific value (11.3-18.0 MJ/kg) (see Figure 7 [30]).

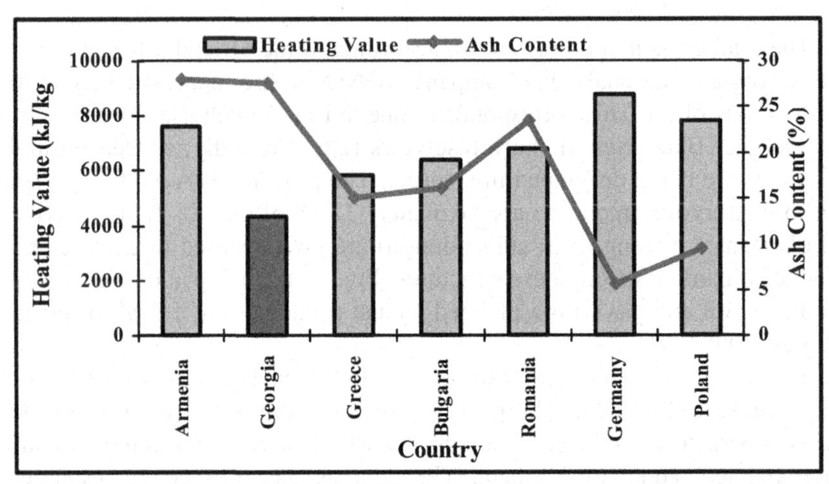

Figure 7. Typical heating values and ash content of Georgian coal compared to other European lignite

The main reserves of coal are located in deeper layers (700-1200 m for sub-bituminous coal at the Tkibuli-Shaori deposit), with few faults and depreciations, making extraction activities very difficult and expensive. Based on international experience coal cannot play a significant role in the formation of the country's energy balance over the foreseeable future.

6.1.2 Renewable Energy Resources

The theoretical potential of Georgian hydropower amounts to 159 TWh p.a. whereas the technical potential is 81 TWh p.a., and the economic potential 32 TWh p.a. (see Figure 8). The installed capacity of existing hydropower plants is 2.7 GW, however, with a current operating capacity of only 1.4 GW. An electric power production of 10 TWh p.a. is planned, whereas the real generation is only 6.5 TWh p.a. Hence, the real and the planned utilization over of the economic hydropower potential equal 20 and 31% respectively. This may, however, increase by adding 1.5-2.5 TWh p.a. through rehabilitation and modernization of existing hydropower plants at reasonable expenses. There are also attractive prospects for new hydro power plants in Georgia. With the completion of the Enguri and Rioni cascades an estimated annual generation capacity of 3-3.5 TWh p.a. can be added. Also some prospects have been proposed for small hydropower plants in Georgia. These are mainly located in mountainous areas separated from the country's high voltage electricity grid.

The realization of hydropower projects, in parallel with the development of regional transmission lines for interstate peak and base loads exchange provide a significant increase of the energy security while minimising the demand for imported fossil fuels.

Some non-traditional resources can also play a role in the Georgian energy supply. The Black Sea basin contains a huge amount of trapped fuel gases such as hydrogen sulphide, methane, propane, and a practically endless (low-temperature) thermal energy reservoir in the sea itself. Although the estimated hydrogen sulphide resources in the Black Sea amount to several dozen Mt, its utilization for energy supply appears to be a very difficult technical problem from the practical point of view.

Wind can be used in Georgia by modern turbines to generate electricity. The technical potential of wind energy in the country is estimated at 530 MW, with an annual generation up to 1.5 TWh. On a short-to-medium term prospective the economic potential of wind power is 500 GWh with an installed capacity of 180 MW [31].

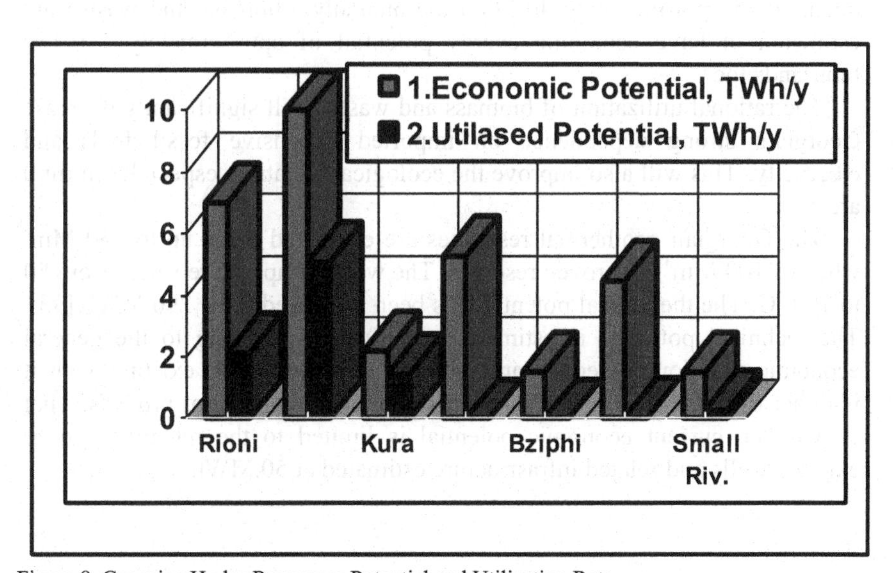

Figure 8. Georgian Hydro Resources Potential and Utilization Rate

The Georgian climate is characterized by long periods of sun radiation. Indeed, the country benefits from 250 to 280 sunny days per year for an average solar radiation of 4200 Wh/m^2 per day. The average annual solar radiation for Georgia at tilt angle 0° amounts to 1525 kWh/m^2, with rather homogenous geographical distribution. But the seasonal variation of solar radiation imposes possibility of simultaneous use of a back-up energy supply system (gas, electricity) to cover permanent needs of consumers.

The technical potential of solar energy for Georgia is around 18 GWh by using solar panels for water heating for niche markets as hotels, hospitals, sport facilities, and single residential buildings. The economic potential is estimated as 3 GWh p.a. [31]. On the medium term basis other niche markets, such as collective residential buildings, agriculture, food processing industry, wood drying etc. may emerge. This may significantly increase the utilization ratio of solar energy.

Biomass comprises material derived from forestry and agricultural operations, industries, commerce, agriculture and household organic wastes which can be used for the production of solid, liquid or gaseous fuels. Nevertheless, the over-cutting of forest resources puts in real danger to the ecosystem. The average renewal of wood in the Georgian forests amounts to 1.82 m^3/year per 1 hectare, and the total wood reserve is approximately 380 Mm^3 [32]. This allows for a annually utilisation of 3-5 Mm^3 of wood. The estimation of the energy utilization potential of biomass and wastes carried out by the Energy Efficiency Centre of Georgia indicates a technical energy potential of approximately 400 000 toe annually. Biomass and wastes are estimated to have economic energy potential of approximately 130-150 thousands toe.

The rational utilization of biomass and wastes will significantly decrease Georgia's strong dependence on imported expensive fossil fuels and electricity. This will also improve the ecological situation, especially in rural areas.

The Georgian geothermal resources are estimated to be up to 250 Mm^3 whereof 100 Mm^3 are proven reserves. The water temperature varies from 50 to 100 °C. The theoretical potential has been estimated to be 290 MWh [33]. The technical potential is estimated at 150 MWh. Owing to the general economic situation in Georgia and with the poor state of the existing wells it is rather difficult to attract investments for new development projects. That is why the current economic potential is limited to the rehabilitation of existing wells and related infrastructure estimated at 50 MWh.

6.2 Energy Resources Consumption

While analysing the structure of primary energy sources consumption in Georgia two periods are examined. During the first period (till 1991) the economy was rather stable as in the entire Soviet Union. The consumption of energy sources over that period reflects the tendencies of the global energy development. This is characterised by a reduced use of liquid and solid fuels and a significant increase in the consumption of natural gas and non-traditional energy sources.

In the former Soviet Union the consumption of primary energy sources per capita was equal to European level and, thus, significantly exceeded the world average. In Georgia by the late 1980-ies the consumption of primary energy sources per capita was 2-2.3 toe (see Figure 9). This rate corresponds to the rate of European countries located in the comparable climatic zones (Portugal, Spain, Italy, Greece, Turkey).

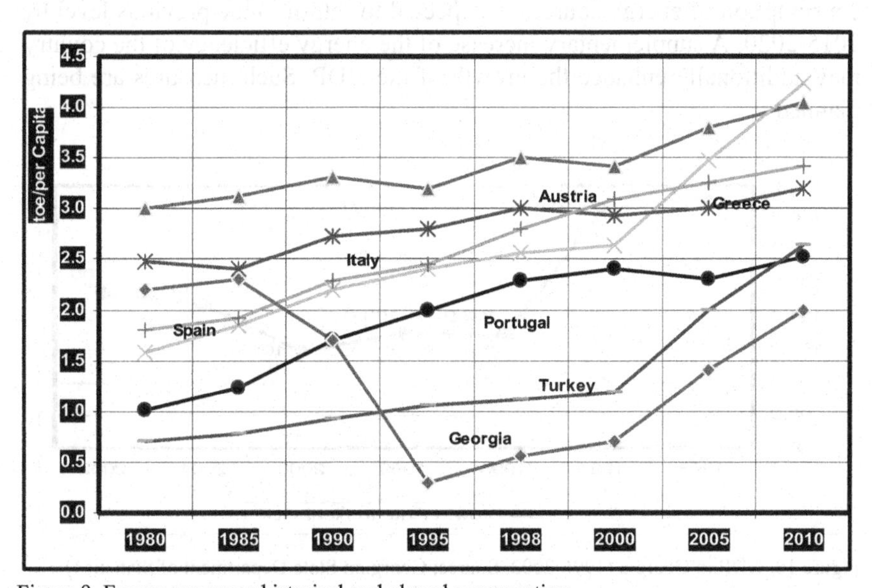

Figure 9. Energy resources historical and planed consumption

The consumption of primary energy sources in Georgia has drastically reduced in the last period. Though, according to official statistics, the current consumption rate is less than 1 toe. This is roughly the same as the world average indices. It should be mentioned that these formal statistics do not include illegal import/export of energy carriers. According to the experts less than 30-40 % of imported petroleum products are registered by the Georgian customs. The official statistics of firewood production and consumption is also doubtful. On the basis of special investigations, undertaken in the framework of a EC TACIS energy project in Georgia [32], the estimated consumption of wood amounts to 450-550 thousands toe per annum. In effect the per capita consumption of the primary energy sources is roughly 25-30 % higher than official statistics [34].

The structure of consumed energy sources also in accordance to the significant changes over the last period. The consumption of expensive imported energy sources (oil products, natural gas, coke) was, however, reduced. At the same time the share of the local hydro resources and firewood were increased. Contrary to the predicted assumptions, the local coal consumption was sharply reduced. A complete substitution of imported fuel would be impossible. This lead to a strong deficit of energy sources that destroyed the country's economy.

Since 1995 the Georgian GDP has been growing, and according to experts it will continue to grow by approximately 5 % per year (as the average of the recent years – see Figure 10). Correspondingly, the per capita consumption of energy sources is expected to return to the previous level by 2015-2020. A supplementary increase of the energy efficiency of the country may additionally enhance the growth of the GDP. Such measures are being planned.

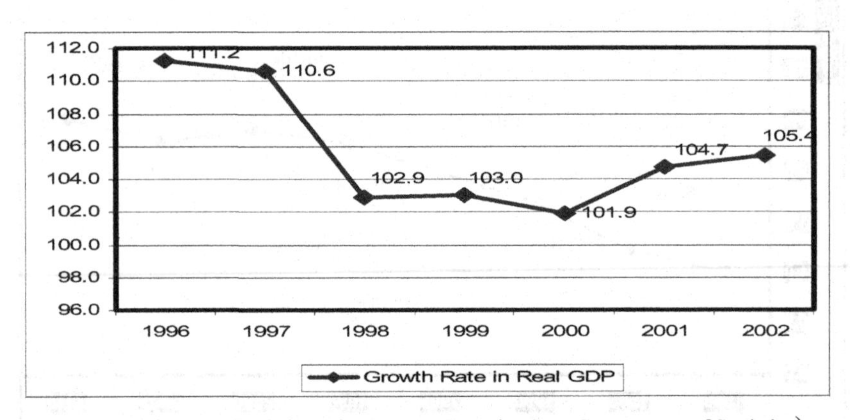

Figure 10. GDP in Georgia (1996-2002, Source: Georgian State Department of Statistics)

By 2010 the planned consumption of the primary energy sources in Georgia is estimated at 4-5 Mtoe (see Table 3). Approximately 75-80 % will be hydrocarbons and up to 20-25 % renewable energy sources. At the same time the planned demand for primary energy sources will amount to 6-7 Mtoe, including 10-12 TWh electricity [35]. The energy structure will be balanced by some 55% natural gas.

Table 3. Energy Consumption Forecasts in Georgia

	2002		2005		2010	
	M toe	TWh	M toe	TWh	M toe	TWh
Demand	**4**	**7.7**	**4.5**	**9-10**	**7**	**10-12**
Local Resources Potential	**0.6**	**7.2**	**1.15**	**9.5**	**2.4**	**10(13)**
Total secured Supply	**0.6**	**7.2**	**1.5**	**9.5**	**3.5**	**10(13)**
Deficit	**-3.4**	**-0.5**	**-3.0**	**-**	**-3.5**	**?**

6.3 Planned Revenues for Natural Gas Transit in Georgia

The planned revenues for the transit of Caspian and Russian gas may have an significant impact on the country's supply potential.

The Statoil-operated South Caucasian Pipeline system is expected to be completed by 2006. This will have access points for gas that according to the Host Government Agreement (HGA). The capacity of the pipeline system will gradually increase according to the development of the gas fields. Hence, the amount of the gas to the transit countries, being proportional to the amount of transit gas, may vary significantly over the seasons, and from year to year. Though, it is designed to transport as much as 30 bcm. However, by 2010 it is not likely to carry more than 10 bcm, mostly because of the lack of demand for gas in Turkey.

Georgia will receive a transit fee corresponding to 5% of the actual amount of the transported gas (i.e. 0.5 bcm for 2010). Under the most optimistic conditions the revenues correspond to the value of 1 bcm of natural gas. This will be the case for 2020 when the throughput of the pipeline is assumed at 22 bcm (including 20 bcm transit volumes for Turkey and Europe). In addition, Georgia will have the right to purchase a maximum amount of 0.5 bcm of gas annually at a discounted price (US$55 per 1000 m^3 reckoned from 2005. The price will be increased by 1.5% per year).[*]

Furthermore, Georgia will receive 10% transit fee for the gas delivered to Armenia by the existing North-South Caucasus gas pipeline operated by the Georgian International Gas Corporation. On the assumption, that Armenia may transport 1.7-2.5 bcm of gas per year in 2005-2010, and up to 5.5 bcm in 2020, the annual Georgian revenues will correspond to 0.17, 0.25 and 0.55 bcm respectively.

The compensation for the gas transport over Georgian territory by these two pipelines amounts to 0.17 bcm by 2005, 0.75 bcm by 2010 and 1.55 bcm by 2020. With additional amounts of Georgian gas production and the Azeri gas contracted at discounted price, this will cover up to 60% of the demand in Georgia.

6.4 Energy Security of Georgia

6.4.1 Natural gas utilization perspectives in the Georgian energy sector

Georgian local energy sources may cover approximately 20 % of the total domestic demand. Therefore Georgia is strongly dependant on imported energy sources. As in most other import countries, there is an increased focus on reliable supply of hydrocarbons to Georgia. The reason is much due to lacking security of the existing infrastructure that is deemed quite detrimental to the country's economic development. Therefore existing and new transit pipelines, and also storage facilities, have become political issues of strategic significance.

At present Russia is the only natural gas supplier to Georgia. This 'one-sided' dependency is de facto a monopoly situation that may impose political pressure on the country, as experienced by Georgia over the last ten years. In order to improve the situation, natural gas supply from the Caspian region should be secured by diversification of the pipeline systems and the supplier countries. In that respect especially the gas supplied by the South Caucasian Pipeline system may be important for the rehabilitation of the country's economy, and may, thereby, have a decisive political impact.

Nevertheless, the rehabilitation of a short segment of the Azeri-Georgian pipeline Kazakhi-Saguramo may provide access to other potential suppliers in Iran, Azerbaijan and Central Asia by offering an alternative route via Azerbaijan. This may significantly diversify supply sources.

The local production of natural gas in Georgia is limited, and the gas supplied from the transit systems has an almost constant output. This represents a major managerial problem as the supply does not correspond to the domestic seasonal demand (see Figure 11). The dependence on seasonal variations shows periods with significant imbalances between supply and demand (see Figure 12) [36]. That's why Georgia has to develop some sound strategy and associated infrastructure to utilize natural gas rationally.

For strategic reasons energy safety will be prioritised, and the social situation will be improved. This implies that the gas will be used first of all for power generation and the household sector respectively.

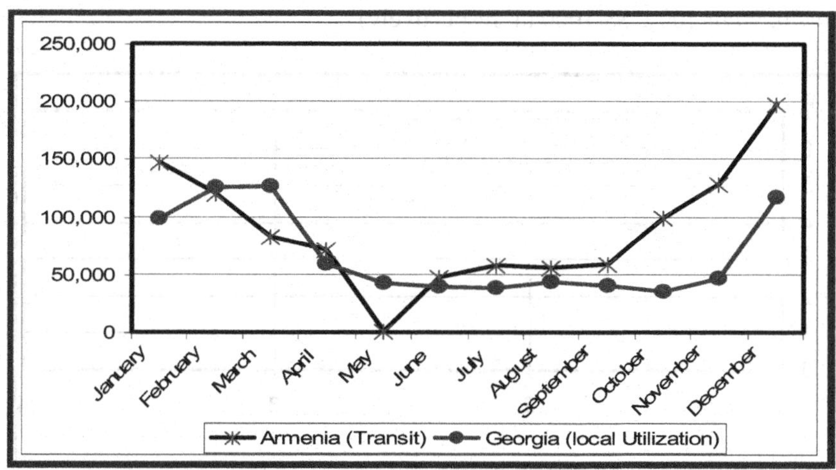

Figure 11. NG consumption monthly fluctuations in Georgia 2002

The main drivers for the utilization of natural gas in power sector is twofold: 1) emission reduction by shifting fuel from oil and/or coal to gas, and 2) the possibility of introducing combined cycle gas turbine (CCGT) technology with high electric efficiency (58 %).

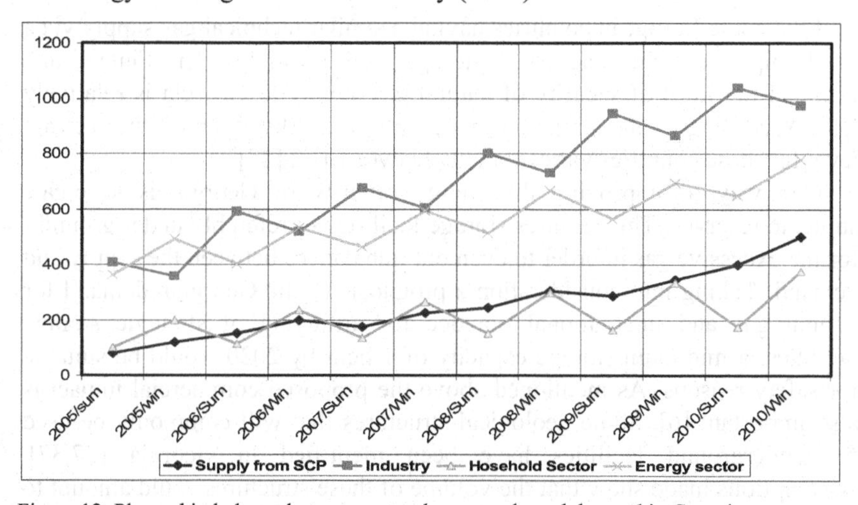

Figure 12. Planned imbalance between natural gas supply and demand in Georgia

This combination significantly decreases the cost of generated electricity. Figure 13 shows the planned cost of generated electricity for Georgian coal (FB technology by indigenous coal production costs 20, 25, 30 or 35 US$/t)

and imported natural gas (CCGT technology and natural gas supply costs pf 50, 70, 90 or 100 US$/1000 m^3 respectively).

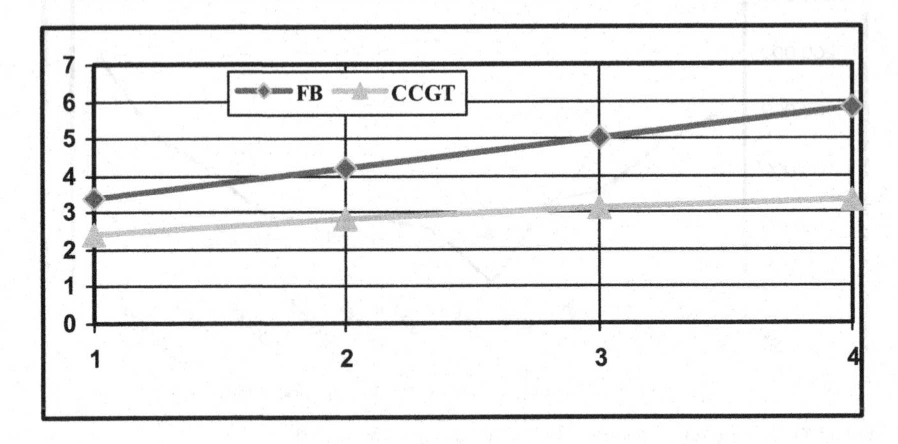

Figure 13. Generated electricity planed cost for coal and natural gas

6.4.2 Energy resources strategic reserves and emerging technologies

Like some European countries having a similar technical gas supply via a single pipeline without access to storage facilities as Sweden, Finland and Portugal the level of security of natural gas supply to Georgia is relatively low. According to an estimation model introduced by Ramboll the security level for these countries varies roughly between 1- 2 [27].

One way of improving the energy security of Georgia is to create strategic reserves. This requires storage facilities - preferably under ground - for the excessive gas in order to even out imbalances between the supply and demand. Taking into consideration a prognosis of the Georgian demand for natural gas and international practice for arranging of strategic storage facilities, a minimum storage capacity of 1 bcm by 2020 would be suitable for safety reasons. As mentioned above the proposed commercial impact is also important [6]. Some geological structures that will commonly be used for underground facilities have been identified in Georgia [27,37]. Assumptions made show that the volume of these structures could amount to 5-6 bcm.

Moreover gas also can be compressed and stored in pressurised tanks, or be liquefied and appear either as natural gas liquids (NGL), or as liquefied natural gas (LNG). The emerging LNG chains may lead natural gas to quite new markets – even those who have no developed infrastructure for natural gas.

Introduction of modern LNG technologies may offer competitive solutions for decentralized LNG production and guaranteed supply of country's remote regions. This, combined with atmospheric, insulated LNG storage tanks suggests a rational solution for a cheaper and simpler storage facility that includes distribution independent from a pipeline system, and fuel for mobile application [38]. It should be mentioned that some 40% of Georgian territory is lacking distribution systems for natural gas. LNG could be an alternative to developing a full coverage pipeline system, the latter being far more expensive. In remote mountainous areas that especially lack biomass LNG would be an adequate solution.

The creation of strategic reserves of liquid fuels for power generation in the existing storage facilities is also highly important, especially in force-majeure situations when gas delivery from a single supplier is expected to be interrupted for technical or other reasons. The total capacity of Georgian fuel oil reservoirs equals to 300 kt. By rational utilization this corresponds to more than 1 TWh electricity (planned efficiency of existing generating units of Georgian thermal power plants is 30 % whereas the current average efficiency is 26 %). The 300 kt capacity includes:

- 40 kt from so-called "receiving reservoirs" that are already rehabilitated (140 GWh electricity potential);
- 140 kt from underground reservoirs (490 GWh electricity potential);
- 120 kt from aboveground reservoirs (420 GWh electricity potential).

The utilization of strategic hydro-energy sources may also play a significant role. Table 4 shows the accumulation potential of Georgian hydropower plants. A rational utilization of the existing potential of hydro resources will result in more than 850 GWh electricity. This will significantly improve the country's energy security situation.

Table 4. The potential of strategic hydro resources

HPP	Installed Capacity, MW	Payload Volume of Reservoir Mm^3	Specific Discharge, m^3/kWh	Potential Storage Capacity, GWh
Enguri	**1300**	**667**	**1.26**	**529.4**
Vardnili	**220**	**37**	**7.02**	**5.3**
Khrami	**113.5**	**195**	**1.51**	**129.1**
Jinvali	**130**	**370**	**3.24**	**114.2**
Shaori	**38.4**	**64.5**	**1.07**	**60.2**
Dzevrula	**80**	**37**	**1.67**	**22.2**
Lajanuri	**111.8**	**12.2**	**3.2**	**3.8**
TOTAL				**864.2**

The following may be summarised in brief. Adequate energy management could significantly improve the Georgian energy security. This includes the utilisation of local energy sources, diversification of imported fuels, introduction of modern energy technologies, and accumulation and rational utilization of strategic reserves. This further requires that:

- On a short-to-medium term basis the main priorities of Georgia pertaining to the energy security strategies are a) diversification of imported natural gas supply and b) rehabilitation and renovation of the hydropower sector;
- The dominating elements of the long-term energy policy of Georgia appropriately reflect the development of renewable energy systems - mainly hydropower projects, and the implementation of high-efficient gas power cycles (CCGT);
- In order to avoid expected interruptions and obstacles special precautions are made for: a) underground gas storage capacity, b) modern small scale technologies for LNG production and utilization, c) strategic planning of water reservoirs and liquid fuels reserves.

7. CONCLUSIONS

1. The export of Caspian hydrocarbon sources to the international market can be secured at minimum expenses through South Caucasus and Turkey.

2. The diversification of natural gas supply and the development of supplementary infrastructure are the major strategic issues for the transit countries. Consequently, the energy security will be increased. This will support the integration of the transition economy of these countries into global structures.

REFERENCES

1. Pierre-Rene Bauquis, La Revue de l'Energie, # 509, September, 1999
2. Survey of Energy Resources, World Energy Council, 2001
3. Natural Gas Information, IEA Statistics, 2001
4. Oil & Gas, World Market Overview, www.tradepartners.gov.uk
5. Warren R. True. Push US natural gas construction plans. Oil & Gas Journal, Sept. 3, 2001
6. Graham Weale, The 2002 European Gas Storage Study, Prepared by The DRI-WEFA Global Energy Practice
7. Natural gas Information, IEA Statistics, Part III, 2002
8. Centre for Global Energy Studies, 2002
9. V.Skokov, Sviazannye odnoy truboy. WWW.Neftegaz.Ru, September 26, 2001
10. James P. Dorian, Oil, Gas in FSU Central Asia, Northwestern China. Oil & Gas Journal, September, 10,2001
11. BP statistical review of world energy, June, 2002
12. Caspian Oil and Gas, The Supply Potential of Central Asia and Trans Caucasus, IEA, 1998
13. Orhan Degermenci, EU Study of Caspian Area oil, gas pipelines compares routes, costs. Oil & Gas Journal, November 5, 2001
14. Platt's Oil gram news, May, 23, 2000
15. M.Foss, G.Gulen, B Shenoy, Caspian Pipeline prospects hinge on transparent. Oil & Gas Journal, August 21, 2000
16. T.Gochitashvili, Natural Gas to Europe XXI Century through Georgia. Center for Strategic Research and Development of Georgia, Bulletin # 13, 2000
17. Russian Energy Strategy up to 2020, www.mte.gov.ru
18. Y.Kristalyov, Statoil confident on sales. Caspian business News, June 9, 2003
19. Production of gas: prognosis and scenarios, October, 2001, WWW.Neftegas.Ru.
20. Cenk Pala, BOTAS, Oil and gas transportation – recent developments and future potential. Materials of International Conference GIOGIE 2003, Tbilisi, March, 2003
21. Study of Observatoire Mediterraneen de L'energie (OME) institution
22. T.Gochitashvili, L.Kurdgelashvili, Comparative analyses for Russian and Caspian natural gas export to Europe, Proceedings of 25th IEEA International Conference, Aberdeen, June, 2002, http://www.iaee.org
23. D. Snieckus, Europe steps on the gas. Oil & Gas Journal, September 4, 2000
24. Black Sea Energy Survey. International Energy Agency, 2000
25. Jan H. Kalicky, Hige stakes hinge on Russian's energy choices. Oil and Gas Journal, March, 19, 2001

26. S.Shatvoryan, Investigations and strategies for security of supplies in Armenia, Proceedings of NATO ARW "Security of Natural Gas Supply through Transit Countries", Tbilisi, May, 2003
27. M.Christensen, Undergro8und gas storage in Georgia, Proceedings of NATO ARW "Security of Natural Gas Supply through Transit Countries", Tbilisi, May, 2003
28. Chitadze N., The Future of the Oil and Gas Industry in Georgia. Georgian Times, February 4, 2002
29. Oil Companies Operating in Georgia, GNOC Information. Materials of Conference "Perspective of Oil Production and Refining", Tbilisi, October, 2000
30. E. Kakaras, G. Skodas, P. Amarantos. Implementation Possibilities of FBC Technologies in Power and Heating Systems in the Caucasus Region. Materials of Workshop "Promotion of Decentralized Cogeneration Units", Tbilisi, 2002
31. The Study of Natural Energy Resources in Georgia, Report, EEC Georgia, 2000
32. Technical Assistance on the Production and Consumption Level, Component 4, Wood Resources, Report, EU TACIS Project GEE/95/92, Tbilisi, 1999
33. Thermal Waters of Georgia. Georgian Geothermal Association. Tbilisi, 1998
34. Capacity Building to Assess Technology Needs.., Report, National Agency on Climate Change, Tbilisi, Georgia, 2002
35. Georgian Ministry of Economy, Black Sea Energy Centre Review, TACIS EEC, 1998
36. Gochitashvili T. Transit Perspectives and Utilization Problems of Caspian Gas in Georgian Energy Sector, Proceedings of GIOGIE 2002
37. David Rogava. Proceedings of NATO ARW "Security of Natural Gas Supply through Transit Countries", Tbilisi, 2003
38. Hetland J., Advantages of Natural Gas Over Other Fossil Fuels. Natural Gas Technologies, Opportunities and Development Aspects, OPET-International Workshop Papers, Vaasa, Finland, May, 2002

POWER SYSTEM SITUATION AUTUMN-WINTER 2002-2003 AND ASPECTS OF ENERGY SECURITY IN GEORGIA

John W A COKER
BP-Statoil

Abstract: This article deals with a planning study of electrical power supply in Georgia for the winter season 2002–2003 carried out by a team set up by the Georgian energy ministry. A mandate was given to specifically address the probable situation and to indicate actions deemed necessary to mitigate the power crisis. Crucial issues and estimates developed in this study are presented herein. As appropriate energy supplies are vitally important for the security of the country it was recommended that the highest level of the Georgian government review the financing programme for energy assistance. A severe and most urgent need for maintenance and repairs in the entire Georgian energy sector was deemed highly critical. Furthermore, the need of securing funding of fuel supplies was addressed. Eventually, as financing appeared to be a major obstacle, a recommendation was given to consolidate the deficit of the Georgian energy system in the state budget. Furthermore, directions were given to establish new policies - backed by Georgian legislation - pertaining to the collection of funds from consumers and even disconnection of non-payers.

Key words: Georgian electrical power, energy deficiency, economic reform

1. INTRODUCTION TO THE GEORGIAN ELECTRICAL POWER SYSTEM

Since the early 1990-ies a serious situation has developed in the fuel and energy sector in Georgia that is worsening due to various reasons. The first one is related to a chronic deficiency of the in-country energy resources, and

J. Hetland and T. Gochitashvili (eds.),
Security of Natural Gas Supply through Transit Countries, 37–45.

secondly the slow pace of economic reform of the electrical power supply system. This has lead to a severe shortage in electrical power supply to the population especially in the winter period. The lack of confidence in the energy sector coupled to a generally bad economical situation in the country has lead to a serious situation with a large number of unpaid electricity bills. This in turn has resulted in the reduced domestic generating capacity through under-investment in the sector, which makes the system more dependent on imported power with the associated risk of the power being cut off. In addition the lack of investment has also meant that essential maintenance and repairs have not be carried out so that units still in operation are approaching the point of total collapse.

The situation was spiralling downwards with actuate cash flow problems the early part of the winter 2002-2003. If the necessary funds needed to cover the cash shortfalls were made available it is believed that the downward spiral could be halted, and the system would start to recover. However, without the appropriate financing the prospect for the energy sector – and thereby the population and the industry – is deemed rather dismal, and the effect on the social structure of the country should not be underestimated.

It was found that the lack of financing for the winter 2002-2003 would also have a serious and negative effect on the recent efforts of the Georgian government - and the donor organizations - in terms of restructuring the economic bases of the power system. In addition to funding - per se - it was deemed necessary to implement a radically different management approach to any such funds in order to ensure that the allocations of money for the subsequent winter (2003-2004) were focused on actions that assumingly would give the best curing effect on the cash flow situation of the energy sector.

It was further recommended that the changes in management structure should cover all areas of the power sector from the government supervision through generation and transmission, to the wholesale market – including the power distribution companies. It was believed that with the correct focus a recovery of the electric power sector could get started. However, without the strongest influence of the highest level of the Georgian government, this is conceived to be unlikely.

2. ASSESSMENT OF THE WINTER 2002 –2003

In order to plan the power supply for the winter 2002–2003 an expert group was set up by the Georgian energy ministry with the mandate of assessing the probable situation and indicating actions that were necessary to mitigate the power crisis. The figures below result from the estimates developed by this group.

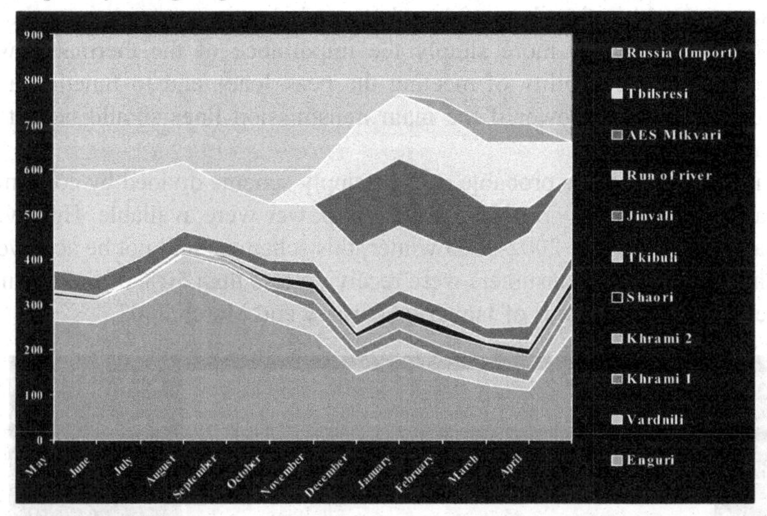

Figure 1 Estimated impact on the power supply in GWh (or million kWh) of possible generating sources and imported power needed to balance the Georgian electrical system and to provide 24 hours of power to Tbilisi and 8 hours to districts outside Tbilisi.

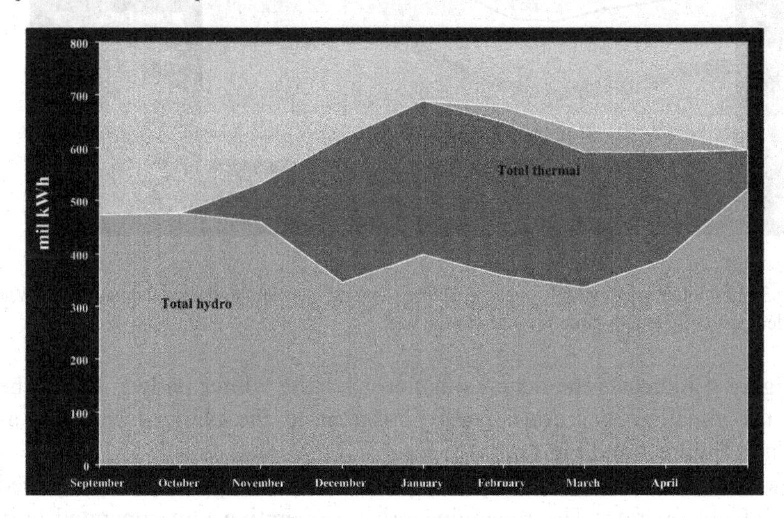

Figure 2 Indications of the importance of the Georgian thermal power plants and their capability of meeting peak loads and functioning as a backup to imported electrical power in cases where the main transmission lines fail.

Figure 1 shows the potential power supply of possible sources of generated and imported power that may contribute to balancing the Georgian electricity system and to provide 24 hours of power to Tbilisi and 8 hours to the districts outside the capital. To achieve this target it was found that the thermal power plants would need extensive refurbishment. Unfortunately, this was not achieved in front of the winter period.

Figure 2 indicates more simply the importance of the thermal power plants and their capability of meeting the peak loads and to function as a backup to imported power if the main transmission lines should be out of operation.

Figure 3 indicates a probable energy supply scheme divided by consumer and assumed demand - provided sufficient power were available. However, as was the case for the 2002-2003 winter, this scheme could not be achieved. On the contrary, the consumers were receiving less than 50% of the planned power in the peak periods of January, February and March 2003.

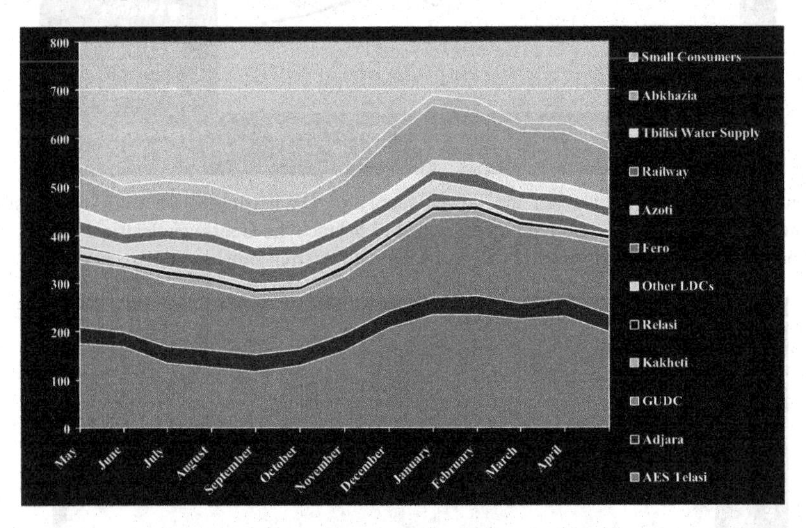

Figure 3 Plausible energy supply scheme divided by end-user and assumed demand (in GWh) provided power were available throughout the year.

Figure 4 indicates the actual situation over the winter period. As can be seen the situation was considerably different to the planned situation as shown in Figure 1 and Figure 2.

As the state-owned thermal power plants were not appropriately repaired they did not operate. The remaining private operative units operated only part time - mainly due to the lack of funds to purchase the fuel gas. The imports from Russia and Armenia were at a maximum during the winter 2002-2003. Hence, considerable energy debts were built up, that needed to be paid during the summer 2003.

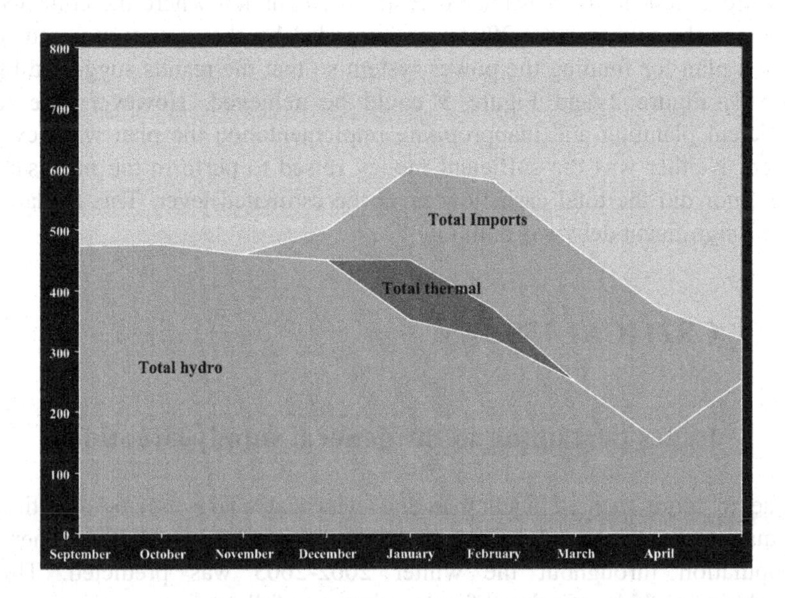

Figure 4 Actual power consumption in GWh over the winter period 2002-2003 divided by source (hydropower, thermal power and imported electricity).

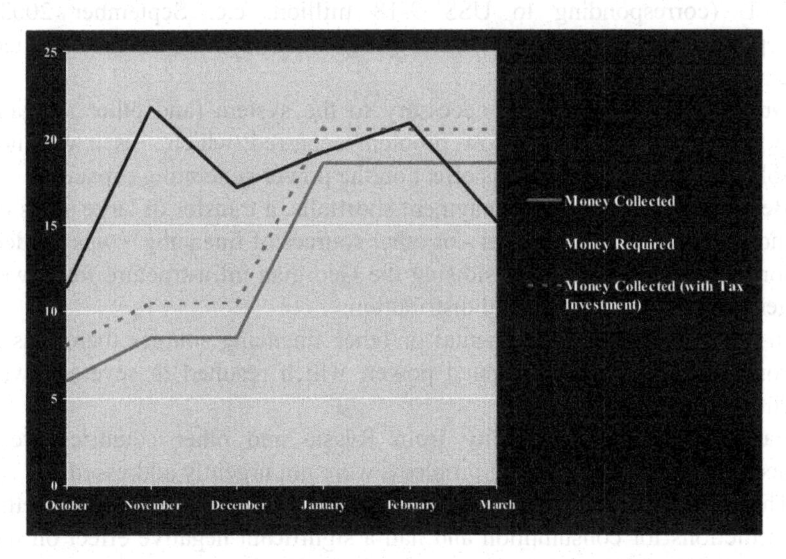

Figure 5 Estimates of the cash flows in GEL deemed necessary for funding the power system and to comply with the results of Figure 1, Figure 2 and Figure 3. (1 GEL corresponds to US$ 0.46 – economic conditions as per September 2003)

Figure 5 indicates the estimated cash flow situation where the collected money for the winter 2002-2003 was exceeded by the money required to pursue a plan for funding the power system so that the results suggested by Figure 1, Figure 2 and Figure 3 could be achieved. However, due to insufficient planning and inappropriate implementation the plan was never realized. Neither was the sufficient money raised to perform the necessary repairs, nor did the total cash flow meet the estimated lever. This explains why the significant debt was built up.

3. CRITICAL ISSUES

3.1 Issues pertaining to the general supply situation

- Due to large cash deficiency in the electrical energy sector a critical situation for the supply of electricity to the industry and the general population throughout the winter 2002-2003 was predicted. This tendency could be stipulated for the winters to follow.
- The estimated yearly deficiency was assumed between 20 and 40 million GEL (corresponding to US$ 9-18 million, e.c. September 2003) depending on the financial situation regarding payment for the supplied electricity.
- Due to the urgent repairs necessary to the system (and other relevant expenses) an acute cash flow problem occurred, which - as it was not solved - resulted in a serious effect on the power generating capacity.
- Because of the anticipated payment shortfalls, a transfer of large sums of money from the state budget - or other sources of financing – was needed for the sole purpose of subsidising the Georgian infrastructure for power generation, transmission and distribution.
- In the absence of governmental or other financing sources there was a considerable lack of generated power, which resulted in severe power cuts and rationing.
- Large imports of electricity from Russia and other countries were necessary, as the generating problems were not urgently addressed.
- The potential lack of generating capacity further reduced all the predictions for consumption and had a significant negative effect on the potential revenues.
- The energy balance that was developed as the basis for this report assumes a full wholesale supply to Tbilisi, and an approximated eight-hour per day supply to the Georgian regions. It was not expected,

though, that 100 percent payment for the wholesale electricity deliveries could be achieved during a reasonable forecast period.

3.2 Technical issues

- In order to perform the minimum of repair work several power plants such as the Gardabani state thermal power plant (Tbilsresi), and most hydropower and transmission systems were in an urgent need of funding. Should money not be raised these plants were deemed incapable to generate and transmit enough electrical power – and especially to respond to the peak load of the coming winters.
- The financing of the repairs before the winter 2002-2003 was stated critical. For the following winter the maintenance work should get started in the summer 2003 in order to be accomplished before the winter 2003-2004 when the demand goes high.
- The amount of electrical power available was reduced by 20% owing to the lacking repair in 2002. As there was no appropriate replacement for the deficient electricity the consequences turned out to become rather severe – not only for the Georgian industry but also for the public.
- Consequently the reliability of the total transmission system was very low. This resulted in partial or total failures of the electrical supply to Tbilisi and the surrounding regions. During the summer 2003 the failure of the transmission system could be taken as an example of ongoing deterioration of the system. The reliability was considered critical as it was expected to worsen until the subsequent winter.

4. LESSENS LEARNED FROM THE WINTER 2002-2003

The winter 2002-2003 can be used as an eminent example to explain the vulnerability of the Georgian power sector. The situation in terms of managerial and operational aspects as well as the general supply situation can easily be extrapolated to the coming winters. As appropriate funds before the winter 2002-2003 were not allocated to the repair of the thermal power plants Georgia suffered severely from power losses – especially when imported power was disrupted. Some important lessons chould be learned from the situation, and the energy sector and the authorities of Georgia should take relevant actions basically pertaining to funding, managerial and policy issues:

- If funds are not allocated to purchase heavy fuel oil the country may continue to suffer from total losses of thermal power generation. This may be rather detrimental to the society in events of disrupted imports of natural gas supplies.
- If necessary repairs are not carried out on the transmission system considerable power outages will occur due to system failure.
- Furthermore, the thermal power plants should be started as planned.
- The hydropower reservoirs should be managed correctly. During the winter 2002-2003 the reservoirs were depleted approximately two months earlier than planned.
- It is important to balance the electrical system. In the winter 2002-2003 the power system was timely out of balance. This resulted in severe deviation in transmission voltages and frequencies that were about 10% under the nominal design level, which in turn made the system unreliable.
- Due to the lack of power supplies, the expected income of the energy sector could not be achieved. And, the extended imports of electrical power are prone to build up debts that may reduce the funding of essential repairs necessary to safeguard a stable operation. Hence, without exceptional efforts during the summer 2003 the situation in the winter 2003-2004 was assumed to become worse than the preceding winter.

5. CONCLUSION

- As appropriate energy supplies to the population and industry are vitally important for the security of the country, it is essential that the highest levels of the Georgian government review the financing programme for energy assistance.
- Financing of the deficiency in the Georgian energy system should be considered consolidated in the state budget – at least on a temporary basis.
- An urgent need to carry out essential repairs in the entire Georgian energy sector has been envisaged.
- Recommendations have been given to establish an energy-monitoring group to supervise the financial actions of the energy sector with reporting to a high level in the government.
- Recommendations were also given regarding a rigorous policy on a maximum collection of funds from consumers – and also the disconnection of non-payers. Successful implementation of this policy would need the full backing of Georgian legislation.
- The management of the Georgian hydropower reservoirs should be rigorously continued to maximize the stored energy.

- Finally, a source for the funding of the Abkhazian energy demand has been deemed necessary, as this particular energy supply amounts to some 14% of the total Georgian energy usage.

ACKNOWLEDGEMENTS

The results, conclusions and recommendations in this paper are based on the information received from

- The Georgian Ministry of Fuel and Energy
- The Georgian Wholesale Energy Market (GWEM)
- AES companies in Georgia
- United Distribution Company of Georgia
- Hydropower stations of Georgia
- Tbilsresi

CURRENT STATUS OF NATURAL GAS SECTOR

Investigations and Strategies for Security of Supplies in Azerbaijan

Ganifa ABDULLAYEV
Nasiraddin Tusi LTD

Abstract: The article addresses the existing gas industry infrastructure of Azerbaijan, current and perspective volume of natural gas supply and demand, consumption and processing. It reviews the current and perspective volumes of the production of natural and associated gas from oil and gas fields developed by foreign oil companies in accordance with production sharing agreements. Perspective of gas export and the assessment of the role of underground gas storages in realization of export potential are presented in this article.

Key words: Pipeline, Gas, Supply, Demand, Transmission system, Transportation, Underground storage, Treatment, Import, Export.

1. HISTORY

Azerbaijan was an early developer of the gas sector, using natural gas in industry of the Absheron area from 1859. The gas industry progressed and extended from industrial to domestic gas supply, first in Absheron then it extended throughout the country as the import and export pipelines network was completed to deliver Iranian gas supplies to the Caucasus and to transport Azeri and Russian gas to Armenia and Georgia.

At the beginning of the 1990's, some 85% of towns and villages in Azerbaijan were supplied with natural gas. In 1991, Azerbaijan's total gas consumption was about 17 bcm p.a., made up as follows:

- Power generation - 5.5 bcm;
- Industrial consumers - 6.0 bcm;
- Communal consumers, including district heat - 3.1 bcm;

J. Hetland and T. Gochitashvili (eds.),
Security of Natural Gas Supply through Transit Countries, 47–60.

- Domestic consumers - 2.4 bcm.

After break-up of the Soviet Union and the ensuing regional conflicts, gas imports from Russia, which superseded those from Iran, were suspended and gas delivery to Armenia and Georgia ceased. During the 1990s, Azeri production declined and there was insufficient gas available to supply distribution consumers outside the Absheron area. Hence the main transmission pipeline network and the unused distribution system fell into disrepair.

2. CURRENT NATURAL GAS DEMAND

Currently, natural gas imported from Russia is mostly consumed for power generation at Ali Bayramli and Mingechevir. The gas market in Baku and the Absheron area is supplied by SOCAR and Production Sharing Agreement (ACG) on gas production.

At the end of 2002, Azerbaijan's daily natural gas consumption (in Mcm/d) was:

- Baku Domestic and Residential 9.3;
- Power Generation 14.0;
- Industry 2.0;
- **Regional gas consumption 4.0;**
- TOTAL 29.3.

3. CURRENT NATURAL GAS SUPPLY

Local Onshore and Offshore gas (about 4.5 bcm p.a.).

Traditionally, onshore gas is consumed on the gas markets, located in or near the production area. Onshore gas is typically fed into the local gas distribution network without any treatment other than separation.

Offshore gas is supplied to consumers either treated or untreated, depending upon its landing point. Higher-pressure gas, about 60% of the total gas supply, is forwarded to the Gas Processing Plant at Karadagh. Lean gas from the GPP is supplied either to:

- Baku and Absheron consumers in winter; or
- The Underground gas storage facilities at Karadagh and Kalmas in summer.

Lower-pressure gas from offshore is forwarded to local collection centres and enters the gas distribution system without any treatment other than separation.

In winter, gas is supplied to the Baku and Absheron consumers and to the Ali Bayramli power plant from the underground gas storage facilities. Some of the gas supplied from storage in winter period is imported gas, which is put into storage in summer.

4. IMPORTED GAS

Gas supplies from Russia were resumed in November 2000 and, at the beginning of March 2002, daily flow is 10.2 Mcm/d, about 4.0 bcm p.a.. Gas imports are expected to increase to 16 Mcm/d in the winter of 2003/2004, as a result of increased use of gas for power generation and resumption of supplies to regional distribution networks.

5. FUTURE NATURAL GAS SUPPLY

5.1 SOCAR production

In general, SOCAR's production of both associated and non-associated natural gas will remain fairly until around 2010, when it will start to decline.

In future, SOCAR will receive natural gas from the Government share and its participating interest in the Azeri-Chirag-Guneshli (ACG) and in the Shah Deniz developments. This is discussed below under PSA Production.

5.2 PSA production

5.2.1 General

SOCAR is a partner in the PSAs for the Azeri-Chirag-Guneshli project and the Shah Deniz Project. As a result of the PSAs, SOCAR will receive gas for its own use or sale. We understand that, in the shorter term, it is proposed that the natural gas will be sold to Azerigaz for use within Azerbaijan. In the longer term, natural gas may be sold to other consumers in Azerbaijan or other countries.

5.2.2 Azeri-Chirag-Guneshli (ACG) Project

The Azeri-Chirag-Guneshli reservoirs contain crude oil with a large volume of associated natural gas. The development of these reserves, have significant implications for the management of Azerbaijan's natural gas resources.

Gas produced by the Early Oil Production (EOP) project is currently transported via Oil Rocks and the Bahar gas field to the Gas Processing Plant at Karadagh. Constraints on the 400 mm pipeline limit the flow of gas to 2.8 Mcm/d, such that approximately 750,000 m^3/day is flared. This will continue until the start of gas deliveries from the ACG Phase 1 development, planned for 2005.

With the commissioning of ACG Phase 1, natural gas will be re-injected to maintain reservoir pressure, with the balance transported via a new 700 mm gas pipeline to Sangachal, where it will be treated to a marketable quality for delivery to SOCAR. In the first quarter of 2005, the gas delivered at Sangachal terminal is anticipated to be some 0.8 Mcm/d, rising to 3.6 Mcm/d by the first quarter of 2006. From the first quarter of 2006, ACG gas will no longer be delivered via 400 mm pipeline to Sangachal terminal, where it will be treated and transferred at a pressure of 39 to 41 bar. All ACG gas delivery will thenceforth be from Sangachal, except in the event of an upset in offshore equipment, when it may alternatively be delivered via 400 mm pipeline to Karadag Gas Processing Plant.

With the commissioning of the ACG Phase 2 development, planned for 2006, gas deliveries to SOCAR will be in the range of 2.9 to 4.3 Mcm/d, rising in Phase 3 to about 19 Mcm/d in 2010 and reducing to about 12.9 Mcm/d in 2015.

The delivery pressures to SOCAR from the ACG Sangachal Terminal are considerably in excess of the presently available operational pressures in the Azerigas system to and from the Karadagh node. SOCAR has a contractual obligation to take gas from the ACG Sangachal Terminal and must have these gas receipt facilities available 3 months in advance of the start of Phase 1 oil production. Thus, the gas receipt facilities must be operational by the end of the third quarter of 2004 and this will become SOCAR project.

Options for transportation of ACG gas are:

- storage in the Karadag and Kalmas underground storage facilities;
- supply to the Baku and Absheron gas system;
- transportation in the transmission system.

Considerable modification and upgrading of the existing gas storage and delivery facilities is necessary to permit the safe and reliable receipt of ACG gas, including safety and control systems.

All the above projected volumes and arrangements are set down in the Gas Protocol, under the terms of the Production Sharing Agreement, but which can be amended through mutual agreement as development plans and options evolve.

Under the terms of the Gas Protocol, all ACG natural gas delivered to SOCAR has priority over all non-associated natural gas, including both Shakh Deniz and imports. This measure is designed to maximize continuing oil production, to the benefit of the partners and the Government, in the event of the loss of the ability to re-inject associated natural gas production offshore. The management of this requirement effectively, implies a need for significant planning and investment.

5.2.3 Shah Deniz project

Shah Deniz is an offshore non-associated gas/condensate field, which is being developed primarily to supply natural gas to export markets in Turkey – and associated liquids to international markets via the crude oil pipeline systems. Production will be brought ashore at Sangachal, where a major gas terminal is to be constructed. In November 2002, the agreement was reached with BOTAS, the Turkish state-owned gas company, for the sale of Shah Deniz gas to Turkey with deliveries commencing in August 2006.

The gas sales agreement for the gas supply to Turkey is presently between BOTAS, as the "buyer" and SOCAR. At the start of gas production SOCAR will transfer its interest to the Shah Deniz Gas Company – which will become the "seller".

The prime customer is BOTAS, the gas and oil pipeline entity of Turkey, which is scheduled for privatisation. Turkey will buy Shah Deniz gas on an 80% annual Take or Pay contract with daily swings at the buyer's option. The sale point for Turkey is at the Turkish frontier.

Shah Deniz gas will be exported via a new 1050 mm (42") diameter pipeline, with an operating pressure of 90 bar. Phase 1 gas flow of 7.1 bcm p.a. will be delivered to Turkey and Georgia, without recompression, at a pressure of about 55 bar – where it will be recompressed by BOTAS to meet the required pipeline operating pressure in Turkey. Recompression will be added in Azerbaijan for the export of Phase 2 volumes of 16 bcm p.a.. The maximum flow potential of the export pipeline is 22 to 24 bcm p.a. with additional recompression.

The Shah Deniz PSA partners do not envisage any other off take points along the export pipeline's route through Azerbaijan.

Georgia will receive gas in lieu of transit royalties and will also have the option to purchase additional gas. The gas delivery point to Georgia is at the

Georgian frontier with Azerbaijan – that is the first connection point after Sangachal.

Azerbaijan, represented by SOCAR, will buy 1.5 bcm p.a. with a further 1 bcm p.a. available as an optional amount following the agreement by both parties. Under the terms of the Shah Deniz-Azerbaijan Gas Sale & Purchase Agreement (SPA):

- the first 1 bcm of the 1.5 bcm is to be delivered at rateable volumes – relatively flat within agreed variations;
- the second 0.5 bcm is for delivery at seller's option;
- the 1.0 bcm, if agreed, would also be for delivery at the seller's option and therefore almost certainly delivered during summer when heating demand is low – thus requiring storage facilities.

The Shah Deniz PSA Partners see the above arrangements to mean that SOCAR will adjust its receipts of natural gas under the SPA, in order to manage:

- the daily swings in deliveries to Turkey; and
- any annual swing on Turkey's 80% Take or Pay.

In addition, it is in the interests of Azerbaijan's balance of trade and fiscal budget for the additional 1.0 bcm of optional natural gas (together with its associated liquids) to be produced and put into storage in summer – and used in winter instead of imported gas or exportable oil products.

While the Shah Deniz PSA partners are clearly concerned about the implications to their funds flow from any failure by SOCAR to manage the swings in natural gas volumes for Azerbaijan, they are reluctant to add a significant investment in the underground storage facilities at Kalmas and Karadag.

6. INFRASTRUCTURE ASSESSMENT & EXPORT INTEGRATION

6.1 Offshore pipelines

The main existing and proposed offshore natural gas pipelines include:

- The 400 mm pipeline currently used to transport ACG EOP natural gas to Oil Rocks and on to the Bahar gas field.
- The pipeline from the Bahar gas field to the Gas Processing Plant at Karadagh.

- The pipeline from the Bulla-Deniz gas field to the Karadagh GPP via Sangachal.
- The pipelines, which deliver natural gas from offshore production to the Absheron peninsula (at Zirya and one other point), where it enters the Baku region distribution system untreated.
- A new 700 mm pipeline, following, the commissioning of ACG Phase 1, which will transport ACG natural gas not re-injected into the reservoir, to Sangachal.
- A new pipeline to transport separated natural gas from the Shah Deniz development to the new processing plant at Sangachal.

6.2 Azerbaijan gas processing plant at Karadagh

The Azerbaijan Gas Processing Plant, located at Karadagh, is a multi-train extraction unit based on the lean oil technology. Of the six extraction units, a number are now in disuse. The fractionation unit facility is currently employed on the batch run basis to produce a stable condensate product and a butane rich LPG.

Currently, two offshore gas pipelines supply the Karadagh facility. One pipeline is from Bahar and Oily Rocks, which links up with the associated gas pipeline from the Chirag Early Oil Project. The other pipeline is from Bulla-Deniz gas field via Sangachal, where other raw gases join the line.

Current supply gas pressure to the Karadag Gas Plant is understood to be of the order of approx 20 bar. Reportedly, the reduction in inlet pressure arises from depletion of the wellhead pressures at various supply fields. The reduced supply pressure has a negative impact on plant extraction efficiency and plant capacity.

Further, the treated gas outlet pressure is limited by the allowable operating pressure (typically 11-15 bar) of the twin transmission lines, which supply Baku from the Karadagh Gas Distribution Station. This station is supplied by pipeline with treated lean gas from the Gas Plant. Gas from this station can also be directed to the storage facilities of Karadag and Kalmas.

The liquid products from the GPP, butane and condensate, are piped to nearby storage and shipping facilities.

In 1996, the World Bank funded a study by John Brown Engineering, to examine the Karadagh GPP. The study recommendations included proposals for a new, two stream facility comprising modern technology, turbo expanders, and feed gas compression.

The replacement plant, 2 x 2.5 bcm p.a. units, had a budget cost estimate of US$120M at 1997 prices. There are no confirmed plans at the present time to proceed with this project.

6.3 Small scale gas treatment

SOCAR now supplies gas to Azerigaz at over 50 discrete locations. In general, the only treatment upstream of these transfer points is physical phase separation of gas and liquids. The Karadag Extraction Plant is the only facility of its kind within the Azeri system, which positively adjusts hydrocarbon dewpoint and extracts gas liquids.

SOCAR has investigated small-scale gas treatment plants, capable of operation at low pressures, and Azerigaz has conducted studies into acquiring 'mobile' treatment units which could be positioned at the more significant of these supply points.

The overall issue of gas quality within the Azerbaijan transportation system will require policy development prior to decisions on infrastructure investment.

6.4 High pressure pipelines

The main transmission pipelines are shown in picture 1.

The gas pipelines from the Russian Border via Siazan to Kazi Magomed; Kazi Magomed to the Georgian Border; and from Alty Agach to Aksu were constructed in the period 1982-86.

Earlier pipelines existing on the Kazi Magomed to Georgia Border segment but Azerigaz no longer view these as main transmission pipelines. The “old” 1000 mm (40”) transmission pipeline from Kazi Magomed to the Georgian border, to which all the regional spur-lines are connected, is now used to transfer natural gas from the main transmission pipeline to the spur-lines.

In view of the lack of gas imports from Russia in the period 1994-2000, the transmission pipelines were largely unused with national gas production generally directed into the Baku/Absheron areas. During this period, the Cathodic Protection (CP) stations on these lines became defunct.

When the transmission pipelines were re-commissioned, following the commencement of gas imports, operation at their original design pressures of 55 barg was not possible. Ruptures and leaks were experienced on the pipelines, and the maximum permitted operating pressure was reduced.

The introduction of imports from Russia, and the progressive incremental flow/pressure requirements on the pipeline system, have necessitated a dedicated review/ refurbishment program. Supply to the Power Plants at Mingechevir and Ali-Bayramli has been the prime objective.

The program is reportedly 90% complete with 100% planned by summer 2002. Progressive increases in import flow/pressure are:

- 7 Mcm/d / 15-20 bar Nov 2001;
- 10 Mcm/d / > 20 bar current Feb 2002;
- 16 Mcm/d / 35 bar planned.

The methodology and highlights of the rehabilitation program are as follows:

- Field Testing including 'Electrometric Testing'.
- Excavation of bellholes and visual inspection at locations deemed problematic as a result of soil aggressiveness, and the interpretation of the test results.
- Repair or replace pipe as necessary, or clean and rewrap.
- Total replacement length of pipe was 5 to 6 kms, at various locations along the refurbished 230km section. New pipe was used in the replacement work.
- CP facilities were rebuilt. 40 stations are to be re-built by March 2002.

For the future, Azerigaz has implemented an on-going rehabilitation review and refurbishment program. Many of the Azeri lines were not constructed in the manner, which permits pigging. Problematic areas include branch design, internal pipe extensions, branch arrangement not suitable for pig passage, certain elbow and bend characteristics, intermediate pipe diameter change, and lack of launchers/receivers. Nonetheless, Azerigaz proposes to implement a program of intelligent pigging where circumstances permit. It is understood that the work program has been outlined with Pipeline Integrity International (PII), but that the participation of a foreign investment partner might facilitate this work.

6.5 Compressor stations

There are 5 compressor stations on the transmission system, at:

- Shirvanovka
- Siyazan
- Kazi Magomed
- Agdash
- Kazakh

These facilities have not been used since 1994, when supplies from Russia were halted.

Although the compressor stations are not required under present throughput plans and pressure profiles, it has been estimated that the budget cost for station reconstruction is US$60 million.

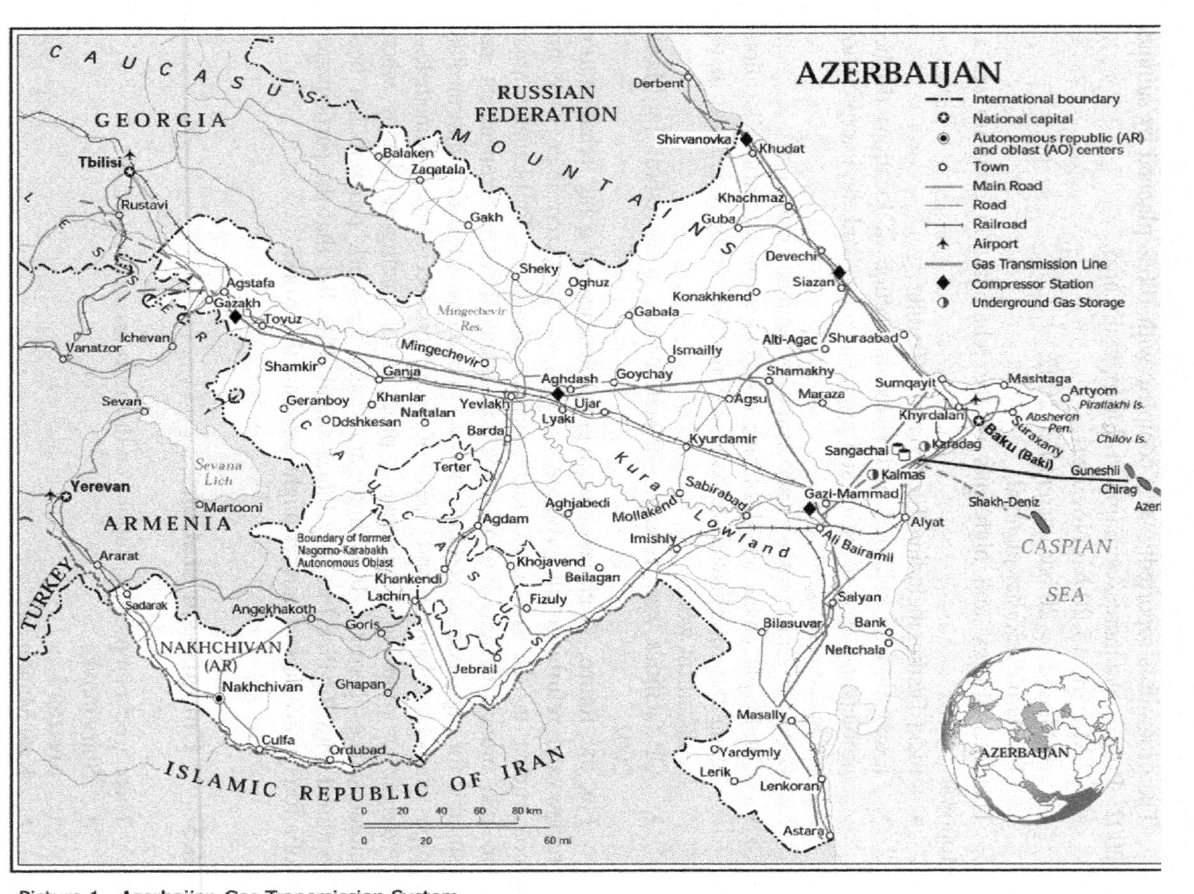

Picture 1. Azerbaijan Gas Transmission System

6.6 SCADA system

A SCADA System for the transmission system has been proposed to enhance operational requirements and safety. The SCADA system has been estimated to cost 15 million Euro.

When ACG and Shah Deniz production commences, the Azerigaz transmission system will be required to manage a number of changing flows and pressures, potentially including:

- rapid and large changes in the volume of natural gas delivery at Sangachal;
- a variety of supply pressures;
- varying pipeline gas supply from Russia;
- the volume dispatched from gas storage facilities; and
- fluctuating off take by major consumers, for instance as CCGT power stations ramp up.

Control facilities, and other projects, will have to be designed and implemented to enable the safe and effective management and operation of future, fluctuating natural gas supply and demand.

6.7 Gas meters - transmission system

The INOGATE Programme is funding the equipment requirements for metering equipment at the Russian and Georgian borders to a value of 2 million Euro.

These 2 metering stations at the Russian and Georgia border have already been constructed.

6.8 Karadagh-Severnaya gas pipeline

Construction of a 400MW power plant is now being completed at Severnaya/Simal, which is located 40km north-east of Baku on the Apsheron Peninsula.

A 150MW unit, built in 1954-60, exists at this site, fuelled by low-pressure gas and mazut. This unit is now due for decommissioning in the next 2-5 years.

The new power plant is financed from a Japanese OECF loan. There are plans to build a second 400MW power plant at the Severnaya site in 2-3 years subject to the availability of finance. It is understood that the new power plant requires fuel gas supply at approximately 26 barg.

Later, the gas will be supplied by a new pipeline from the Karadagh area, which is under construction. The concept of this new pipeline also provides for improving the reliability and efficiency of gas transportation from Karadagh to the Baku network, and for future, replacement supply to industry and a potential power plant at Sumgait. This new Severnaya Pipeline system, has a budgeted cost US$118 million, and includes the following elements:

- A Compressor Station located near the Karadag Gas Distribution Station. Two compressors will be installed initially, with space allowed for a further 4 machinery sets.
- Discharge pressure at Karadag of 38 barg.
- Approximately 45km, 40" underground pipeline from Karadagh to Digah. From Digah to Severnaya the pipeline size would reduce to 20" for the 45km distance.
- Connection point provided for future gas supply to Sumgait, where reconstruction of electrical generating facilities are under planning. Alternative locations of this connection are under consideration.
- Digah Connection - New line from Digah to Sumgait, under review with design conditions as the new Karadag - Severnaya pipeline.
- Southern Connection - providing a link via a pressure reduction station into the existing Karadag - Sumgait pipeline system (twin lines 500 and 700mm diameter constructed in the 1960s) and operating at approx 12 barg. The interconnecting pipeline length for this option would be approximately 3km.
- Five offtake stations to supply gas into the Baku Gas distribution system.
- Additional future offtake facilities are under consideration.

7. NATURAL GAS STORAGE

There are existing, depleted field, natural gas storage facilities at Karadagh and at Kalmas. Both facilities require significant investment to operate at their full potential. Studies have been conducted on these systems, and budget costs developed.

Details of active gas enhancement, and budget prices are included in Table 1 below. The total required investment is of the order of US$272M for both sites.

Under the contractual arrangements for the ACG and Shah Deniz field developments, SOCAR has assumed the major responsibility for managing the volumes of natural gas delivered to the Azerbaijan market, including the potential for significant, short notice variations in the volume. In effect, this responsibility extends to include the ability of Azerigaz' transmission system

Table 1 Main Technical-Economical Indicators for Reconstructing Kalmas and Karadag Underground Gas Storage Facilities

Characteristics	Unit of measure	Kalmas UGS			Karadag UGS				
		Before Reconstruction		After Reconstruction	Before Reconstruction		After Reconstruction		
		Original Design	Current State		Original Design	Current State	Total	Including 1st Stage	Including 2nd Stage
1. Start year of Operation	Year	1976			1986				
2. Total volume of gas reservoir - including volume of active gas	Billion m^3 Billion m^3	2.6 1.3	1.0 0.3	2.5 1.5	1.7 1.0	1.0 0.2	5.0 3.0	3.2 1.3	1.8 1.7
3. Number of wells	Item	101	101	101	35	35	50	35	15
4. Compressor Station - Number of Units - Type - Capacity	Item MW	16 Piston 17.6	10 Piston 11	3/8 Turbine 51	4 Turbine 30.7	5 Piston 5.5	5 Turbine	2 Turbine	3 Turbine
5. Daily Capacity - While extracting - While injecting to storage	Million m^3 Million m^3	10.0 8.0	2.5 2.0	15.0 12.0	10.1 8.3	1.2 1.0	28 21	14 10	14 11
6. Investment	US$ million			81.0			191.4	80.1	111.3

to continue to deliver large volumes of natural gas to the power stations at Ali Bayramli and Mingechevir. Any failure by SOCAR to manage the volumes of natural gas is likely to translate into reduced production and hence one or more of:

- increased gas imports;
- reduced gas exports;
- increased use of exportable liquid fuels;
- lost PSA Partner revenues, which are presumably provided for contractually; and
- most importantly, lost revenue to the Azerbaijan State.

There is an urgent need for SOCAR to conduct a risk analysis of its gas management responsibilities and, as a result, evaluate what it should spend on measures, including the use of storage, to mitigate the risks of reduced production.

8. CONCLUSION

a) The analyses of current supply and demand of natural gas in Azerbaijan are presented in the article.
b) Future sources and volumes of natural gas production are determined.
c) Current state of Azerbaijan natural gas industry and infrastructure of the main transmission pipeline are assessed.
d) The analysis of perspective of natural gas export given in the article.
e) Roles of underground gas facilities in securing and supply of local gas consumers with natural gas and its export are determined

ITALIAN EXPERIENCE PERTAINING TO NATURAL GAS IN TRANSPORTATION AND DEVELOPMENT; RATIONALE FOR THE ITALIAN INVOLVEMENT IN A TRANSIT COUNTRY

Monitoring and Protection of Gasline. An Italian-Georgian Collaboration

Mauro PICCOLO
EUREKOS srl Via Leopardi, 13 30026 Portogruaro-Italy mpiccolo@eurekos.it

Abstract: This article discusses the Italian experience on natural gas transportation and development and the rationale for transferring knowledge and involvement in transit countries. Some aspects of the pipe protections and monitoring are presented with a description of some innovative monitoring techniques based on thermal anomalies detection and the use of fibre optic as sensors.

Key words: Thermography, fibre optic, thermal anomaly, leak, natural gas, directional drilling

1. INTRODUCTION

Italy is one of the largest European gas consumers with a widespread network of main and regional distribution network covering almost the whole territory with the largest concentration in the Northern part.

The security of supply is guarantee by the diversification of sources. The safety of the distribution network is obtained by a constant improvement of the monitoring system and passive protection in difficult areas such as river and mountain crossings.

J. Hetland and T. Gochitashvili (eds.),
Security of Natural Gas Supply through Transit Countries, 61–71.

The Italian technologies in the gas sector are very advanced and of high interest for gas transit Countries; with this spirit and in the frame of the law 212 for the Internationalisation of Enterprises, and in order to foster the links between Italy and former Soviet Union the Italian Ministry for Productive Activities co-funded the Project G.E.ORG.I.A. SYS –"Gas and Electricity Organisation by Innovative Advanced System" that with a duration of 18 months will provide the Georgian Partners with training, technological innovation and advanced equipment.

2. THE ITALIAN MARKET

From 1996 to 2000, the natural gas consumption in Italy grew from around 56.2 billion cubic metres to 70.4 billion, a rise that derived primarily from growth in the power generation sector, in which consumption doubled. In the same period, natural gas was the primary energy source that grew most rapidly, increasing at an average annual rate of 5.8%.

In 2001, Italy was ranked number three in Europe in terms of gas demand with 71 billion cubic meters. The demand is expected to grow at a rate that is higher than in other European countries as the demand will increase by 22 billion cubic meters from 2000 to 2010.

2.1 EU suppliers

In 2002, ENI's Gas & Power division supplied 62.33 billion cubic meters of natural gas. Volumes of natural gas from domestic production accounted for 20% of total supplies (23.3% in 2001) with a decrease of approximately 2 billion cubic meters over 2001. Natural gas volumes supplied outside Italy represented 79.7% of the total supplies (76.6% in 2001) with an increase of 1.87 billion cubic meters over 2001, up 3.9%, in particular due to higher volumes purchased from Norway (3.73 billion cubic meters) and the Netherlands (0.55 billion cubic meters). In 2002 a total of 1.43 billion cubic meters of natural gas were put into underground storage mostly by STOGIT s.p.a. .

2.2 Regulation of transportation

The natural gas sector has been the object of new regulations at national and community level within the framework of the sector's liberalisation policies and of the creation of a sole European market for natural gas.

The regulation process was initiated at European level through the Gas Directive (Directive 98/30 CE dated 22nd June 1998), which was assimilated in Italy by the "Letta" Decree (Executive Order 23rd May 2000 N. 164). On the basis of this Decree, the activities of transport and dispatch of gas are declared to be of public interest and are regulated. The companies that carry out such activities are obliged to allow all requesting users to have access to their networks.

Furthermore, starting 1st January 2002, the transport and dispatch of gas must be carried out by subjects who do not cover any other activities in the gas sector, with exception to the activity of stocking, for which a separate accounting administration must be kept.

The system comprises also, reduction and mixing plants and other facilities necessary for the transport and dispatch of gas. In accordance with resolution number 120/01 ("Definition of criteria for the determination of rates for natural gas transport and dispatch and for the use of LNG terminals and the reservation of capacity"), the Snam Rete Gas system is divided into two parts: National Gas Pipeline Network (7,896 km), and Regional Gas Pipeline Network (21,711 km) as shown on Figure 1.

Figure 1 The Snam Rete Gas transmission system consisting of 11 compression plants, 29,600 km of natural gas pipelines with diameters from 25 to 1,200 mm operating at pressures between 0.5 and 75 bar.

2.3 International connections

To manage this huge amount of gas deliveries from different suppliers requires a well-developed network of lines crossing Europe from North to South and from East to the West, and also the South-North connections with Algeria. The Italian network is consequently connected to foreign suppliers as explained below:

- TENP pipeline: transporting natural gas from the Netherlands through Germany, to Wallbach at the German-Swiss border. The pipeline is 845-kilometer long (i.e. a 500-kilometer long simple line and a 345-kilometer long doubling line) with four compression stations.
- Transitgas pipeline: 282-kilometer long, with one compression station. The pipeline is stretched from the Netherlands and from Norway via Switzerland with its 167-kilometer long main line plus 60-kilometer doubling line, from Wallbach (TENP) to Gries Pass at the Italian border.
- TAG pipeline: 2 lines each 380-kilometer long and a third partial 185 km line with 3 compression stations from Russia via Austria from Baumgarten (border of Austria and Slovakia) to Tarvisio).
- TTPC pipeline: made up of two lines each 370-kilometer long, with three compression stations. The pipeline goes from Algeria across Tunisia from Oued Saf Saf at the Algerian border to Cap Bon on the Mediterranean coast where it links with the TMPC pipeline.
- TMPC pipeline: for the import of Algerian gas, made up of two lines each 156-kilometer long and three lines each 155-kilometer crossing underwater the Sicily Channel from Cap Bon to Mazara del Vallo in Sicily, the entry point to Italy.
- A new 55-kilometer long line for Norvegian gas from Rodersdorf at the French-Swiss border to Lostorf (interconnection point with the line coming from Wallbach)

3. ENTRANCE POINTS

Entrance points are considered all points from where the natural gas can enter the distribution network; as a consequence they apply to all connections to the international supply lines entering the Italian territory or to internal production fields or to the connection with underground storage or re-gassification plants.

There are 3 entrance points at the intersections with the import pipelines:

- Near the boarder of the Italian territorial waters (gasline TMCP) for the import from Algeria
- Gries Pass for the import from the Netherlands and Norway
- Tarvisio/Gorizia for the import from Russia and Slovenia

There are several entrance points that correspond to the internal supply:

- N° 1 entrance point at the GNL re-gasification plant of Panigaglia.
- N° 10 virtual points in correspondence with the main national production fields, from treatment plants or storages centres (i.e Minerbio Storage Point)
- N° 2 virtual entrance points "hub" one per Storage Operator (ENI , Divisione AGIP and Edison Gas).

3.1 Pressures and diameters

By the Ministerial Decree of 24th November 1984: "Fire-prevention Safety Standards for the transportation, distribution, storage and use of natural gas of a density no greater than 0.8 and subsequent modifications" the natural gas transportation and distribution pipelines are classified into 7 categories, on the basis of the different maximum operating pressures.

Pipeline Category	*1st*	*2nd*	*3rd*	*4th*	*5th*	*6th*	*7th*
Maximum operating pressure (bar)	**>24 bar**	**24>p> 12**	**12>p> 5**	**5>p> 1.5**	**1.5>p> 0.5**	**0.5>p> 0.04**	**>0.04**

Most pipelines operated by Snam Rete Gas are of the 1st, 2nd or 3rd category. There are only a limited number of examples of the 4th category of pipelines and even fewer of the remaining categories: these latter are in fact typical of local distribution networks. For pipelines of the 1st, 2nd or 3rd category the operating authorisation - up to a pressure determined by the CPI (Certificate of Fire Prevention) - is released by the appropriate authority. For the other categories this value is fixed according to the project's pipeline pressure:

Underground pipelines with maximum 48 inch pipe diameter is in the range 24 <p< 75 bar. Likewise, for 20-26 inch underwater pipelines the pressure is higher or equal to 115 bar.

According to the regulation the import from Algeria are formed by 2 lines with a 48/42-inch diameter, each approximately 1,500-kilometer long, including the smaller pipes that cross underwater the Messina strait.

The imports from Russia (and countries of the former Soviet Union) are formed by 3 lines with 42/36/34-inch diameters extending for a total length of 900 kilometres.

The import from the Netherlands and Norway uses one line with 48/34-inch diameter, 177-kilometer long.

3.2 Determination of tariffs

In the past, transport and dispatch tariffs were determined through freely negotiated agreements between Snam, Italian farmers or Shippers.

On delegation of the "Letta" Decree and with the successive Resolutions, the Authority disciplines the tariff system, establishing the criteria of tariff determination for diverse periods of regulation (for example, the first period, of a 4 year duration, began on the 1st October 2001 and will end on the 30th September 2005).

The proceeds are determined on the basis of the following elements:

- the known cost of the invested capital (or Regulatory Asset Base - RAB)
- a rate of return on RAB
- rates of depreciation and operating costs
- annual revision of income through the formula "Revenue Cap"
- revision of income per volume unit or "Price Cap"
- additional income in function of investments on the network.

The resolutions have introduced an incentive system for new investments destined to the development of the network and of the liquid natural gas (LNG) terminals. At the end of the first period of regulation, the increase in assets deriving from the new investments carried out during the first period will be considered part of the new RAB. The Authority will have the responsibility to ensure a fair yielding of the invested capital. The tariff system is therefore studied to reward efficiency and supports the development of investments.

4. SAFETY

Security, environment and safety of workers are key points for the Italian companies in the gas sector. Security is obtained by:

1. Monitoring of the pipelines, (inspections, pigging, corrosion, cathodic protection , helicopter survey, etc.)
2. Passive protection of the pipe (tunnels)
3. Advanced monitoring methodologies (gas detectors, infrared monitoring, thickness measurements by eddy current and magnetic methods)

The first point represents a group of common methodologies for the inspections of pipelines, with the only difference that helicopter surveys are a pretty common procedure in case of potential damages. This type of survey is also used for preventive monitoring in case of environmental problems such as flooding, landslides, fires, earthquakes, etc.

The second point represents a very safe procedure for the crossing of mountain areas where risk of landslides, avalanches, erosion and other environmental issues could represent a menace for the pipe safety.

The crossing of the Italian Austrian mountainous region, interested by strong erosion, narrow and steep valleys and canyons, is obtained by many tunnels with a diameter of more than 5 metres that guarantee the safety of the pipes in every conditions.

In general the tunnels are dug by a tunnel-boring machine that in rocks like limestone can provide an advancement of tens of meter per day. The cost of tunnelling could appear high compared to normal cost of laying a gas pipeline, but in a middle-to-long perspective the costs are compensated by the advantages in term of maintenance savings, cheaper inspections in remote areas and strong improvement of the safety.

Directional boring is another advanced technique that is used for river crossing. This permits a safer and faster preparation of the structures needed to protect the pipe. Avoiding the digging of the trench diminish the impact on the riverbed and special shoulders minimise the possibility to trigger erosion or to determine the weakening of the river banks.

5. THERMOGRAPHIC GAS LEAK DETECTION

The third point is the monitoring of leaks. This aspect is very important because of the potential risk of explosions but also because the precise identification of small leaks will permit to avoid larger spills that could have dramatic consequences. Many techniques can be applied but on long distances airborne thermography is the faster and more cost-effective. It is based on a very sensitive thermo camera able to detect variation of temperatures in the order of 1/100 °C, thus the leak can be identified because of the cold thermal signature determined on the soil above the pipe. An ultra light aircraft or a small helicopter can easily carry out the coverage of the pipeline in a very short period of time. The image processing procedure and interpretation is not complicate and is, indeed, very fast and allows a precise determination of the position and presence of leaks.

The same technique can be used in pumping stations for the fast determination of the point of leak (valves, joints, connections, etc.). In this case it is based on the temperature difference of the metallic part involved in the leak. By analysing the images of the structures and equipment it is usually not difficult to determine the parts that are responsible for the leak.

The picture below shows two examples of the thermal effect determined by a leak: the thermal anomaly on the ground recorded by an helicopter, and the extreme effect of a leak of a block valve (Figure 2 and Figure 3, respectively).

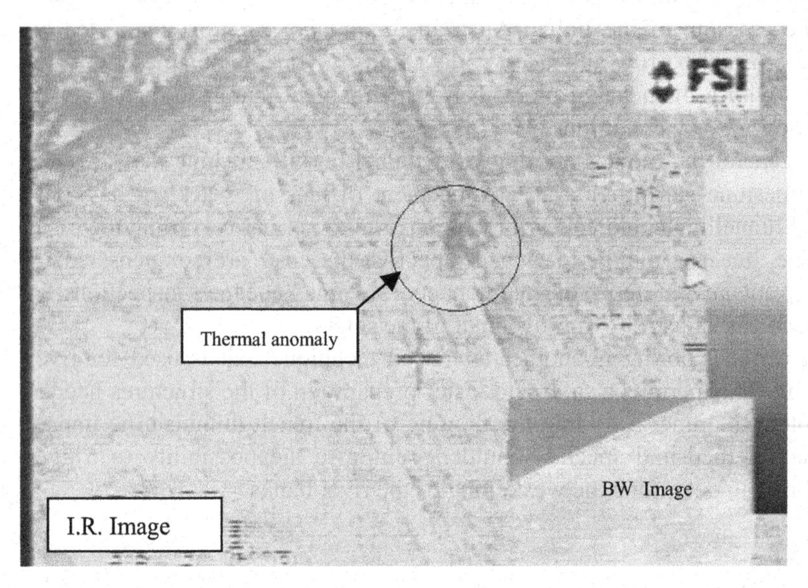

Figure 2 Examples of thermal effects determined by gas leaks:
the thermal anomaly from an underground leak.

In association with all the above methodologies, the thickness of exposed pipes is controlled with high precision by using magnetic or eddy current methods that can provide a fast and precise identification of internal, external corrosion or blistering inside the metal thickness.

Figure 3 Thermal effect of a gas leak, a frozen valve

Those methodologies can guarantee a very high level of safety of the distribution network and permit also the planning of an adequate predictive maintenance programme that can strongly diminish the risk of failures.

6. REMOTE MONITORING BY FIBRE OPTICS

The methodologies described above require the presence of an operator and cannot be performed automatically 24 hours per day. In some situations or some special structure there is a need for a continuous monitoring or the activation of an alarm when a determined parameter is reaching a certain level. In other cases the location could be remote or dangerous; in this case the remote monitoring by fibre optics is a methodology that can be applied over long distances for the monitoring of a series of physical parameters such as inclination, vibrations, shocks, elongations, bending etc.

The continuous monitoring of can be obtained by using a fibre optic cable attached to the pipe itself or to the structural elements connected to the pipe. The fibre can act as a simple data transmitter (insensitive fibre) or be itself a sensor for its total length or for only selected sector (sensitive fibre). The different possibilities are presented in Figure 4.

By studying the behaviour of a laser signal into the fibre it is possible to monitor pipe movements, vibration, temperature variations, inclination, bending and many other geometric and physical parameters. By setting the appropriate thresholds it will be possible to implement automatic alarms from remote locations. The applications are very interesting mainly in areas where landslides, earthquakes, mechanical stress, elongation and bending could affect the pipe and determine a menace for its integrity.

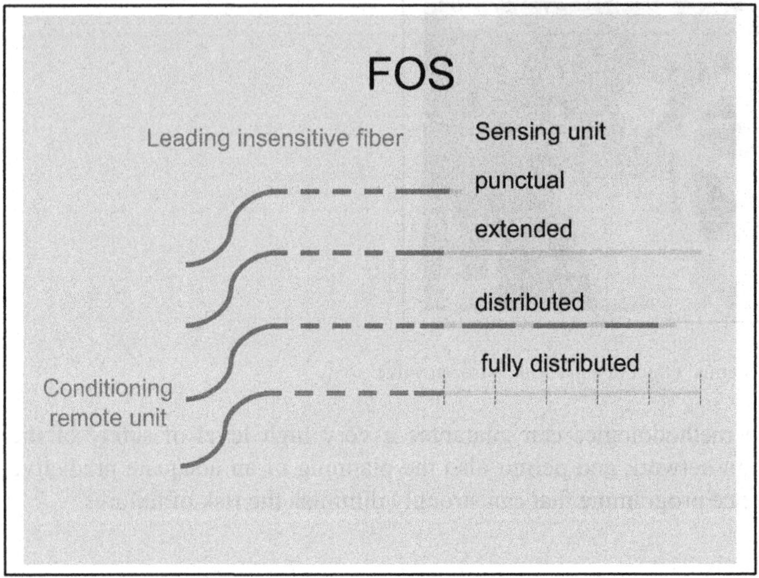

Figure 4 Different partition of sensitive elements from fibre optical sensors.

The main advantages of the use of fibre optic as sensors are the following:

- Wide range of physical parameters
- High sensitivity and a wide dynamic range
- Immunity from electromagnetic interferences
- High reliability and robustness
- Flexibility of the sensing length
- Distributed and continuous detection
- Remote reading by fibres that are insensitive to disturbance
- Possibility to deploy sensor networks
- Possibility of embedding the fibre sensors into structures without compromising the integrity of either the sensor or the structure
- Lower cost compared to normal sensor network
- Long life and poor vulnerability

Intrinsic sensor are used for:

- Strain
- Temperature
- Pressure
- Angular Rotation

Extrinsic sensor are used for:

- Positioning
- Vibration
- Acceleration
- Fluid Velocity

7. CONCLUSIONS

Italy can be considered a leading country in the security of gas supply and safety of gas lines. The Italian market is well developed and the new regulations permit to invest in safety and research to improve both the efficiency and safety of the network system. Very advanced technologies are used to guarantee the safe supply of gas to the industry.

The Italian Ministry for Productive Activities has financed a cooperation and pilot Project "Gas and Electricity ORGanisation by Innovative Advanced SYStems" (Acronym: G.E.ORG.I.A. SYS) in Georgia for the improvement of both the gas and electric networks by application of modern technologies and training of Georgian personnel. This financial support provided in the frame of the Italian law 212 for the support for the Countries belonging to the former Soviet Union is intended to provide modern methodologies and training for the Georgian personnel.

Georgia is a gas transit Country that needs a fast development toward modern standards of safety for its pipeline network; the access to information, modern technologies and equipment will permit to fill the gap with the western Countries. This technological exchange will provide a framework for the development of a safe and effective management of the gas resources and transportation in Georgia and in the Caucasus region.

PART II: MODERN TECHNOLOGIES

STATE OF THE ART IN LIQUEFACTION TECHNOLOGIES FOR NATURAL GAS

Ways of providing LNG for Transit Countries

Einar BRENDENG, Jens HETLAND
SINTEF Energy Research, Dept. of Energy Processes, Kolbjorn Hejes vei 1A, N-7465 Trondheim, NORWAY

Abstract: LNG has been in practical use for more than fifty years - especially in the United States and Japan - and is attracting increased interest from the European energy sector. Owing to substantial improvements in large-scale liquefaction technology over the last twenty years, LNG is now gaining acceptance as an alternative energy carrier for a variety of applications. At the same time the production cost has decreased considerably, and so have the expenses for transport and storage. In several events the distribution of condensed natural gas - LNG - is being conceived as an economical alternative to developing new firm pipeline capacity.

Transit countries are usually given the option of receiving compensation for their land lease in kind. A relevant question is how the option gas should be used and distributed, and how infrastructure that should be developed. In this context LNG offers some advantages over natural gas that should be further looked at.

This paper deals with modern technologies pertaining to the production of LNG from natural gas. Experience is drawn from large scale LNG production to applications of medium to small scale. Special attempts are made to describe actual technologies in some detail in order to show the true diversity in modern LNG production.

Key words: LNG, natural gas, liquefaction, technology

J. Hetland and T. Gochitashvili (eds.),
Security of Natural Gas Supply through Transit Countries, 75–102.

1. INTRODUCTION

Despite Norway is a major exporter of natural oil and gas it has no infrastructure for domestic gas distribution. The reason is partly the large extent of hydroelectricity (99.4%) and its remote population. However, since the 1980-ies until present time, specific knowledge on liquefied natural gas (LNG) has been developed owing to substantial targeted research. In 1984 SINTEF/NTNU[1] entered a strategic co-operation with Statoil on liquefaction. Under the State R&D Program for the Utilization of Natural Gas, SPUNG (1987-1993), the thermodynamic properties of natural gas mixtures were subjected to fundamental and experimental research financed by the Norwegian Research Council [1].

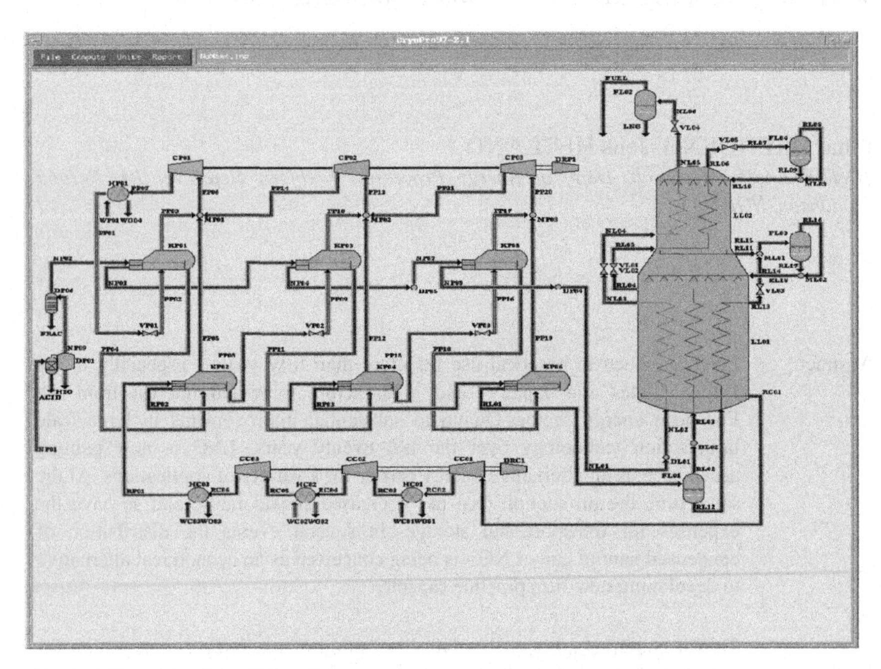

Figure 1 User-interface of CryoPro.

Long-term strategic R&D has enabled SINTEF and NTNU to develop advanced tools like CryoPro for LNG plants design and evaluation. This includes propane pre-cooling and the dual-mixed process, and advanced correlations for heat transfer and pressure drop. It also incorporates the corresponding state equations for thermodynamic properties, multi-variable optimisation of refrigerant composition, pressure and cryogenic heat exchanger layout, and furthermore, the user-interface system as shown on Figure 1. Extensive laboratory experiments were carried out on thermodynamic properties and spiral-wound heat exchangers. Beyond 13 Ph.D. and 45 M.Sc. theses [2,3,4,5,6,7] the most tangible result of this work

[1] The Norwegian University of Science and Technology, Trondheim

is the construction of a novel 4.2 MTPA liquefaction plant designated the Statoil/Linde MFCP[2]. Decided in 2002 the plant is being set up in the northernmost city of the world, Hammerfest (Norway) - scheduled to go on stream in 2006.

2. LIQUEFIED NATURAL GAS (LNG)

LNG is a compact energy carrier that is easy to transport over seas, on rail and road, independent on gas pipelines. By lowering the temperature of natural gas to sufficiently low temperatures the gas will condense into liquid phase. At atmospheric pressure natural gas turns liquid at minus 162 °C – the boiling point of methane.

LNG is simple to re-gasify in order to deliver almost pure methane at the end users. In contrast to natural gas that contains typically around 90% methane, and some ethane, propane and heavier hydrocarbons, the liquefaction process involves pre-treatment of the gas in order to remove carbon dioxide, sulphur compounds, water, and petroleum gases with carbon number higher than one (butane, propane etc.). This is done in order to avoid formation of solids in the cold heat exchangers. The presence of nitrogen is usually limited at about 1% (refer Table 2 on page 81).

Liquefaction means a volume reduction of approximately 600 times. This corresponds to an energy density that is 600 times higher than natural gas at atmospheric conditions. Since LNG has no odour, leak detection requires special instrumentation. Furthermore, LNG is colourless and non-toxic, and non-carcinogenic. The specific weight of LNG is 45% of water. In vaporised phase - mixed with air - LNG can only ignite if the concentration of gaseous methane is in the range 5-15%. Neither LNG, nor its vapour can explode in an unconfined environment [8].

3. STATE OF THE ART IN LNG TECHNOLOGY

The main challenge pertaining to modern cryogenic technology relates basically to suitable refrigerants and low temperatures. Although mixtures of refrigerants were suggested in the nineteenth century, multi-component refrigerants did not appear until the 1930-ies, and were further developed and used for the liquefaction of natural gas in the 1960-ies. The first base-load LNG plants came on stream in Arzew, Algeria, in 1963, and in Kenai, Alaska, in 1969. They were fairly conventional cascade plants, with propane, ethylene and methane as refrigerants [9]. Today LNG plants represent the largest refrigeration plants in the world.

The state of the art of base-load LNG plant design is reflected by four different technologies:

[2] Mixed Fluid Cascade Process

1 The C3/MR concept
 ⇨ Mixed refrigerant with propane pre-cooling
2 The optimised cascade concept
 ⇨ Combined propane, ethylene, methane refrigerant system
3 The DMR concept
 ⇨ Dual mixed refrigerant process
4 The MFCP concept.
 ⇨ Mixed fluid cascade process

So far the C3/MR process is the most significant process in the market as it accounts for nearly 80% of the number of liquefaction units and almost 90% of the global LNG-production capacity. Table 1 provides a non-extensive list of LNG plants around the world that are in either operation, under construction or in the planning. The status of main liquefaction technologies is as follows (2003) [10]:

- C3/MR: 57 trains in operation plus 7 under construction
- Optimised cascade: 3 trains in operation (Trinidad and Tobago) plus 1 under construction (Idku, Egypt)
- DMR: 2 trains in operation (Sakhalin)
- MFCP: 1 train in construction (Snøhvit, Norway)

Table 1 Overview of LNG plants and projects (non-extensive list)

Project	Place	In operation	Cycle	MTA	kWh/t
CAMEL [11]	Arzew Algeria	1963	Cascade	0.36	509
Phillips Marathon [11]	Kenai Alaska	1969	Cascade	1.15	440
Esso [11]	Marsa el Brega Libya	1970	MR	0.69	529
Brunei LNG [11]	Brunei	1973	C3/MR	1.08	457
Skikda 1,2,3 Tealarc [11]	Algeria	1974	MR Two pressure	1.03	610
Skikda 5,6 Prico [11]	Algeria	1981	SMR	1.70	349
Karratha I Karratha II Karratha III	Australia	1989 1993 2004	C3/MR C3/MR C3/MR	2x2 1x2 1x4.2	
Point Fortin, ALNG, Trinidad [12]	Trinidad	1999	C3/MR	2.9	
Ras Laffan, RasGas, Qatar [11,12]	Qatar	1996	C3/MR	2x3.2	374
Ras Laffan, Qatargas, Qatar [12]	Qatar	1999	C3/MR	2.30	
Malaysia Sdn. Bitulu LNG-1 Dua LNG-2 [11] Tiga LNG-3 [12]	Malaysia	 1983 1996 2002	 C3/MR C3/MR C3/MR	 3x2 3x2.6 1x3.8	 328
Qalhat, OLNG, Oman [12]	Oman	2000	C3/MR	2x3.3	
Bonny LNG Train 1,2 [12] Train 3	Nigeria	 1999 2003	 C3/MR C3/MR	 2x2.95 3.7	 296?

Bonny LNG Extension Train 4,5	Nigeria	2005	C3/MR	2x4	
Atlantic LNG ConocoPhillips	Trinidad/ Tobago Train 1,2, Train 3 Train 4.	2003 2006	Cascade Cascade	(In total more than 15) 2x3.4 3.4 5.2	
Shell [12]	Oman	2000	SGSI C3/MR	2x3.3	
Shell	Sakhalin, Russia	2007	SGSI DMR	2x4.8	
Statoil/Linde	Snøhvit, Norway	2006	MFCP	1x4.2	
Fenosa SEGAS 1	Egypt	2004	C3/Split MR	1x5.0	
ConocoPhillips	Idku, Egypt Train 1. Option for Train 2	2005	Cascade	1x3.6 Optional 1x3.6	
Linde	Shan Shan Xinjiang, China	2003	SMR N_2, CH_4, C_2H_4, C_3H_8, C_4H_{10}	0.43 54 t/h Spiral wound	
YLNG	Yemen		C3/MR		
Shell	Mariscal Sucre Venezuela	2007/8	C3/MR?	4.7	
North Slope	Alaska		Cascade C3/MR?		
	Papua New Guinea		C3/MR? Alternative?		

C3/MR: Propane pre-cooled mixed refrigerant
DMR: Dual mixed refrigerant
SMR: Single mixed refrigerant
MFCP: Mixed refluid cascade process

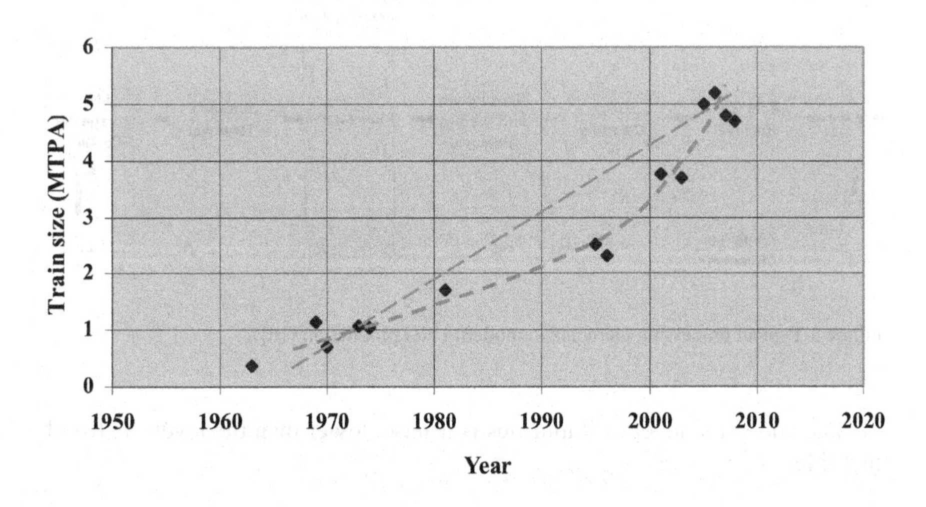

Figure 2 LNG train size development versus time.

Today the largest unit in order, SEGAS LNG, Egypt, has a capacity of 5 MTPA, scheduled for commissioning in 2004. The plant comprises two spiral wound heat exchangers in parallel.

4. LARGE-SCALE LNG FACILITIES

Today base-load trains have production capacities of up around 5 million ton of LNG per annum, or about 15 000 ton per day. Usually they receive natural gas from the well stream - or a pipeline - and delivers LNG in bulk tanks – most often carried by ships. Base-load plants are in permanent operation during most of the year. The processing equipment may vary from plant to plant depending on the composition and the purity of the gas. Vitally important to large LNG plants is efficiency. To obtain a high efficiency, complex refrigeration processes with specialized equipment are applied. The investment costs are high. However, owing to large production volumes the relative cost per produced unit of LNG is likely to become competitive. For very small LNG plants the relative investment cost may constitute a major challenge, especially if efficiencies similar to large base-load plants are to be achieved.

Grouped according to duty the typical elements of a base-load plant are as shown in Figure 3: reception, acid gas removal, CO_2 cleaning, dehydration and mercury removal, liquefaction and fractionation, nitrogen removal, LNG storage and loading system for shipment as shown in [13,10].

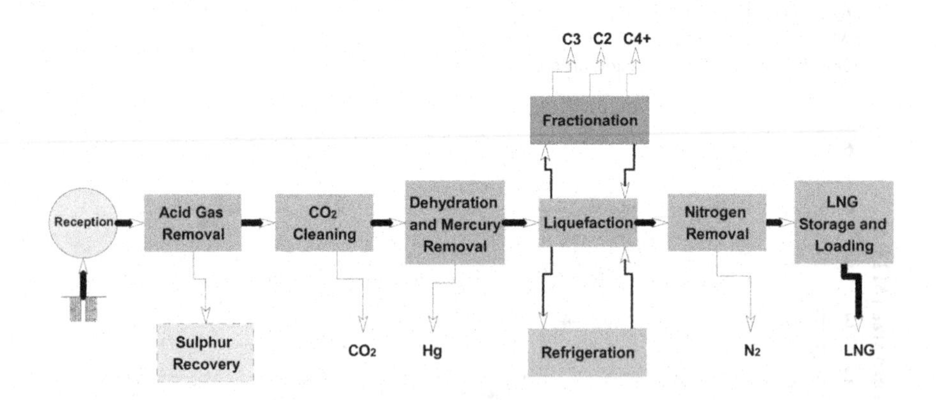

Figure 3 Typical processing elements in modern LNG production [10]

The allowable level of impurities is usually lower than the levels as listed in Table 2.

Table 2 LNG Specification [14]

Impurity	***Limit***	***Cause***
CO_2	50-100 ppm	A
H_2O	0.1-1 ppm	A
Hg	<10 ng/Nm^3	A
$HgCH_3$		A
Aromatics	< 2 ppm	A
C_6H_{14}	<250 ppm	A
C_7H_{16}	<90 ppm	A
C_8H_{18}	<0.6 ppm	A
H_2S	3.5 mg/Nm^3 (4 ppm)	B
COS		B
CS_2		B
Mercaptans		B
Total sulphur	10-50 mg/Nm^3	B
N_2	0.5-1.5 mole%	B
A: Solubility limit B: Product specification		

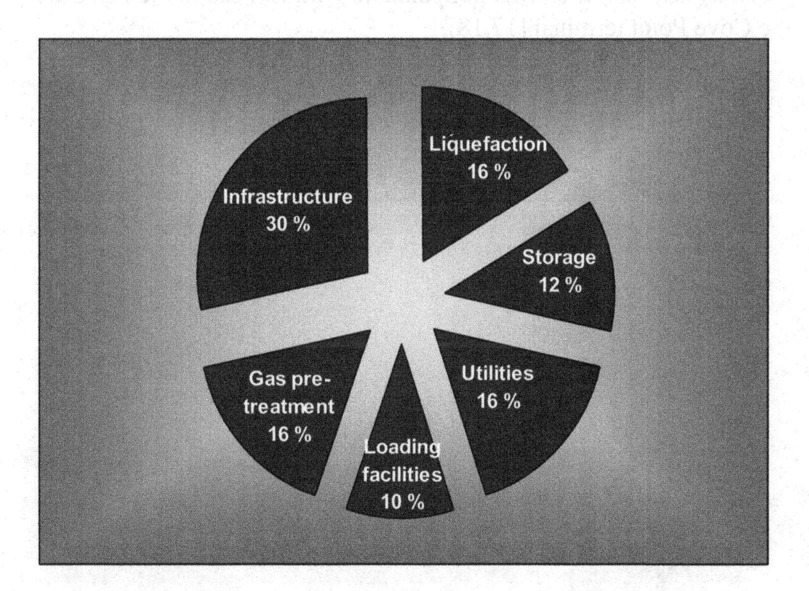

Figure 4 Cost breakdown of a modern base-load LNG plant [15,16]

4.1 Main components of LNG plants

The breakdown of the investment cost for large-scale LNG plants may vary depending on the plant specification. A representative scope for a typical greenfield plant located to the Middle East for offshore production with an onshore LNG plant having two liquefaction trains shows that the LNG plant per se would account for roughly 50% of the total project, almost

40% for the shipping, and about 10% for the terminal [16]. Furthermore, the cost of the LNG plant could be broken down as shown in Figure 4. In base-load LNG plants the predominant heat exchanger type in is the spiral-wound type as shown on Figure 5.

4.2 The Norwegian Snøhvit development project

The development project Snøhvit is scheduled to go on stream in 2006. It includes three offshore gas fields - Snøhvit, Albatross and Askeladd in the Barents Sea, 140km off Hammerfest in the northern part of Norway. The development project also includes an onshore liquefaction plant on Melkøya (Figure 6). The offshore field contains mainly natural gas with some condensate totalling more that 193 billion cubic metre of natural gas and 113 million barrels of condensate [15]. Statoil operates the license on behalf of a consortium that includes Petoro, TotalFinaElf, Gaz de France, Norsk Hydro, Amerada Hess, RWE Dea and Svenska Petroleum Exploration. The annual export prospect is 5.75 billion Sm^3 LNG, 747 000 ton condensate and 247 000 ton liquefied petroleum gases (LPG). Long-term contracts have already been signed with Iberdrola in Spain and with El Paso in the USA, for rent of the Cove Point terminal [17,18].

Figure 5 LNG heat exchanger at fabrication workshop, MFCP.

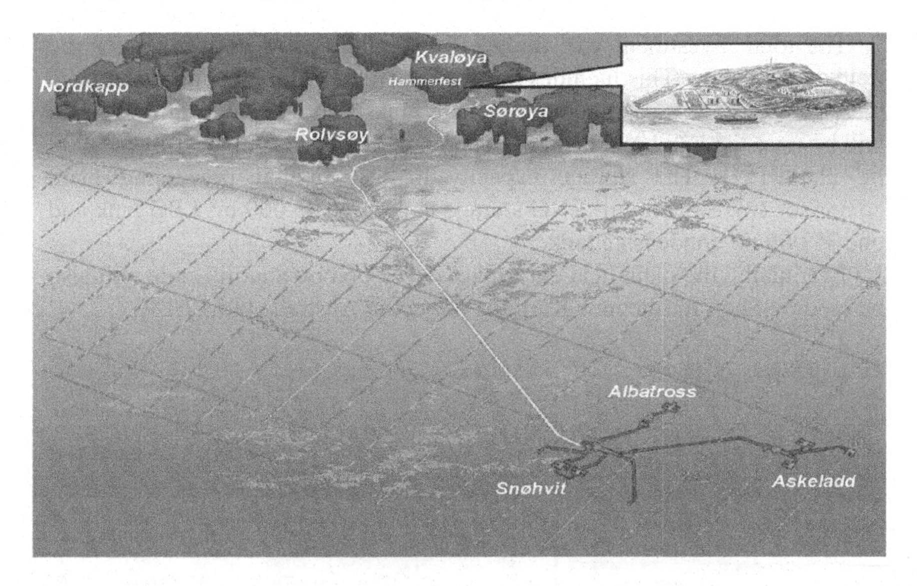

Figure 6 The Snøhvit development project featuring sub sea templates down to 320 m water depth, well stream to shore, and hydrate control by MEG. The system is closed so that there will be no emissions offshore. The project comprises CO_2 re-injection and remote power system and control facilitated from the shore at a distance of 160 km [17].

The liquefaction train is the novel mixed fluid cascade process (MFCP) patented by Statoil/Linde. It is the first LNG base-load plant ever to employ electric drive for the compressors. Today the Oman plant (Shell) is claimed to be world's most efficient LNG plant as it consumes only 8% of the gas for its operation. When Snøhvit goes on stream (2006) it is expected to become even more efficient as it will consume only 5% of the feed gas for the liquefaction. This means that Snøhvit will constitute the most efficient LNG plant ever. The high efficiency is attainable much owing to the low temperature of the ambient air and the cooling water in the artic region.

Snøhvit will also be the first major development on the Norwegian continental shelf without support from a floating unit. A sub sea production system on the seabed will feed the land-based LNG plant via a 0.68 m ID, 160 km two-phase pipeline. In addition two chemical lines, an umbilical and a separate pipeline for the transport of carbon dioxide will be included. The unprocessed well stream from Snøhvit is separated and the gas is treated and cooled to a temperature below the boiling point (-162 °C) to form LNG. This operation employs a large cold box (40 m high and a footprint of 15x17 meter).

The purity of the well stream is an important factor in the design of LNG plants. As the Snøhvit gas contains 5-8 % carbon dioxide CO_2 will be captured from the inlet gas stream by a MDEA process and returned via a separate line for off-shore re-injection beneath the seabed close to the wells [17,18].

The total investment will include 3600 million Euro allocated for the field development. This includes the sub sea well-stream pipeline system and the onshore processing plant and storage tanks. An other 700 million Euro will be invested in a LNG tanker fleet comprising 4 ships in the 140-145 000 m^3 class [17]. The storage capacity of the Snøhvit processing plant comprises two 125 000 m^3 LNG storage tanks, one 45 000 m^3 LPG tank and one 75 000 m^3 condensate tank.

Eventually, the main heat exchanger developed by Statoil/Linde will also be used by Shell for the new Karratha project in Australia [15,17,18].

Figure 7 The spiral-wound heat exchanger (in front) to be built for the Snøhvit LNG processing plant. The cold box measures 15x17 meter (footprint) having a height of 40 meter. The Statoil/Linde heat exchanger is the same type as the heat exchanger used in the Mossel-bay LNG plant in South Africa [17]. The lower tower (right) contains a plate fin heat exchanger for pre-cooling.

5. SMALL-SCALE LNG PRODUCTION

Land based small-scale LNG plants can be divided into two main categories:

- Peak shaving plants
- Small-scale LNG plants for decentralised LNG production

The production capacities for these types of plants are usually in the range of 10 – 500 tonnes of LNG per day. Table 3.

In peak shaving applications large LNG volumes are stored for winter operations. When the gas demand exceeds the normal gas volume in a localized area, LNG is pumped from the storage, vaporized and distributed in gas phase to the customer. The volume is then replenished during the off-peak periods.

The latest global market is for transportation fuels. Liquefaction facilities for vehicle usage may essentially be the same as the peak shaving units; although, LNG storage will be much smaller, and the LNG will be loaded to the vehicle tank in liquid phase rather than vaporized gas.

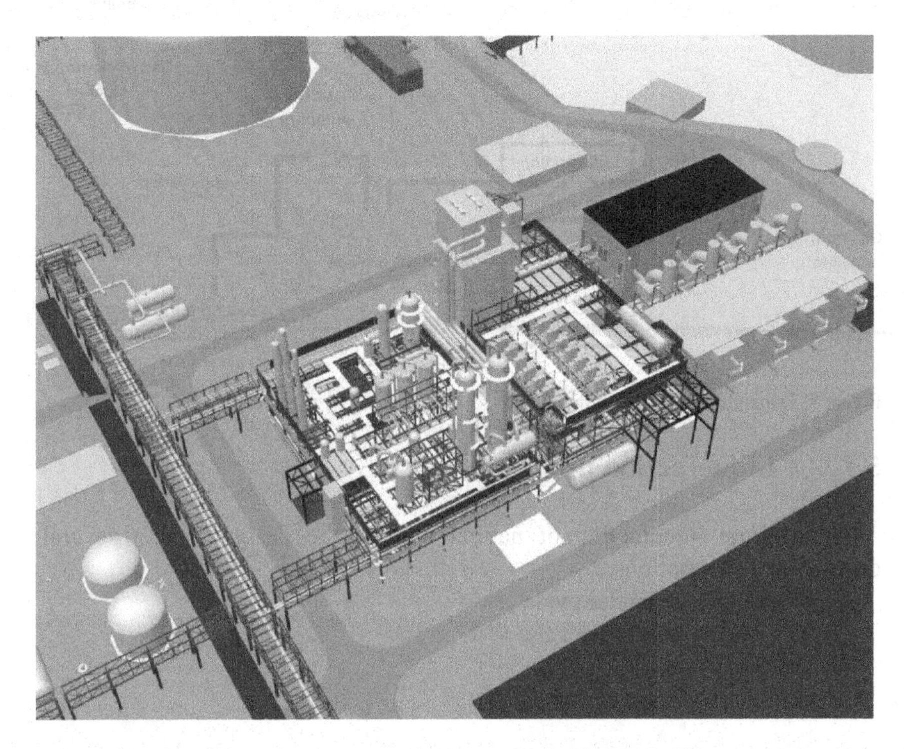

Figure 8 Bird's-eye impression of the 4.2 MTPA Snøhvit liquefaction plant at Hammerfest, Norway. The plant is scheduled to go on stream in 2006 [17]

Table 3 LNG plant characteristics [19]

Plant characteristics	Peak shaving	Vehicle fuel
Liquefaction, ton/day	100-500	10-400
Operating period, days/year	150-200	365
Storage, m^3 (days production)	50 000-100 000 (150-200)	5 000-10 000 (5-10)
Sendout	Vapour	Liquid
Sendout rate (relative to liquefaction rate)	10-20 times	2-3 times
Sendout type	Pipeline	Truck/rail

In Figure 9 the principle for a peak shaving plant with combined delivery of gas to pipeline and truck or rail loading is shown. Different process cycles are developed for such applications.

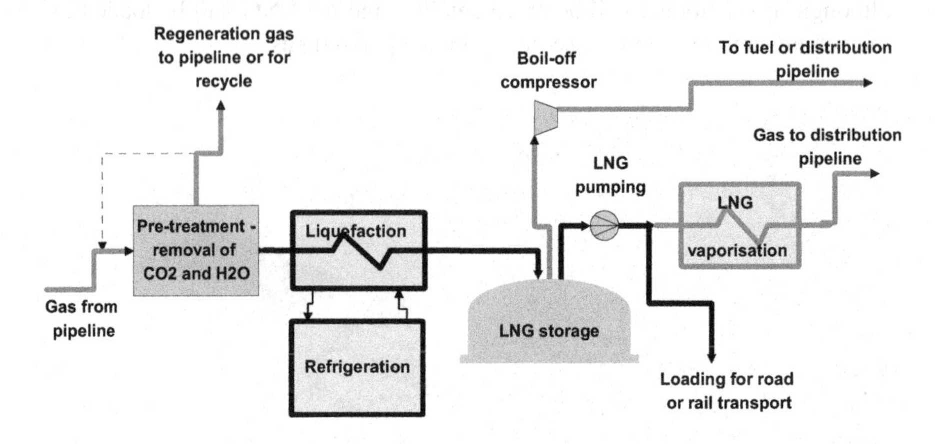

Figure 9 Principle for peak shaving plant.

5.1 Liquefaction unit

Small-scale liquefaction of natural gas can be designed with several different process options.

1: Natural gas expander cycle
2: Refrigerant expander cycle
3: Cascade cycle
4: Mixed refrigerant cycle

In situations where high-pressure pipeline gas is reduced to the pressure of a distribution system, some LNG can be produced without the use of a compressor. This can be achieved simply by using the Joule-Thomsen effect of the gas for liquefying a certain fraction of the natural gas.

5.1.1 Natural gas expander cycle

Natural gas expander cycles can be utilized at places with a high-pressure main gas pipeline and a low-pressure distribution line. The natural gas is expanded in a turbine, thereby cooling the exit gas. Dependent on the inlet conditions, the exit temperature will be so low that a part of the gas is condensed, typically around 12 to 30 %. The liquid is separated from the gas flow in a liquid-vapour separator as shown in Figure 11.

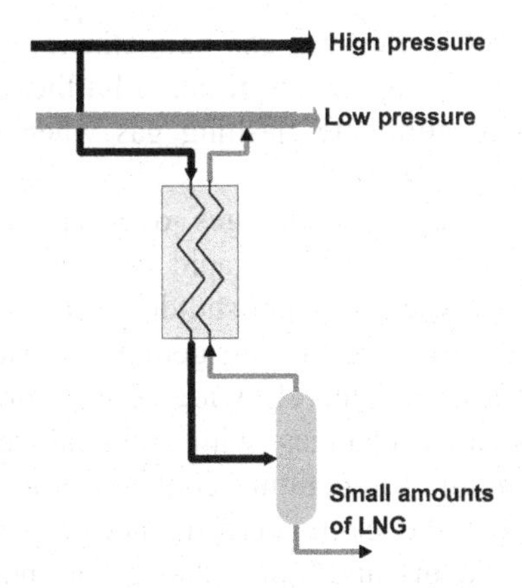

Figure 10 Using pressure drop between the high-pressure pipe and the low-pressure pipe to produce some LNG without the use of a compressor.

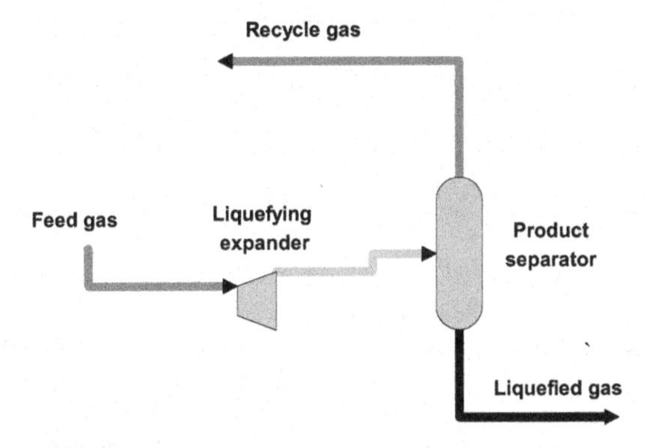

Figure 11 Principle for natural gas expansion process.

The cold recycle gas from the separator is fed to the pre-cooling heat exchanger and is recompressed to the pressure in the distribution line. The work from the expander may be utilized in a compressor for the recompression. Dependent on the conditions, additional recompression may be needed.

5.1.2 Refrigerant expander cycle.

Refrigeration can also be provided by compression and expansion of a single-component gas stream. High-pressure gas is cooled in a counter-current heat exchanger with a cold recycle gas. At an appropriate temperature, the cycle gas is expanded with high isentropic efficiency through an expansion turbine, and the gas temperature will drop. The useful work that is created is normally recovered through a booster compressor, supplementing the main cycle compressor. The cold, low-pressure gas stream from the expander is then returned through various stages of heat exchange that are cooling the incoming natural gas and the incoming refrigerant cycle gas. The refrigerant gas can be methane or nitrogen, but a mixture of these gases may also been used. Further sub-cooling to temperatures low enough to eliminate flashing gas when the LNG is let down is possible.

Expander cycles have several advantages over cascade and mixed-refrigerant cycles. They enable relatively rapid and simple start-ups and shutdowns. This is important when frequent shutdowns are anticipated, such as with peak-shaving plants. Because the refrigerant is always gaseous and the heat exchanger operates with relatively wide temperature differences, it tolerates feed gas composition changes with minimal requirements for changing the refrigerant circuit. Temperature control is not as crucial as for mixed-refrigerant plants and the cycle performance is more stable. The problem of distributing the vapour and liquid phases into the heat exchanger is eliminated because the cycle fluid is maintained in a gaseous phase.

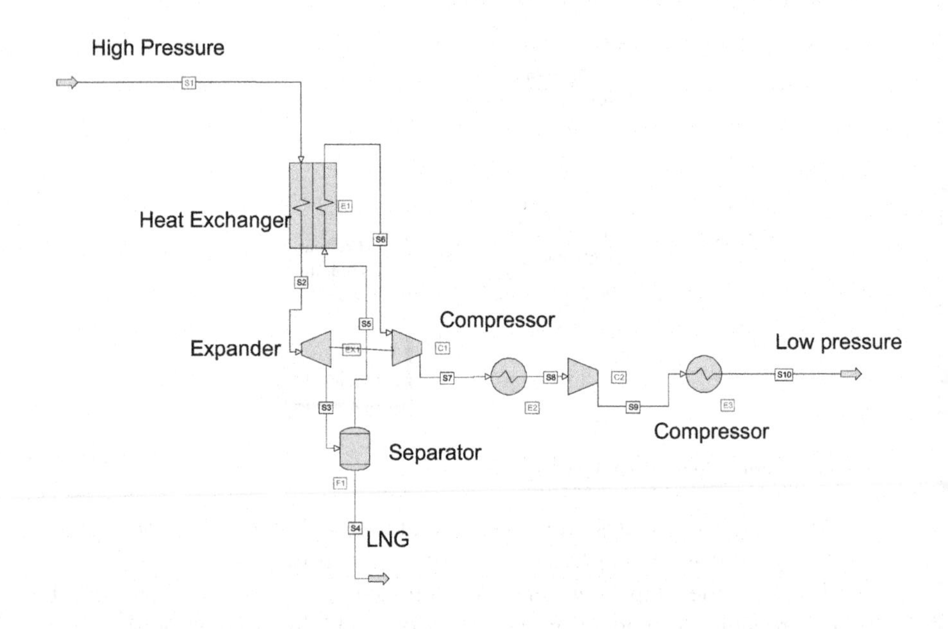

Figure 12 Natural gas expander cycle.

A major disadvantage of the expander cycle is its relatively high power consumption, compared with the cascade and mixed-refrigerant cycle. Changes can be made, however, to the single expander cycle that will increase its efficiency. Power consumption can, thus, be reduced by about 20% when using natural gas pre-cooling with a conventional vapour compression cycle. A further alterative to pre-cooling is to use two expanders that operate over different temperature levels. Two expanders allow closer matching of the warming and cooling curves than with a single expander. In this manner the effective driving temperature difference is reduced and, hence, the thermodynamic efficiency improved. A single expander cycle for this purpose is shown in Figure 13.

With three or more stages for each refrigerant, the power consumption for the cascade cycle is found to compete quite favourable with other liquefaction processes, especially in arctic conditions. This is mainly because of the low flow rate of refrigerant. The cascade cycle is also more flexible in operation, since each circuit of refrigerant can be separately controlled. For the classical cascade, however, an overlap may occur of refrigerants like methane and ethylene. Thus, as methane condenses at an elevated pressure, this will inherently cause some throttling losses.

Table 4 Comparison of efficiencies for different cycles at tropical conditions [20,21].

Cycle	C3/MR	Cascade	DMR	SMR	N2
kWh/ton LNG [20]	293	338	300	348	375
Power % [21]	100	102.4	99.7	110.3	N/A

Economics of scale show that the cascade cycle is most suited to large train sizes. The main disadvantage of the cascade cycle is the relatively high capital cost due to the number of compressors in the refrigerant circuits, as each circuit requires a compressor and a refrigerant storage. Maintenance and spare equipment costs tend to be relatively high due to the large number of mechanical equipment. The principle for the cascade cycle, with only one stage for each refrigerant, is shown in Figure 14.

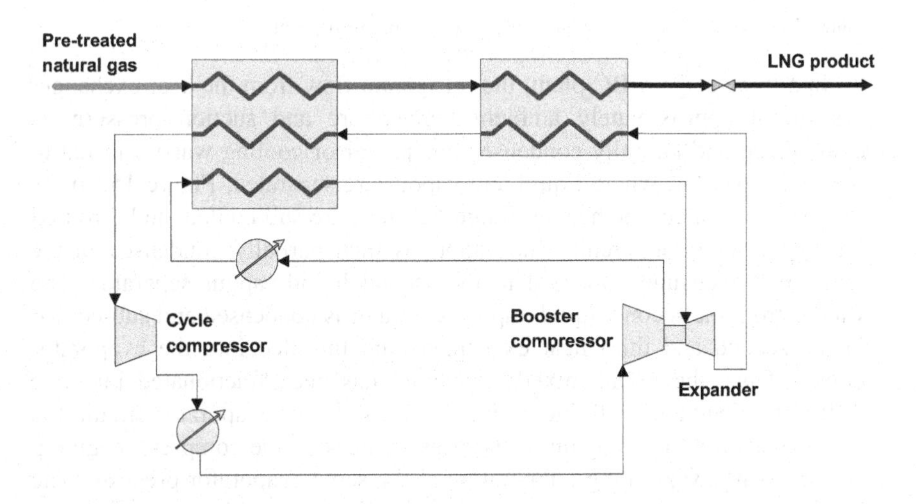

Figure 13 Single expander cycle.

5.1.3 Mixed refrigerant cycle

The mixed refrigerant cycle (MRC) uses a single mixed refrigerant instead of multiple pure refrigerants as the cascade cycle. The mixed refrigerant consists normally of nitrogen, ethane, propane, butane and pentane. The mixed composition is specified so that the liquid refrigerant evaporates over a temperature range similar to that of the natural gas to be liquefied. Small driving temperature differences give operation nearer to reversibility; leading to a higher thermodynamic efficiency. Simultaneously, the power requirement will be lower, and the entire machinery smaller.

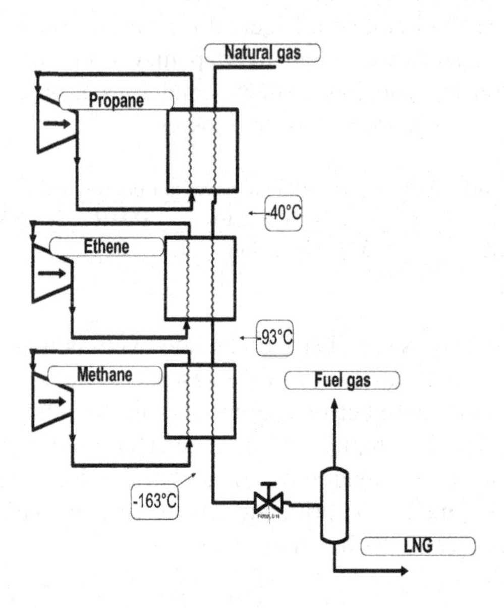

Figure 14 Cascade refrigerant cycle, only one stage per refrigerant.

In a three-stage MRC plant, the refrigerant flow from the heat exchanger system, at approximately ambient temperature and suction pressure, is compressed and partially condensed against air or cooling water and fed to the first separator, where liquid and vapour are separated, Figure 15. In the first heat exchanger the heavier liquid fractions are sub-cooled, and throttled into the evaporator circuit. The vapour is then partially condensed in the same heat exchanger, and fed to the second liquid-vapour separator. The vapour from the second liquid-vapour separator is condensed and sub-cooled in the second and third heat exchanger, and throttled into the evaporator circuit. Thus, the initial mixed refrigerant has been fractionated into two different refrigerants with increasing volatility. In the evaporator circuit, the components are mixed again in the passage back to the compressor, giving an increasing evaporating temperature at the same evaporator pressure. The initial composition of the mixed refrigerant should be optimised in terms of compressor discharge and suction pressure, and also the temperatures in the separators, and to the pressure and composition of the natural gas. With increasing temperature the evaporation curve will better match with the cooling curves of the refrigerant and the natural gas.

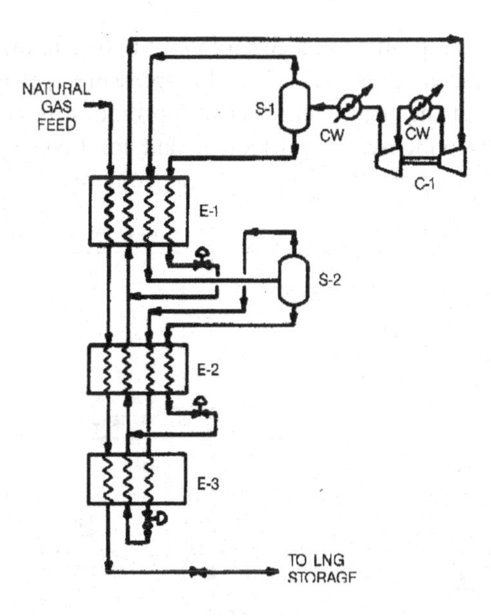

Figure 15 Conventional three stage mixed refrigerant cycle.

The refrigeration duty must be sufficient to sub-cool the incoming refrigerant liquid phase, partially condense the inlet refrigerant vapour and, thus, cool and liquefy the natural gas. The refrigeration duty required to condense and sub-cool the refrigerant is normally a large portion of the total duty.

The optimum number of stages for partial condensation, separation and throttling depends on the relative importance of capital cost, operating complexity and operating costs. The more stages the higher the energy efficiency and complexity. When comparing processes with increased complexity, the reduced power demand by the compressors can roughly be assumed as shown in Table 4.

Many peak shaving plants utilize a single mixed refrigeration system - designated Prico® - as an efficient and simple way to produce LNG, as shown in Figure 16. Pre-treated gas is fed to the main heat exchanger where it initially cooles to between -45°C and -73°C. Gas and heavy hydrocarbons, which might solidify at LNG temperatures, are removed from the exchanger at this point and sent to a separator. Cold gas is then returned to the main heat exchanger where it is totally liquefied and sub-cooled. LNG exits the heat exchanger usually at temperatures between minus 150°C and 160°C as it is sent to storage at near atmospheric pressure.

The closed loop refrigeration circuit compresses and cools the high-pressure refrigerant, which is partly gas and liquid. The liquid is pressurised by pumping and mixed with the vapour at the inlet of the main heat exchanger. This two-phase mixture passes the heat exchanger and leaves the exchanger in liquid phase at essentially the same temperature as the LNG.

The refrigerant is then throttled in a control valve before returning to the heat exchanger at reduced pressure. The low-pressure refrigerant stream vaporizes up-flow in the heat exchanger and provides all the refrigeration that is required to condense the refrigerant and for LNG production. The loop is complete as the refrigerant returns to the compressor.

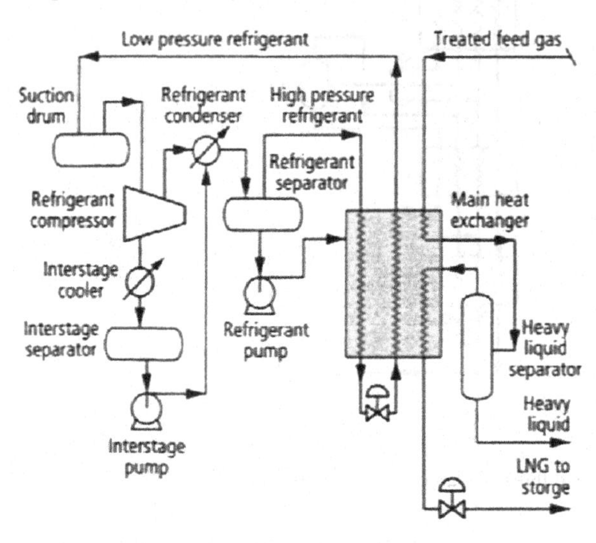

Figure 16 Single mixed refrigerant cycle. (Prico) [19]

5.2 Comparison of power requirements

A comparison of power requirements of different cycles can only be approximate because of the difficulty in defining equivalent operating conditions for the various cycles. Table 4 and Table 5 present values for different cycles, and for different versions of the mixed refrigerant cycle.

Table 5 Comparison of efficiencies for mixed refrigerant cycles. [22]

Number of stages	Approximate power consumption relative to one stage process
1 (refer Figure 16)	1
2	0.93
3 (refer Figure 15)	0.9
4	0.88
5	0.87

6. THE ECONOMICS FOR SINGLE MIXED LIQUEFACTION PLANTS [19]

6.1 Compressor drive selection

The refrigeration compressor is usually the most expensive equipment of small-scale liquefaction plants. Smaller units may employ screw or reciprocating compressors, whereas centrifugal compressors always are used in larger units. Driver selection is a key decision point in project development, and can be either turbine, gas engine or electric motor.

Electric motor drives are preferred if the power cost is low, since electric packages are less costly than turbine drives, and motors generally have lower maintenance requirements. Turbine drives are preferred if the fuel gas cost is more attractive than the cost of electric power. Turbines can also provide process heat for the pre-treatment system and may, thus, reduce the need for heating equipment. In many installations, gas turbine fuel is provided from flash gas in the main process.

6.2 Storage and send-out

LNG plants for peak shaving and for vehicle fuel production are basically similar in terms of pre-treatment and liquefaction. However, these plants differ greatly in terms of LNG storage and send-out facilities as shown on Table 6. Typical of peak shaving units is that they involve large atmospheric LNG storage tanks, LNG pumping to high pressure and LNG vaporization to a pipeline, in contrast to plain liquefaction that supply gas for end-usage at low or moderate pressure. LNG fuel plants typically store LNG for five to ten days. Storage may be low pressure or medium pressure tanks rated at 8-10 bar.

Liquefied natural gas is loaded to truck or rail cars and sent to market as a liquid. LNG fuel plants are usually running year round, while the peak shaving units run only in off-peak periods - typically in the spring to autumn season - to the extent that storage is complete for the high demand winter season. This distinction, however, results in large variations in the capital cost, and should be carefully considered in the context of transit countries.

6.3 Project cost

As turbine-driven plants generally cost 2-4 million Euro more than similar plants driven by electric motors the selection of drive will reflect the cost of the plant versus capacity as indicated on Figure 17 showing the regression line for LNG plant cost development versus time (year). Likewise, Figure 18 shows the cost development versus the development of LNG train size [10]. The investment may also reflect various requirements for the liquefaction train and necessary support facilities.

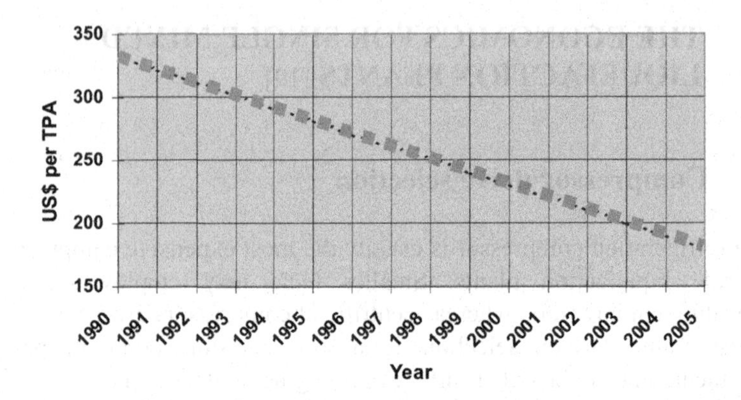

Figure 17 Typical LNG plant cost development versus year [10].

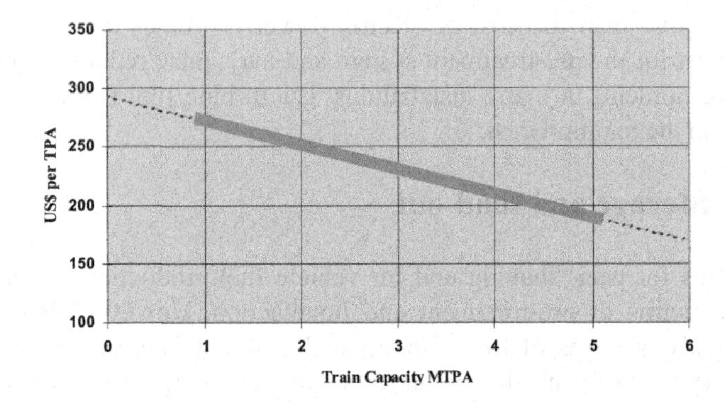

Figure 18 LNG plant cost versus train capacity [10]

Figure 19 indicates the relative plant cost versus LNG production capacity of small-scale LNG plants relative to the unit cost in large-scale production.

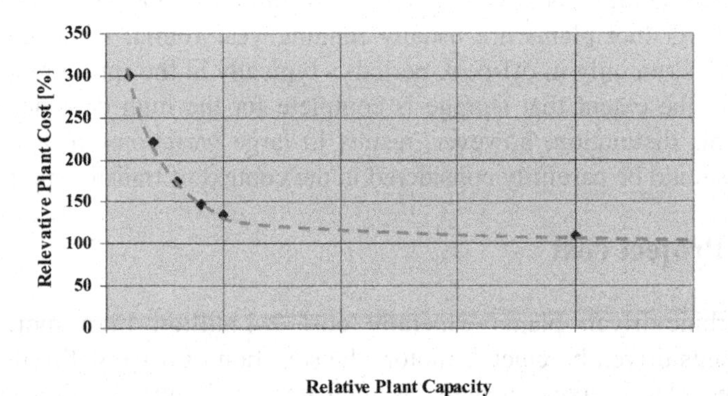

Figure 19 Relative plant cost versus production capacity of small-scale LNG plants relative to state-of-the-art base-load LNG plants.

Table 6 Comparative project cost [19]

	Peak shaving		Vehicle fuel	
	300 ton/day liquefaction		300 ton/day liquefaction	
	100 000 m^3 storage		7 000 m^3 storage	
	4 000 ton/day send-out		600 ton/day send-out	
	200 day/year operation		350 day/year operation	
Drives	**Motor**	**Turbine**	**Motor**	**Turbine**
Power, US¢/kWh	3	5	3	5
Fuel, USD/GJ	3	2	3	2
Capital, million USD	39	43	23	27
Operating cost, USD/ton	23.5	19.5	23.5	19.5
Capital, USD/ton	78	56	25.5	30
LNG to tank, USD/ton	101.5	105.5	49	49.5

7. SMALL-SCALE NATURAL GAS LIQUEFIERS

Small-scale natural gas liquefiers are developed at several organizations, mainly for the LNG vehicle market. The capacity for these liquefiers may be as low as 75-225 kg/h. Gas Technology Institute in USA is working on a single mixed refrigeration system with an ordinary oil-lubricated screw compressor and a plate-fin heat exchanger. The main purpose of the project is to avoid the conventional negative scaling effect, and to achieve an LNG price that can compete with the large production units [23].

SINTEF Energy Research has taken a similar course, and is aiming at a unit with a maximum LNG capacity about 400 kg/h. Oil-lubricated screw compressors and copper-brazed plate heat exchangers are used, so far in a pilot plant with a production rate of 50 kg/h LNG as shown on Figure 20 [24]

For small-scale LNG production, the cost of pre-treatment of the natural gas, i.e. drying and CO_2 removal, may constitute a larger portion of the total cost. Cryogenic CO_2 removal has been used for peak-shaving plants by Chicago Bridge & Iron Inc. This feature is also included in the small-scale LNG concept by Idaho National Engineering and Environmental Laboratory, in a letdown cycle [25] as shown on Figure 21.

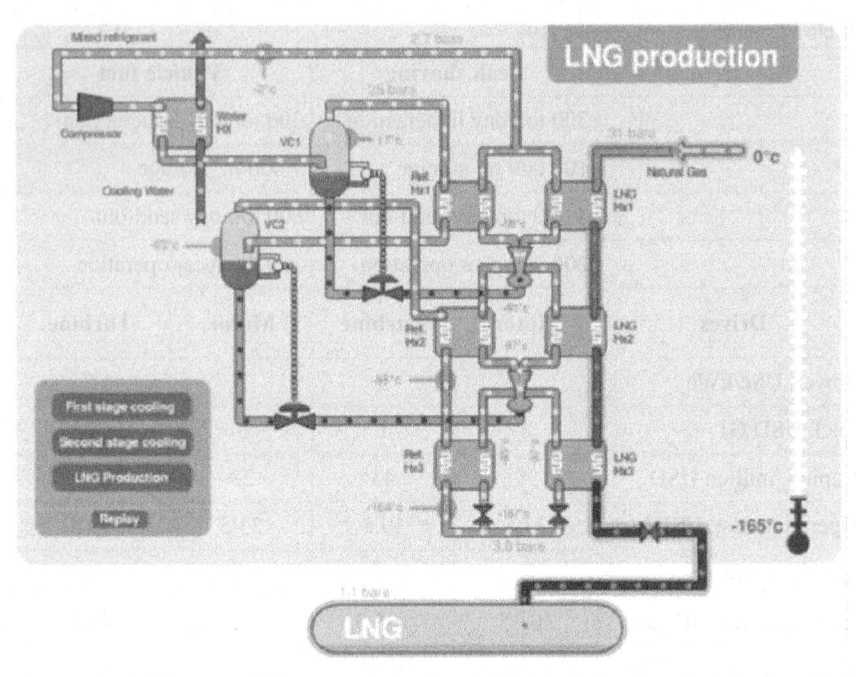

Figure 20 Small-scale LNG concept developed by SINTEF using basically standard refrigeration equipment such as copper-brazed heat exchangers and oil-lubricated screw compressors in order to suppress investment cost. The cycle uses a mixed refrigerant (N_2, C_1, C_2, C_3, C_4), and will require about 0.6 kWh per kg LNG.

A different approach has been taken in the development of the LNG Micro-Cell Process at Curtin University of Technology as shown on Figure 22 [26,27]. The capacity of the unit is 1200 l LNG/day, corresponding to 25 kg/h. Reportedly, the production rate can be increased by a factor of 10. As used in the mini LNG module the process is based on a combination of the conventional nitrogen and compression cycle. The system makes use of a mixture of three refrigerants in order to achieve continuous refrigeration at temperatures between –80 °C and –110 °C, and then turns to liquid nitrogen down to –161 °C. The nitrogen is liquefied under controlled pressure and temperature using the compression refrigeration cycle, then recycling it through a natural gas heat exchanger chamber. Liquid nitrogen is continuously cycled between the compressor cycle and the LNG production chamber for maximum efficiency. The CO_2 in the natural gas is frozen out as a solid in the LNG chamber and removed in a cyclone.

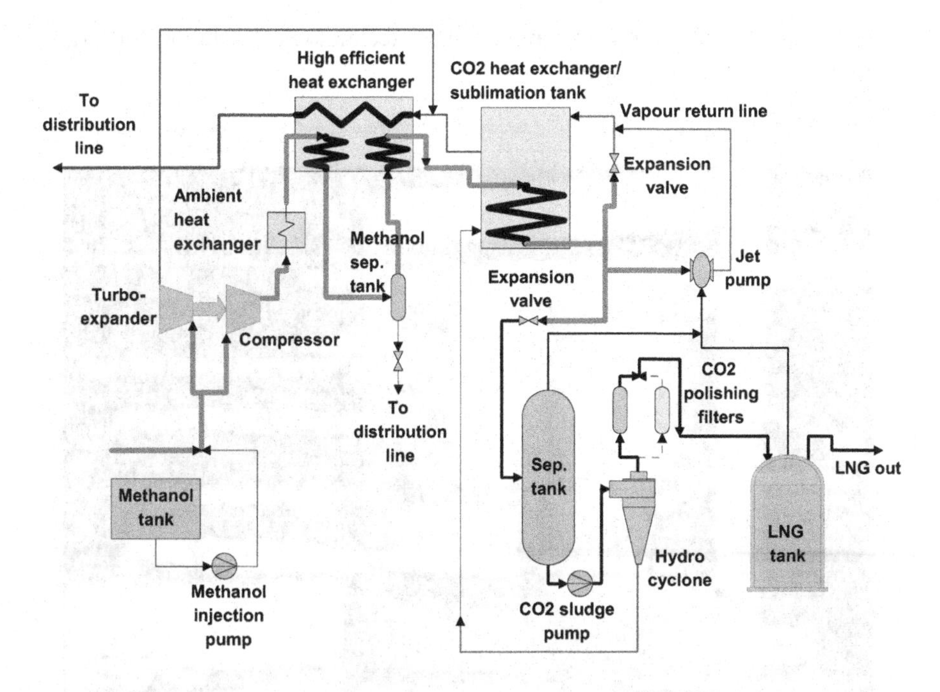

Figure 21 Natural gas expander cycle with CO_2 freeze out included.

8. NON-TRADITIONAL TECHNOLOGIES FOR NATURAL GAS LIQUEFACTION

Cryogenic refrigeration with compressors, refrigerants and eventually expanders is the common method for LNG production. Several new methods are under development.

8.1 Thermo-acoustic heat engines and refrigerators [28]

By combining two conversion technologies a cryogenic refrigeration duty can be provided without moving parts at temperatures as low as –240 °C. A thermo-acoustic heat engine and refrigerator concept consists of a collection of heat exchangers arranged within a network of piping containing pressurized helium gas. In the engine, one heat exchanger is heated up to roughly 700 °C, a second heat exchanger is held at ambient temperature, whereas a third heat exchanger - operating in-between the two - is thermally floating. Owing to the heat input the temperature gradient across the heat exchangers causes an oscillating pressure wave in the helium gas. This oscillating wave drives the pulse tube refrigerator and produces refrigeration power at cryogenic temperatures. It is assumed that the simplicity of the system could result in low manufacturing cost and high reliability, but the underlying physics is rather complex and a technological solution for larger systems is not yet developed.

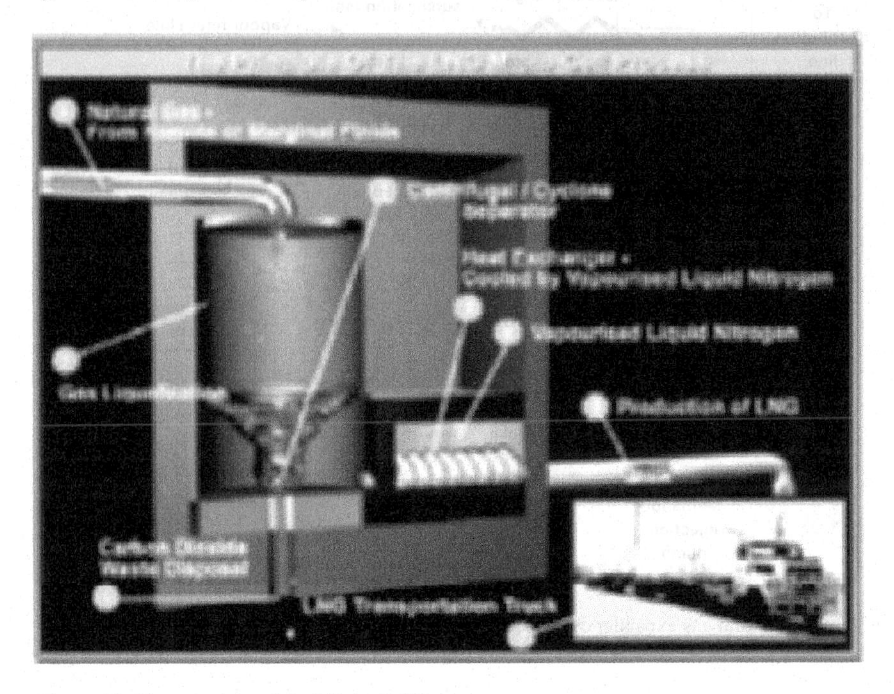

Figure 22 The principle of the Micro-Cell Process.

The engineering challenges include heat exchanger design, performance and accommodation of high pressures, temperatures and thermal stresses. If successfully developed the technology could be applied in the liquefaction of natural gas to provide a low-cost alternative to diesel fuel. So far one unit is reported built having a liquefaction capacity of about 35 kg/h. In this unit, 30% of the input natural gas stream was consumed as heat input, with a 70% yield of LNG. A future system with a capacity of about 700 kg/h LNG and with a projected liquefaction rate of 85 % of the input gas stream is under development.

8.2 Magnetic Refrigeration [29]

Magnetic refrigeration technology takes advantage of the magneto-caloric effect – namely the ability of a magnetic material to increase its temperature in the presence of a magnetic field, and vice-versa to loose temperature as the field vanish. The Curie point is known as the material's ordering temperature where most of the change in magnetic entropy occurs. This is a characteristic point as the material changes from being ferromagnetic to paramagnetic, and the farther away from the Curie point the weaker is the magneto-caloric effect.

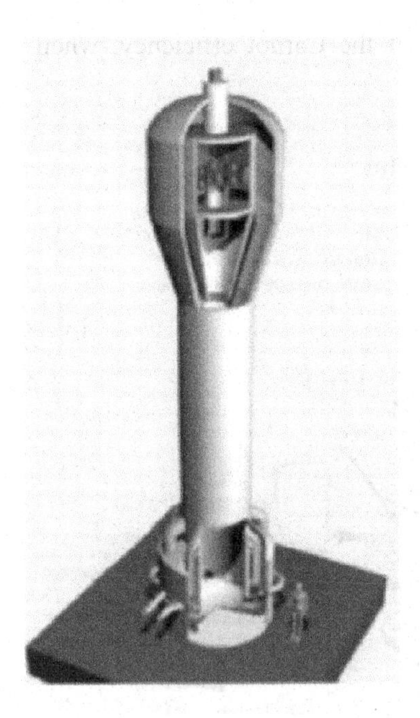

Figure 23 Conceptual configuration, 700 kg/h TASHE-OPTR.

The useful portion of the magneto-caloric effect is typically about 25°C on either side of the material's Curie temperature. Therefore, in order to span over a wide temperature range, a refrigerator must contain several different coolants arranged according to their differing ordering temperatures.

Magneto-caloric materials store heat energy in the way that atoms vibrate and in the way in which electrons spin within each atom. When a strong magnetic field is applied to the coolant magnetic material, the magnetic moments of its atoms become aligned, making the system more ordered. As the ordered materials have lower entropy they are prone to compensate for the entropy loss by increasing the temperature. When a strong magnetic field is removed, the magnetic moments return to their random directions. Hence, the entropy increases and the temperature of the material drops. Typically, the temperature of a material can drop by about 10 to 15 °C, depending on the strength of the magnetic field.

Research work has revealed that a germanium-silicon based material exhibits a fairly strong magneto-caloric effect in the range from 30 Kelvin to near room temperature. This can be obtained by adjusting the ratio of silicon to germanium. Excellent heat transfer between the circulating fluid and the magnetic material is essential, and the magnetic material should be porous in order to enhance the surface area. Active magnetic regeneration (AMR) characterises magnetic cycles where the magnetic material can simultaneously act as the refrigerant and the regenerator.

Magnetic refrigeration offers high thermodynamic efficiencies due to the reversibility of the magneto-caloric effect. For active magnetic regeneration the efficiency is expected to reach 60 % of the Carnot efficiency, when working between 300 K and 30 K.

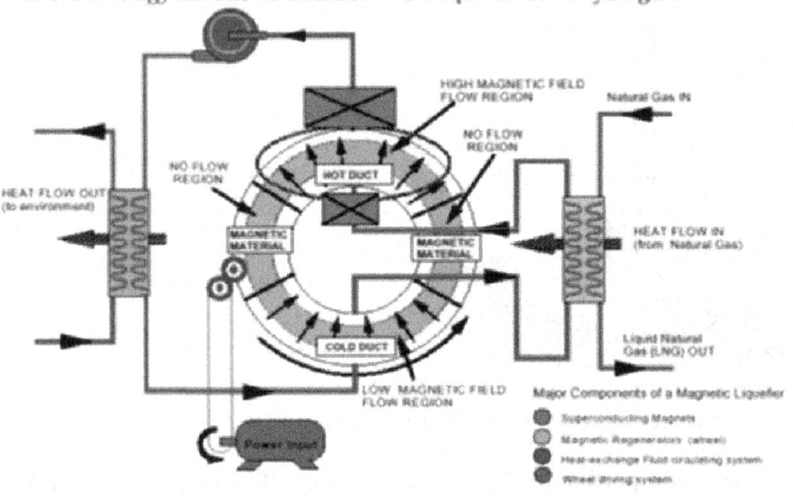
A Rotary AMR Liquefier
The Cryofuel Systems Group at UVic is developing an AMR refrigerator for the purpose of liquefying natural gas. A rotary configuration is used to move magnetic material into and out of a superconducting magnet. This technology can also be extended to the liquefaction of hydrogen.
HIGH MAGNETIC FIELD FLOW REGION
Natural Gas IN
NO FLOW REGION
NO FLOW REGION
HOT DUCT
HEAT FLOW OUT
MAGNETIC MATERIAL
MAGNETIC MATERIAL
HEAT FLOW IN (from Natural Gas)
COLD DUCT
Liquid Natural Gas (LNG) OUT
LOW MAGNETIC FIELD FLOW REGION
Major Components of a Magnetic Liquefier

Figure 24 Example of an active magnetic regeneration concept (AMR) used in a cooling and refrigeration system for LNG production.

9. CONCLUSION

For several reasons LNG may answer directly to the needs of transit countries for the use of their option gas. Firstly, because of the high energy density that facilitates storage and transport in cryogenic tanks – either for peak shaving or as a fuel. Secondly, because LNG offers a true option to pipeline distribution infrastructure for natural gas - as would be the case in some transit countries. Thirdly, LNG is gaining acceptance as an alternative energy carrier for various applications. The reason is the increased competition in the market combined with technological progress over the latest years that have reduced the cost of the entire LNG supply chain including production, transport and storage.

As LNG is 600 times more compact than natural gas – and the energy density is 600 times higher - LNG can efficiently be stored and transported on road, rail and sea in cryogenic tanks. LNG is easy to re-gasify and can, thus, serve as an easy extension of the general gas supply system. LNG can further be used as vehicle fuel and for peak shaving. To serve these purposes small-scale LNG units could be deployed at the transit pipeline to process the option gas year round. Peak shaving would require a storage capacity large enough to receive the liquefied gas over the low-demand season, and a high-pressure discharge system in order to supply the gas over the high-demand season.

LNG production involves cryogenic technologies that include concepts, refrigerants and process design. A state-of-the-art review of liquefaction concepts – especially on smaller scale – have been given. Cost level and relative pricing have been indicated in order to provide rough estimates.

REFERENCES

1. Brendeng, E.; Grini, P. G.; Jørstad, O.; Melaaen, I. S.; Maehlum, H.S.; Owren, G.A.; Puntervold, S.: *'Measurement and Calculation of Thermodynamic Properties of Natural Gas Mixtures'*, NTNU-SINTEF 1993.
2. Hetland, J.: *'Norwegian Frontier Technologies for Liquefaction and Deep-Water Exploitation of Natural Gas'*, Presented at the OPET – International Workshop on Natural Gas; Technologies, Opportunities and Development Aspects; Vaasa, Finland, 31 May, 2002
3. Brendeng E.; Neeraas B. O.; Owren G.: *'Research in Norway at NTNU and Sintef'*, LNG Journal, July/Aug 1999
4. Fredheim A.; Jørstad O.; Owren G.; Vist S., Neeraas, B. O.: *'Coil, a model for simulation of spiral wound LNG heat exchangers'*, World Gas Conference, Nice, 2000
5. Fredheim, A.O.: *'Thermal design of coil-wound LNG heat exchangers, shell-side heat transfer and pressure drop'*, Dr.ing. thesis, NTH, 1994
6. Neeraas, B.O.: *'Condensation of hydrocarbon mixtures in coil-wound LNG heat exchangers, tube-side heat transfer and pressure drop'*, Dr.ing. thesis, NTH, 1993
7. Aunan, B.: *'Thermodynamic analysis of small and medium scale natural gas liquefaction cycles'*. Public lecture for the degree of doctor ingeniør, The Norwegian University for Science and Technology, 1999
8. *'Answers LNG FAQS. Frequently Asked LNG Questions'*. Northstar Industries, Methuen, MA, USA. http://northsaring.com/lngfaqs.html, 2003-04-15
9. Brendeng, E.: *'Multicomponent Refrigerants – a Giant Success for an Old Idea'*, Refrigeration, Energy and Environment, International Sympositum on the 40th anniversary of NTH Refrigeration Engineering, June 22-24, Trondheim (Norway), 1992
10. Technic-Coflexip: 'LNG Plant Pricing Considerations', http://www.technip-coflexip.com/pdf/Morgan_Stanley_Oil.pdf
11. Ohishi et al: *'Availability of Refrigeration Process of Base-Load Plant'*, AICHE 2002, New Orleans
12. Shell information, http://www.shellglobalsolutions.com/news_room/press_cuttings/benchmarking03.htm

13. Litzke, W-L.; Wegrzyn, J.: *'Natural gas as a Future Fuel for Heavy-Duty Vehicles'*, The Engineering Society For Advancing Mobility Land Sea Air and Space International, Government/Industry Meeting Washington, D.C. May 14-16, 2001, SAE Technical Paper Series 2001-01-2067
14. Melaaen, I.S.: *'Pretreatment in LNG plants. Accepted Processes and Possible Future Developments'*, Public lecture for the degree of Doktor Ingeniør, The Norwegian University of Science and Technology, Trondheim, 1993
15. Owren, G.; Moger, J.; Neraas, B.O.: Plant-specific information on LNG plants and storage provided by Statoil, 29 April 2003
16. Stone, J.B.: *'Applying New Technology to Lower LNG Cost'* 4th Doha Conference on Natural Gas, Exxon Mobile, Quatar, 2000
17. *'Snøhvit Gas Field, Barents Sea, Norway',* the website for the offshore oil & gas industry, http://www.offshore-technology.com/projects/snohvit/
18. *'Snøhvit The World's Northernmost LNG project'.* Statoil website. https://www.statoil.com/STATOILCOM/snohvit/svg02699.nsf?OpenDatabase&lang=en
19. Price, B.C.: *'Small-scale LNG facility development'.* Hydrocarbon Processing, January 2003
20. Vink, K.L.; Nagelvoort, R.K.: 'Comparison of Baseload Liquefaction Processes', LNG 12, Perth, Australia, 4-7 May 1998
21. Yu Nan Liu, Baugherty, T.L.; Bronienbrenner, J.C.: 'LNG Liquefier Cycle Efficiency' LNG 12, Perth, Australia, 4-7 May 1998
22. Finn, A.J.; Johnson, G.L.; Tomlinson, T.R.: 'Developments in natural gas liquefaction', Hydrocarbon Processing, April 1999, Vol. 78, No. 4
23. Wurm, J.; Kountz, K.K.; Liss, W.E.: *'Small-Scale Natural Gas Liquefier Development',* IIR Cryogenics 2002, Prague, April 23-26 2002
24. *'Flytendegjøring av naturgass'*, Norwegian Patent PCT/NO01/ 00048
25. Wilding, B.: *'Sacramento Small Scale Liquefier Plant'*, NGVTF Technical Committee Meeting January 28-29, 2003
26. *'Kurdish Professor Innovation in Australia'*, SUN, No. (4) June, 2002.
27. 'Small Scale LNG Unit, Large Scale Potential'. Australian Energy News, Issue 22, December 2001
28. Wollan, J.J.; Swift, G.W.; Backhaus, S.; Gardner, D.L.: *'Development of a Thermoacoustic Natural Gas Liquefier',* 2002 AIChE New Orleans Meeting, March 11-14
29. Hall, J.L., J.A. Barclay, *'Analyzing Magnetic Refrigeration Efficiency: A Rotary AMR-Reverse Brayton Case Study',* Advances in Cryogenic Engineering, V. 43B, pp. 1719-1728, 1998

ON THE RELEVANCE OF INTEGRATING LNG WITH THE ENERGY SUPPLY SYSTEMS OF TRANSIT COUNTRIES

LNG Chains and Storage Facilities

Einar BRENDENG, Jens HETLAND
SINTEF Energy Research, Dept. of Energy Processes, Kolbjorn Hejes vei 1A, N-7465 Trondheim, NORWAY

Abstract: Usually security of energy supply means diversified provisions and infrastructure development. Sometimes special precautions are required, as for instance when stretching a gas pipelines through a transit country. In this event security of supply would include pre-emptive actions to ensure stability on a long-term basis. One initiative is to enable the transit country to prosper from the option gas that it receives as compensation for land lease.

This article addresses LNG as an alternative to pipeline distribution. In contrast to natural gas LNG requires no firm pipeline infrastructure, as it is transported as a bulk commodity in cryogenic tanks on road, rail and sea. The article suggests that countries that lack a well-developed gas distribution system may draw upon recent experience from countries that employ LNG, as for instance Norway. Although Norway is among the largest exporters of natural gas in the world, the domestic outlook is that natural gas will be introduced to the Norwegian energy system basically in liquid phase.

The paper describes vital elements pertaining to the LNG chain. This includes the provision and distribution of LNG, principles for satellite systems, peak shaving, and suggestions on how LNG can be used as a fuel for prime movers in ships and vehicles. As cryogenic tanks are integral parts of any LNG concept, some tank concepts and containers are presented. Eventually, the combination of small-scale LNG provision and large tank volumes may offer a viable alternative to underground storage of natural gas (UGS).

Key words: LNG, LNG usage, peak shaving, LNG storage, LNG distribution, cryogenic tanks

J. Hetland and T. Gochitashvili (eds.),
Security of Natural Gas Supply through Transit Countries, 103–133.

1. INTRODUCTION

A crucial question is how to provide clean energy in the future? As energy management closely relates to the environmental concern, there is a switch from coal and oil to natural gas. From 1980 to 1999 the world's energy demand increased annually by 1.5%, and, eventually, only by 1.2% per year since 1990 [1]. The recordings of CO_2 emission are interesting in the post Kyoto debate since they are useful in explaining to which extent industrialised countries reduce their greenhouse gas emissions – especially in power generation and transport. The share of CO_2 from these two sectors reached a total of 58% in 1998 against 48% in 1980. From 1990 to 2001 the global emissions of CO_2 increased by 6% whereas the European Union seemingly stabilised its emissions over the same time span, much owing to a massive fuel switching from coal and oil to natural gas (Figure 1).

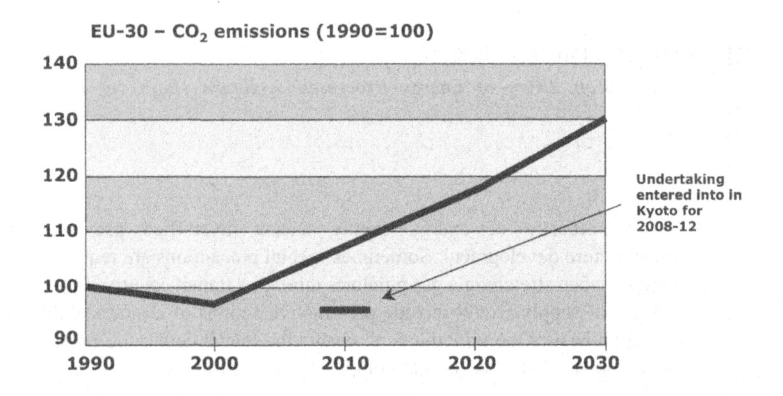

Figure 1 Prognosis of the greenhouse gas emissions caused by the European Union up to 2030 [2]. De-facto emission so far showing a reduction with a minimum in the year 2000 – mainly owing to fuel switching from coal and oil to natural gas. Beyond 2000 up to 2030 a substantial increase is expected – requiring new measures and initiatives.

However, Norway has taken an extreme position regarding natural gas: Despite its position as one of the world's largest gas exporters Norway has no tradition for using natural gas. Neither is fuel switching an issue because of the predominance of Norwegian hydropower (99.4%). So far only 1% of its produced gas is consumed domestically - mainly LNG produced by a methanol plant at Tjeldbergodden (refer section 6.3). The outlook is that only limited amounts of natural gas will be introduced to the Norwegian infrastructure - basically LNG.

For more than fifty years LNG has had a practical role to play especially in the United States and Japan. Owing to the progress in large-scale liquefaction technology, LNG is gradually gaining acceptance as an alternative energy carrier for various applications. The growing interest is much due to a substantial cost reduction over the entire LNG chain from production via transport and storage.

In several cases LNG is prone to be regarded as an economical alternative to developing new pipeline capacity. Another important aspect relates to the security of the energy supply pertaining to the industrialised world. Technically, LNG carriers are capable of diverting LNG to most countries around the world that are connected to the sea - independent of a rigid infrastructure. Hence, the supply of LNG is by nature less vulnerable, and the safety records of transportation are outstanding. However, as the number of LNG-importing countries is still rather limited 48.5% of the global LNG production is for Japan [3].

2. NATURAL GAS ON A MACRO-SCALE

Natural gas is widely used for heating and cooking purposes, but has recently gained increased interest for power generation for two reasons: 1) fuel switching from coal or oil to gas in order to reduce the greenhouse gas emission, and 2) the introduction of combined power cycles. The latter owes to the ability of natural gas to provide so high firing temperature in modern gas turbines that the exhaust gas can be used to generate steam for a steam turbine bottoming cycle. This combination offers electric efficiencies up to 58% (current state of the art).

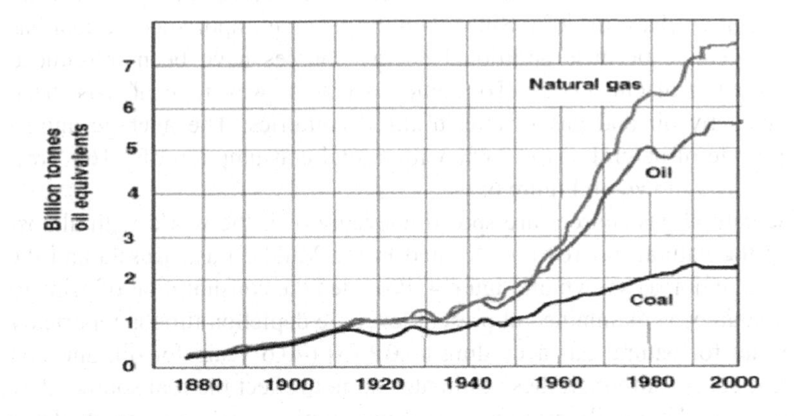

Figure 2 Global use of primary energy since 1875 until present time.

Natural gas is strategically important as it accounts for 24.1% of the global market for fossil fuels (2002) almost the same level as coal [3]. In 2002 the global annual consumption of fossil fuels reached 8.2 billion tons oil equivalents (TOE), or 350 EJ (i.e. 10^{18} Joule). Today fossil fuels stand for 87% of the global usage of primary energy. Figure 2 shows the historic development of the global usage of primary energy since the 1870-ies up to present time. According to BP the world consumption of natural gas grew by 2.8% in 2002 on the back of a 3.9% increase in the US consumption, and more than 7% in non-OECD Asia Pacific. However, coal was reported to be the far most increasing fuel in 2002 with 28% increase in China and 7% on the world basis.

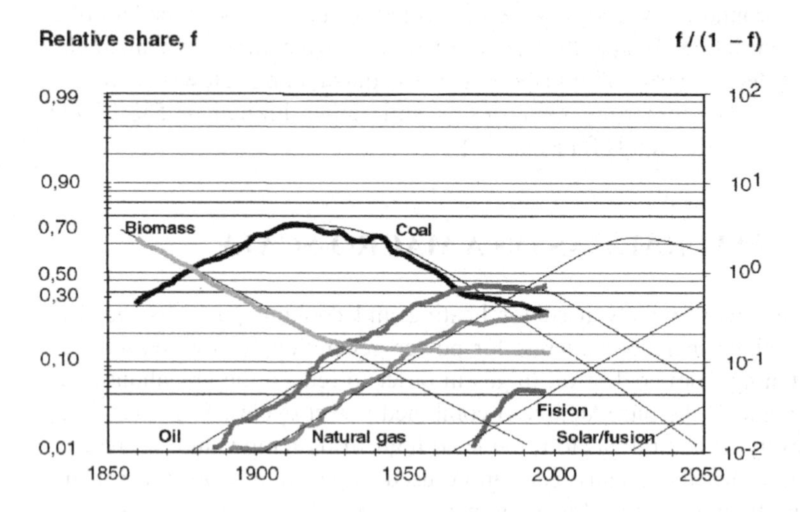

Figure 3 showing the world primary energy substitutions since 1850 until present time.

As indicated in Figure 3 the relative share of coal reached a peak in 1900 with a relative share of 70%. Since then the relative importance of coal has been on decline because additional energy sources have been introduced rather than replacing coal. To some extent, however, coal has been substituted by oil and gas in industrialised countries. The average annual increase rate of natural gas is 2.6% with a total consumption of 2.164 Tm_n^3 p.a. (2000) as shown on Figure 6.

The natural gas sources are spread unevenly over the world with almost 70% of the natural gas reserves located in the Middle East, Russia and the Caucasus region as shown on Figure 4. Provided the consumption of primary energy sources is maintained at present level the depletion time is reportedly 60.7 years for natural gas according to BP [3] (40.6 years for oil, and 204 years for coal)[1]. Although these estimates do not reflect the real source, they confirm that the fossil resources are rather limited and may be depleted within a few generations.

[1] Defined as the estimated quantum of the reserves to the annual production rate In this context it should be mentioned that *reserves* only reflects sources that are discovered and deemed economically exploitable. This means that there is no direct correlation between *reserves* and *resources* – as new sources are added to the *reserves* each year – especially natural gas.

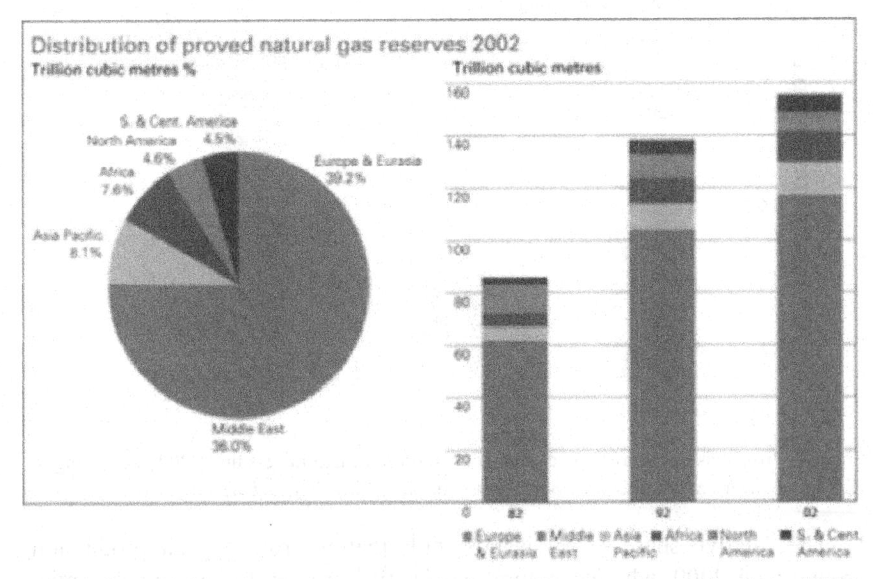

Figure 4 Distribution of proved natural gas reserves 2002 according to BP statistics

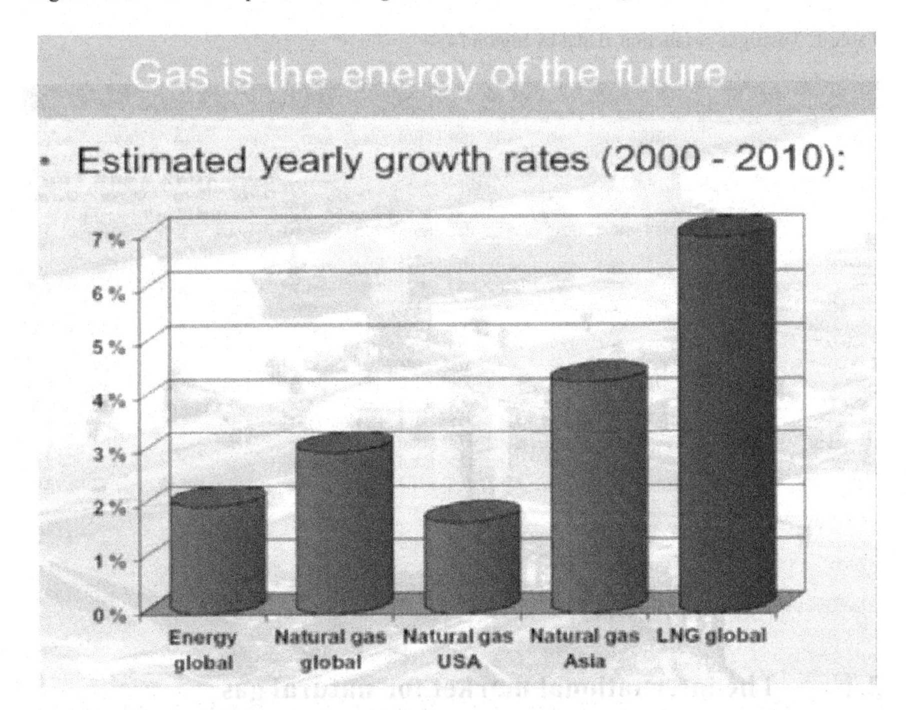

Figure 5 LNG Is Growing Fastest - around 7-8% aai over the last 20 years. In 2000: 137 million Sm^3 = 100 million tons LNG

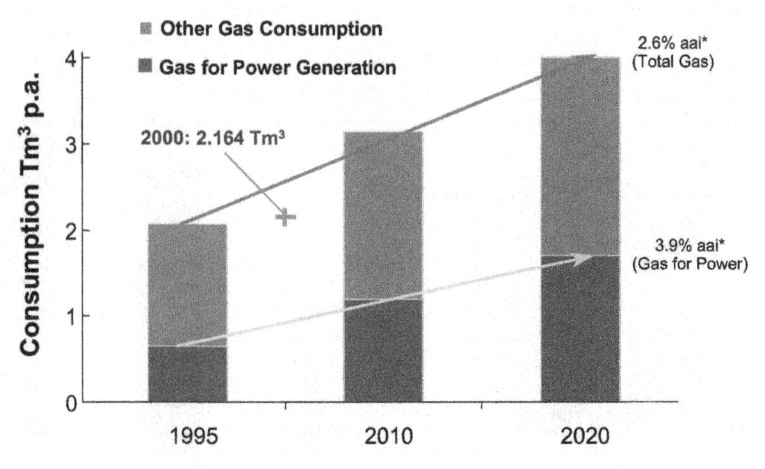

Figure 6 Prognosis of the average annual consumption of natural gas up to 2020 according to IEA. (aai* means average annual increase. 1 $Tm_n^3 \equiv 1012m^3 \equiv 37.3$ Tcf)

The relative share of oil is now culminating from a peak production around year 2000, whereas natural gas is still foreseen to have a substantial relative growth potential. However, owing to the limited reserves of fossil fuels the role of natural gas can only be seen as a *bridge to the future.*

Table 1: Total gas production: Total by region [4]

Mtoe	1980	1985	1990	1995	1997	1998	1999	85/80	90/85	97/90	98/97	99/98
									Annual % Change	Annual % Change	Annual % Change	Annual % Change
								Annual % Change				
World	1245,4	1429,1	1703,9	1818,1	1910,8	1953,1	1998,1	2,8%	3,6%	1,7%	2,2%	2,3%
								-	-	-	-	-
Western Europe	156,1	155,3	157,0	194,8	223,1	222,8	229,6	-0,1%	0,2%	5,1%	-0,1%	3,0%
European Union	*133,3*	*131,9*	*132,9*	*166,6*	*182,1*	*181,5*	*185,5*	-0,2%	0,2%	4,6%	-0,4%	2,2%
EFTA	*22,8*	*23,4*	*24,1*	*28,3*	*41,0*	*41,3*	*44,1*	0,6%	0,6%	7,9%	0,8%	6,7%
Rest of OECD	549,9	496,2	553,5	621,2	640,8	649,4	647,7	-2,0%	2,2%	2,1%	1,3%	-0,3%
NAFTA	*539,7*	*480,0*	*530,5*	*590,3*	*608,3*	*616,2*	*613,2*	-2,3%	2,0%	2,0%	1,3%	-0,5%
OECD Pacific	*10,2*	*16,2*	*22,8*	*30,8*	*32,3*	*32,8*	*33,9*	9,7%	7,1%	5,1%	1,5%	3,5%
Central and Eastern Europe	43,6	44,2	32,0	24,3	20,9	19,6	18,6	0,3%	-6,2%	-5,9%	-5,8%	-5,4%
CIS (1)	359,6	520,1	656,3	569,0	541,4	554,4	560,6	7,7%	4,8%	-2,7%	2,4%	1,1%
Africa	20,3	42,4	61,8	75,2	89,7	94,9	105,7	15,9%	7,8%	5,5%	5,8%	11,4%
Middle East	36,1	54,3	82,2	124,4	147,9	156,6	163,0	8,5%	8,7%	8,8%	5,9%	4,1%
Asia	48,3	75,4	109,4	145,7	172,2	175,0	187,4	9,3%	7,7%	6,7%	1,6%	7,1%
Latin America	31,6	41,3	51,7	63,5	74,8	80,4	85,6	5,5%	4,6%	5,4%	7,5%	6,4%
of which (%)												
European Union	10,7	9,2	7,8	9,2	9,5	9,3	9,3	-2,9%	-3,3%	2,9%	-2,5%	-0,1%
OECD	56,7	45,6	41,7	44,9	45,2	44,6	43,9	-4,3%	-1,8%	1,2%	-1,3%	-1,7%

(1) Including Baltic countries for statistical reasons

2.1 The international market for natural gas

In contrast to oil and coal, the traditional gas export has basically been linked to pipeline construction and infrastructure development. In a future perspective new solutions may eventually change this picture. Typical of the international market for natural gas is that it is divided into several regional markets, and that trading links between these markets are fairly rare. On the other hand the volume of international gas trading is expanding. This underlines the relative importance of natural gas in modern societies. Still, however, natural gas is mostly consumed in the vicinity of its origin, and only limited amounts are exported.

Natural gas constitutes an important energy source as it provides the lowest emission of CO_2 among fossil fuels. In addition natural gas is widely used as feedstock for the petrochemical industry, such as methanol and ammonia production – and may on a future perspective also have a potential for bio-protein production. The only disadvantage of natural gas seems to be a fairly costly storage – either as under ground (UGS) or compressed or in condensed phase (CNG, LNG).

3. TRANSIT COUNTRIES

The export of natural gas usually involves infrastructure that requires access to land that affects third parties. Eventually, this may represent a potential security problem. Therefore, gas pipelines - besides being a transport device – may also be conceived as a geopolitical issue - especially the new export pipeline from Baku to Turkey (with further perspectives to Europe). The transit countries are remunerated in kind reckoned as a certain percentage of the gas flow (option gas[2]). As several Caucasian societies are in recession the option gas represents an important stimulus that may give boost to increased prosperity and industrial growth, and is important for the restructuring of these societies. Immediate questions are: How may the *option gas* fit to local energy plans, and b) how can the social stability of the region be improved?

The South-Caucasian Gas Pipeline System (SCP) from Azerbaijan (Shah Deniz) to Turkey is now being planned. It is scheduled to go on stream in 2005. Its capacity will gradually increase as the Caspian gas fields develop. Hence, the amount of the *option gas* for the transit countries may vary a) over the seasons, and b) from year to year, similar to the fluctuating demand.

In Georgia, for instance, the energy dependency is twice as high as the average international level [5]. The prevailing practice among European importers of fuel is a reserve requirement corresponding to approximately 50 days of supply in average. This is seen as necessary in order to sustain critical situations.

[2] Georgia will receive 5% of the gas that passes its territory.

4. LIQUEFIED NATURAL GAS (LNG)

By cooling the natural gas to very low temperatures it will condense to liquid phase when reaching minus 162 °C at atmospheric pressure – the boiling point of methane. LNG is compact and easy to transport by sea, rail and road – independent on gas pipelines. LNG is also simple to re-gasify for the delivery of purified methane. In contrast to natural gas that contains typically more than 90% methane, and some ethane, propane and heavier hydrocarbons, the liquefaction process involves pre-treatment of the gas in order to remove oxygen, carbon dioxide, sulphur compounds, and water. Impurities like other petroleum gases as butane and propane are also removed in the liquefaction process, and the presence of nitrogen is usually limited to about 1%. Liquefaction results in a volume reduction of approximately 600 times. This corresponds to an energy density that is 600 times higher than natural gas at atmospheric conditions. Since LNG has no odour, leak detection requires special instrumentation. Furthermore, LNG is colourless, non-toxic, and non-carcinogenic, and its specific weight is 45% of water. LNG – or natural gas - is less flammable than gasoline, and is indeed rather difficult to ignite. It has the narrowest range in which it ignites: In vaporised phase - mixed with air – the gas can only ignite if the concentration of methane is 5-15%. Neither LNG, nor its vapour can explode in an unconfined environment [6,7].

The current, global LNG production is almost 120 MTPA provided by 69 liquefaction trains. Furthermore, there are 120 satellite facilities. The distribution is 39 satellite and 55 liquefaction facilities in the United States, and 81 satellites and 14 liquefaction facilities in the remaining countries. Owing to the deregulation of the natural gas industry, the construction of LNG facilities in the United States has increased, and the prevailing role of LNG facilities is changing from peak shaving to base-load for the supply of natural gas year round [7,8].

4.1 Competing technologies

The concept map shown in Figure 7 indicates the viability of LNG versus the distance from the liquefaction plant to the end-user and the annual bulk quantities, with comparison to optional technologies. There are windows or areas in which technologies appear to be viable. The windows for CNG and small-scale LNG facilities are interesting – particularly in regards of gas supply to regions that lack infrastructure and distribution pipelines.

New LNG technology has made LNG cheaper and simpler to handle. This implies that LNG could constitute an ample alternative to a full-coverage pipeline system – as the latter will become far more expensive in regions with sparsely population (i.e. extreme radials). In Georgia only 60% of the territory is likely to be supplied by natural gas through pipelines (Figure 8). For remote mountainous areas, especially those that lack biomass, LNG would offer an adequate solution. Another important aspect is the high energy density of LNG that makes it useful as vehicle fuel.

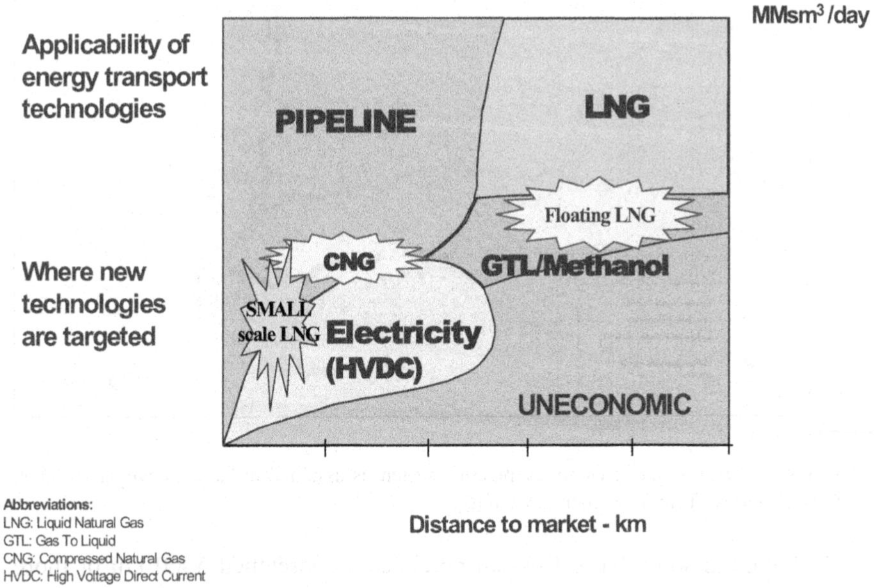

Figure 7 Concept map showing the applicability of energy transport technologies with practical and economical limitations originating from the utilisation of natural gas sources. The distance given is indicative, and may to some extent vary depending on local conditions.

Figure 8 The prospective Georgian pipeline system is assumed to have a coverage of about 60%. (Courtesy Teimuraz Gochitashvili [5])

Although some large Russian pipelines are stretched 5000 km or more, the economical viability of pipelines tends to be limited around 3000 km as indicated on the concept map presented in Figure 7. Provided sufficiently large volumes, however, only liquefaction appears to be viable on very long distances. For instance does the largest gas-fired power plant of the world, the Japanese Kawagoe plant as shown on Figure 9, receive LNG basically supplied from Indonesia and the Middle East by large LNG carriers (140 000 m^3 LNG). Eventually, there are few options to LNG for the use of natural gas in Japan, as it would hardly be technically feasible for a pipeline to cross the sea to supply Japan with piped natural gas.

Figure 10 shows the established worldwide trade routes for natural gas. There is however, a strong-growing trade over the Atlantic to the United States of America, where especially Trinidad, but also Nigeria, are exporting LNG to the United States of America. And by the year 2006 a new corridor will be opened for overseas export of LNG from the Northern part of Norway to basically the United States and Spain as shown on Figure 11.

Forecasts point to a doubling of LNG trade volumes over the next decade. The new LNG exporters will include, among others, Egypt, Iran and Venezuela, while on the import front, China will almost certainly overtake India as the most exciting prospect. This could have important ramifications for the Chinese shipbuilding industry [9].

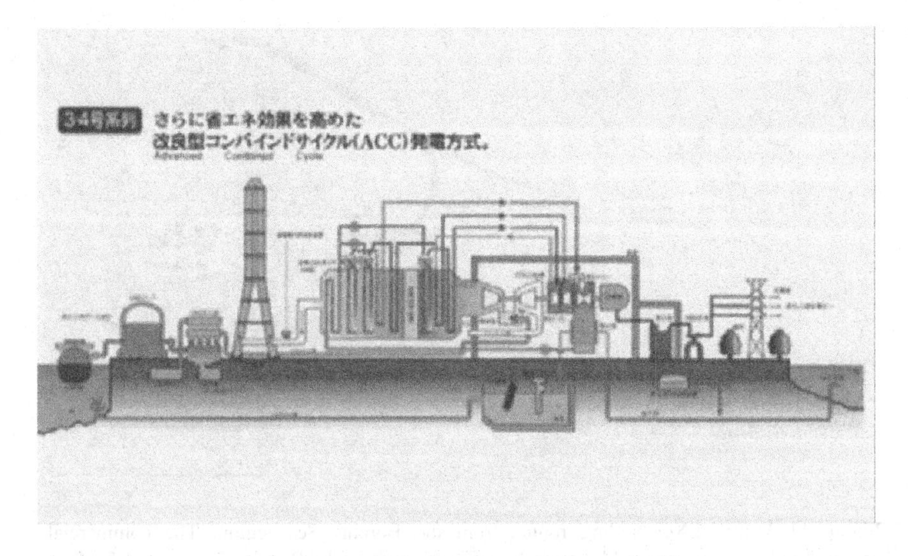

Figure 9 Kawagoe 3&4 Ultra Supercritical, Double Reheat Power Generation System fired with LNG from the Middle East and Indonesia. The largest gas power plant in the world with a rated power of 4.7 GW electric - having a LNG storage tank capacity of 120 000 m3 (Courtesy Kawagoe plant, Japan 2002)

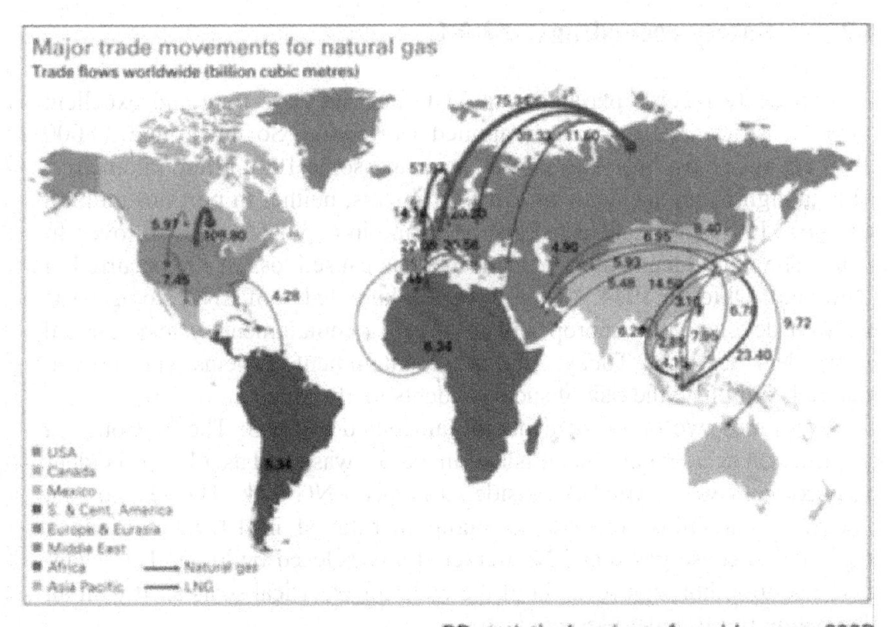

Figure 10 Current major trade movements for natural gas and LNG according to BP statistical review of world energy 2003 [3].

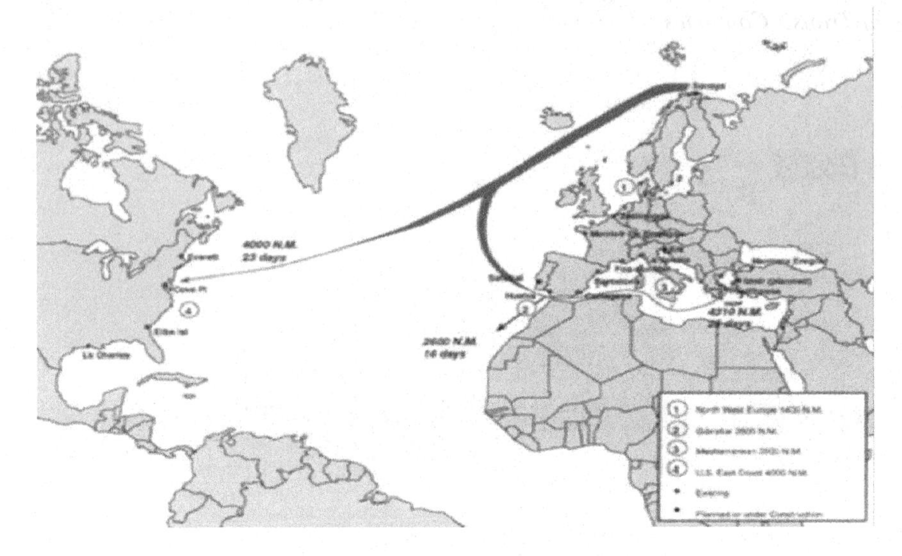

Figure 11 Future LNG trading routes from the Barents Sea region. The commercial exploitation of a new gas field off the Norwegian coast in the Barents Sea, named *Snøhvit*, scheduled to go on stream in 2006. The on-shore based 4.2 MTPA LNG plant will process 5.6 billion standard cubic metres of gas per year. The LNG will basically be shipped to Spain and the United States.

4.2 Safety recordings of LNG

The safety records pertaining to LNG activities are long and excellent owing to strict safety standards applied worldwide. So far around 33 000 LNG voyages have been carried out, covering some 100 million kilometres, without significant accidents or safety problems, neither in port nor on high seas [10]. LNG terminals have been operating in populated areas for over 40 years. The only documented incident that has caused loss of life occurred in Cleveland, Ohio, in 1944. At that time the knowledge of LNG storage was not well developed as improper materials of a containment system resulted in a LNG leakage. Today, multiple containment systems and proven materials eliminate the risk of such accidents to almost zero.

Sometimes two other incidents are attributed to LNG: The first one - a construction accident on Staten Island in 1973 - was cited as a LNG accident because work was carried out inside an empty LNG tank. The second - an electrical seal failure on a LNG pump that failed in 1979 - caused an explosion because gas (not LNG) entered an enclosed building. This event gave rise to a thorough revision of the code for electrical seals used with all flammable fluids under pressure [10].

5. LNG CHAINS

5.1 The Structure of LNG satellite plant

LNG is usually shipped overseas, or transported by rail or road, and generally fed to a LNG satellite facility as shown on Figure 12. The purpose of the satellite facility is to offer local distribution of LNG throughout the year at relatively low cost and low risk within a territory. The satellite plant would generally comprise an unloading system to discharge the ship, storage tanks, pressure-control and send-out facility. Depending on the operating conditions a satellite facility may include various equipment as indicated on Figure 13 [11].

Typical of an installation for re-gasification of LNG is that it includes a vaporiser and one or more LNG storage tanks - either vertical or horizontal type. The vaporiser receives heat as required to evaporate the LNG either from the surrounding air or from hot water. In the latter case, hot water is usually supplied via a gas-fired burner that consumes about 2% of the corresponding LNG stream to be vaporised [12]. Hot-water-based evaporation is more expensive than the ambient-based evaporation, but is less dependent on the ambient conditions. The re-gasification installation will furthermore include gas controllers and an odoriser. Since natural gas does not smell odour should be injected to the gas flow for safety reasons. Except for the filling of the LNG tanks from time to time - and also the unloading LNG carriers - such facilities could be unmanned.

6. LNG USAGE

To some extent LNG is conceived as a portable pipeline because the LNG tanks can easily be transported by road, rail and sea fully equipped with a vaporiser and control systems. In this manner gas can also be used to temporarily feed the local distribution pipelines in situations of interrupted supply of natural gas.

Besides base-load LNG production in quantities up to 5 MTPA – mainly for overseas gas export, some LNG is produced in decentralised units with a small to medium production capacity. Land based small-scale LNG plants can be divided into two main categories:

- Peak shaving plants
- Small-scale LNG plants for decentralized LNG production

The production capacities for these types of plants may be in the range of 10 – 500 tonnes of LNG per day as shown in Table 2.

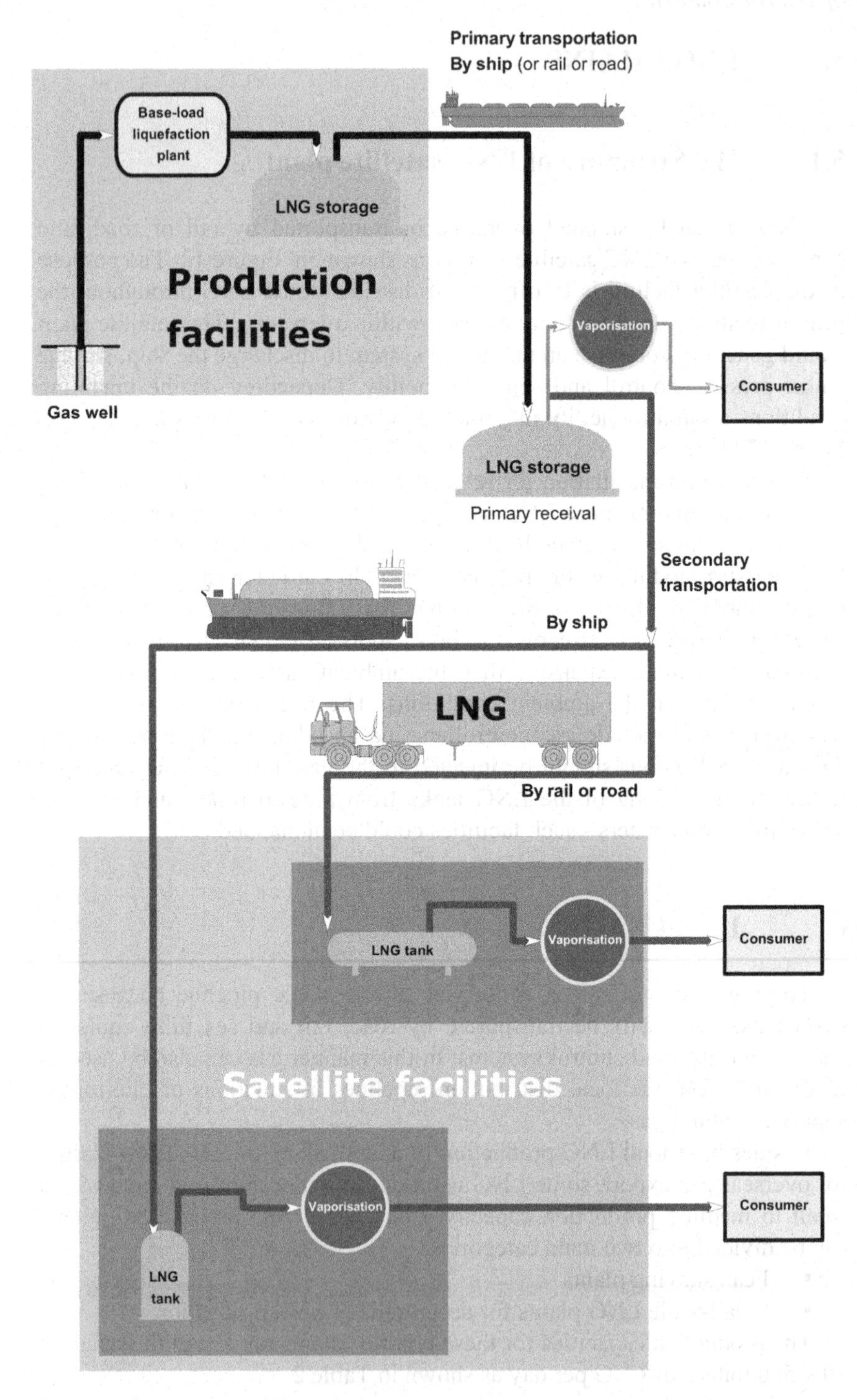

export terminal and LNG transportation in large gas tankers for delivery at end-users and remote satellite LNG plants. The transportation may alternatively be by road or rail.

Table 2 LNG plant characteristics [13]

Plant characteristics	Peak shaving	Vehicle fuel
Liquefaction, tones/day	100-500	10-400
Operating period, days p.a.	150-200	365
Storage, m^3 (days of production)	50 000-100 000 (150-200)	5 000-10 000 (5-10)
Sendout	Vapour	Liquid
Sendout rate (relative to liquefaction rate)	10-20 times	2-3 times
Sendout type	Pipeline	Truck/rail

6.1 Peak shaving

In peak shaving applications a large volume LNG is stored for winter operations.

LNG has been used for peak shaving purposes for many years. The idea of peak shaving is to liquefy natural gas and store LNG during the summer season, and vaporise the gas during periods of high demand (winter). When gas volume is needed in a localized area, LNG is pumped from storage, vaporized and sent to the customer. Volume is replenished during off-peak periods. Numerous local gas distribution companies are operating their own liquefaction plants in this modus [14]. Some LNG peak shaving plants do not have on–site liquefaction, but receive the LNG via tanks from other liquefaction plants prior to the heating season.

In contrast to base-load operation peak shaving differs from the large-scale LNG trains that are generally driven by cost of the LNG as a commodity in comparison with alternative fuels available in the market. Peak shaving on the other hand can be justified by the cost of omitted pipeline charges and the reduced capital cost pertaining to pipeline infrastructure [7].

Pine Needle, one of the larger LNG peak-shaving plants in the United States, has a capacity of 145 MTPA (17 tonnes per hour). It is designed to fill two 95 000 m^3 LNG storage tanks during a 200 day liquefaction season. The tanks have a double steel wall, designed for 0.1 bar overpressure, 55 m/s wind impact and site-specific earthquake requirements. The send-out capacity is 34 tonnes per hour corresponding to a send-out time of 10 days. Usually the discharge pump operates at 40-50 bar pressure [15].

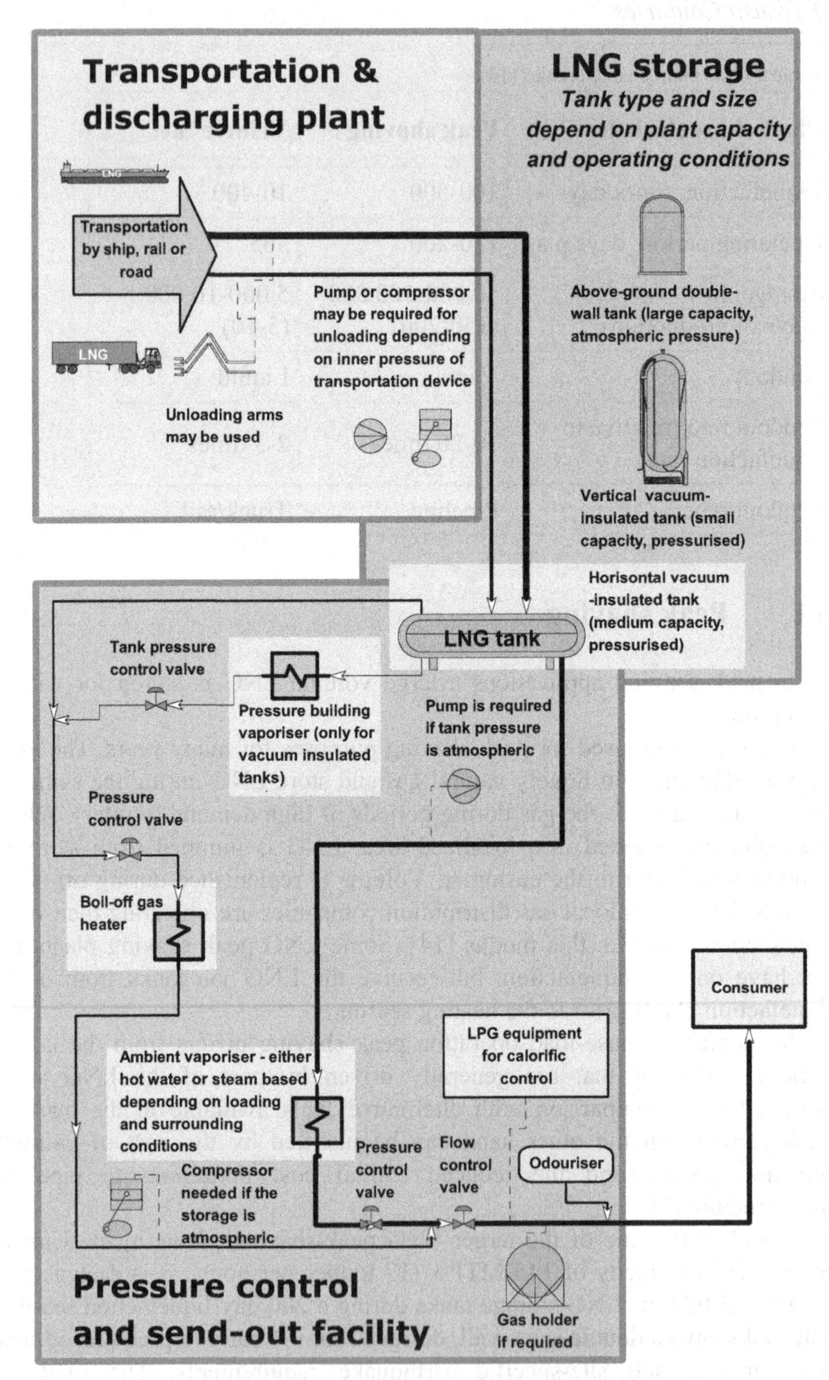

Figure 13 Typical equipment and control devices of a LNG satellite facility.

6.2 LNG for mobile applications

A novel global LNG market is for vehicle fuels. Liquefaction facilities for vehicle usage may essentially be the same as peak shaving units; however, LNG storage is much smaller and is prone to load out the fuel as a liquid rather than vaporized gas. Different process cycles are developed.

The use of natural gas as a vehicle fuel has been known for many years – however, basically as compressed natural gas (CNG). The volume of natural gas consumed by transportation vehicles is projected to rise from 5300 TJ in 1996 to around 620 000 TJ in 2015. Expectedly 50% of this gas is going to be consumed by the heavy-duty trucks [16]. It is further assumed that targeting of a fleet of vehicles with high fuel demand would make LNG utilisation more economical. This may also be required in order to justify the incremental cost of the vehicles and the infrastructure.

The rationale of introducing natural gas in road traffic is primarily to reduce emissions. Several factors should be considered when making a decision on putting up an alternative fuel system. LNG is prone to be preferred by heavy vehicles only when payload is important – as is the case for large distribution fleets like buses and trucks that operate within reasonable distance from the filling station. Whereas CNG turns better out for small, light vehicles when the routes are not fixed, and payload is no issue.

Often LNG offers the most flexible method for the use and delivery of natural gas as vehicle fuel [6]. Furthermore, since life cycle cost analyses usually are in favour of gas rather than diesel fuel, some incentives could be justified to give boost to the development of the infrastructure pertaining to gas (CNG and LNG).

The vast majority of vehicles that burn substantial quantities of diesel or gasoline cannot - from a practical standpoint - carry enough fuel as CNG, whereas LNG is prone to meet these needs [17]. The concept of using LNG as vehicles fuel – in contrast to CNG - is based on a system for charging the engine with LNG from the car's tank via a vaporiser that is integrated with the engine-cooling system. Improvements in on-board LNG storage vessels and vaporiser equipment have encouraged the use of LNG as vehicle fuel [12]. In general terms this would imply improved safety as the tank is atmospheric, and enhanced range because of the energy density of LNG that is substantially higher than the energy density of CNG.

Table 3 shows that the energy density of LNG per volume (MJ/l) is higher than any alternative gas-based fuels, and offers a range that is roughly 2.5 times the range of CNG. The table also shows that LNG has almost 60% the energy density of diesel fuel.

Table 3 Energy density of common automotive fuels compared with natural gas in various states [6]

Fuel	*MJ/kg*	*MJ/l*
Gasoline	42.5	32.7
Diesel	42.5	37.7
LPG	48.0	24.4
Methane	50.0	0.035
Gaseous methane at 248 bar, CNG	50.0	8.7
Liquid methane at –162 °C, LNG	50.0	21.6
Hydrogen at 248 bar	120.0	2.5
Hydrogen at –250 °C	120.0	8.5

6.3 LNG in Ship propulsion

In conjunction to one of the world's largest methanol plants at Tjeldbergodden, Norway, a 10 kTPA LNG plant has been on stream since 1997. Two more small-scale plants will go on stream in 2003 in Norway: a 18 kTPA LNG plant at Haugesund, and a 42 kTPA LNG plant at Bergen. Due to the Norwegian topography the distribution of the LNG will be mainly by a coastal tanker, and some by road. The provision of LNG has, however, given rise to some innovative application projects – not least in the marine sector; the roro-passenger ferry M/F Glutra (2000) and the offshore supply vessel M/V Viking Energy (2003).

By 28 September 2000 the Norwegian built roro-passenger ferry M/F Glutra - owned by Møre og Romsdal Fylkesbåtar (MRF) - received the "Ship of the Year" award at the international shipping exhibition SMM in Hamburg [18]. The ferry is the first LNG powered commercial ship in the world. The planning and building of the ferry without having adequate rules and regulations for the gas system was a major challenge. The estimated cost of the ferry was approximately about 17.6 million Euro - some 30% higher than the cost of a comparable diesel-electric ferry - much owing to the prototype effect [18,19]. The price difference would probably drop by about 10% for the next LNG ferry project.

6.3.1 M/F Glutra

Glutra has provided excellent experience. Owing to the gas engines, the noise level is low, and the emission of NOx and CO_2 is reportedly reduced by as much as 90% and 30% respectively. Hence, Glutra is conceived as the most environmentally benign marine vessel operated by air-breathing engines. As the decision of building the ship was basically political, the additional cost is justified by referring to the cost of any alternative initiative required to meet the Norwegian NOx reduction target.

Figure 14 The first LNG operated ship, M/S Glutra, having a payload capacity of 100 cars and 300 passengers. The propulsion is gas-electric provided by four Mitsubishi GS12R-PTK ultra lean burn natural gas engines, each rated at 675 kW running at 1500 RPM. The fuel storage capacity is two vacuum-perlite insulated tanks totalling 55 m^3 LNG. The engines are mounted at the upper deck, and the LNG tanks at the water line section. The overall length is 94.8 m, breadth 15.7 m, and depth 5.15 m. The ship was built by Langstein Yard of Tomrefjord, Norway, and delivered by January 2000.

6.3.2 M/V Viking Energy

Viking Energy is the first cargo/offshore support vessel to be fuelled with LNG. The celebrated project has attracted much attention for its efforts and innovative approach lead by the Norwegian shipping company Eidesvik AS[3]. The quoted emissions reduction of NOx and CO_2 is 84% and 20% respectively as compared with conventional marine diesel operation. Furthermore, the sulphurous gases are practically zero [20].

The project includes two similar LNG vessels. Eidesvik AS and Simon Møkster Shipping AS will own and operate one ship each to enter a ten-year charter for Statoil. The price tag per ship is about 38.2 million Euro.

[3] Eidesvik has been nominated for the Thor Heyerdahl International Environmental Award 2003 for its work on M/V Viking Energy.

Figure 15 M/V Viking Energy, a new generation platform support vessel operated on LNG taken over by Eidesvik of Norway by April 2003. The ship has four main engines, Wärtsila 6L32DF with a rating of 4x2010 kW. The fuel tank has a capacity of 230 m^3 LNG. Quoted emission reductions are 84% less NOx and 20% less CO_2 as compared with conventional diesel (MDO). The overall length is 94.9 m, breadth 20.4 m, and the maximum draft 7.9 m. The vessel was built at Kleven Yard, Ulsteinvik, Norway, and delivered by April 2003. (Courtesey Eidesvik AS, N-5443 Bømlo, Norway)

6.4 LNG used in filling stations

Compressed natural gas from LNG (LCNG) is provided by pumping the LNG to typically 200-250 bar prior to vaporisation and tank loading. In this concept odour will usually be added to the gas. Alternatively natural gas can be pressurised to the same level by means of a compressor in order to charge the car's pressurised tank. Filling stations can be equipped to offer both LNG and CNG (LCNG) as shown on Figure 16. The advantage of CNG over LNG is the infinite storage time as there always will be some heat in-leak that inevitably causes boil-off.

The advantage of LNG over CNG is its high energy density and the avoidance of high-pressure tanks. Drawbacks of gas fuel stations are the higher investment in additional equipment and the many regulations that apply to gas filling stations, but not to the traditional fuel stations. And since the fuel is either a cryogenic fluid or it is under very high pressure, the filling is by nature slightly more complicated, and special safety precautions are required. Eventually, local regulations may affect the complexity of the system such as whether a spill containment area is required or not.

A LNG filling station can be located almost anywhere, as it only requires a bulk storage tank. During offloading of LNG from a transport tank to a bulk tank at a fuel station, the typical saturation pressure will be about 1.5 bar with a corresponding density of 400 kg/m^3. The pressure will, however, be raised to 8 bar to supply the engine at ambient temperature.

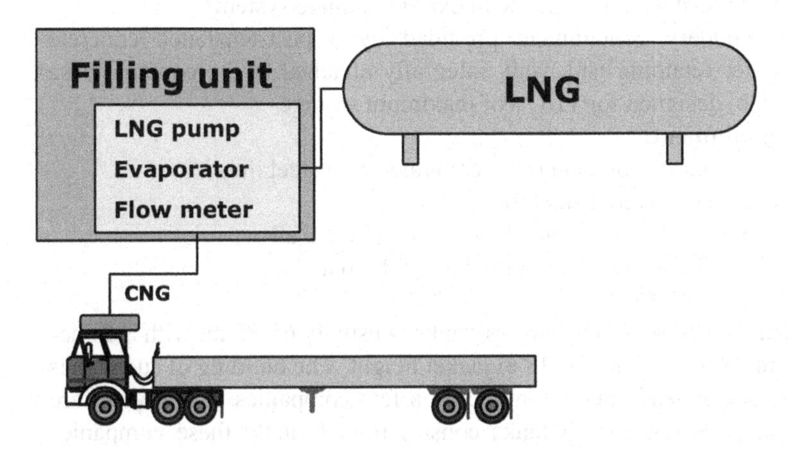

Figure 16 LNG-to-CNG (LCNG) as applied to tank-filling stations. The LNG is pressurised to CNG pressure by pumping (200-300 bar) before vaporisation and filling of the car's tank. Alternatively when LNG is used as mobile fuel the pump and vaporising unit will become part of the car's fuel supply system.

7. LNG STORAGE

LNG is stored in cryogenic tanks and distributed in special containers by road, rail and sea. Typical of LNG tanks are:

1. **Large-scale LNG storage tanks** are usually constructed as freestanding tanks, membrane tanks or buried tanks. Some characteristics and details pertaining to freestanding tanks are given below. Such tanks are flat-bottomed, vertical, cylindrical tanks for low-temperature services according to BS 7777 and API 620 [8,21]. Basically three tank design principles prevail as shown on Figure 17:
   - ⇨ a): single containment,
   - ⇨ b): double containment
   - ⇨ c): full containment.

   The main features of these concepts are; freestanding inner tank and bottom in 9% nickel steel and aluminium suspended deck, with all penetrations from the tank roof. Pump wells are either supported from the tank base (as for type a+b) or from the roof (type c). The insulation material is usually perlite. Further specific features:

Single containment:

- Carbon steel outer tank vapour and insulation container
- Carbon steel dome roof with external deluge system
- Secondary containment provided by composite sand core and crushed rock dikes designed for 110 % tank storage capacity

Double containment:

- Carbon steel outer tank vapour and insulation container
- Carbon steel dome roof with external deluge system
- Secondary containment provided by a post-tensioned concrete outer retaining tank wall, integrally attached to the concrete base slab, designed for 110 % of maximum storage

Full containment:

- Pre-stressed concrete outer container with steel liner
- Concrete covered steel roof
- Liquid spill area limited by outer concrete wall
- Normal operating pressure up to 250 mbar

The diameter of base-load storage tanks is usually 65-85 m, with a dome-height up to 48 meters and 37-38 m jacket height. The building of such tanks is a highly specialised undertaking. Only a few companies in the world are involved in large-scale LNG tanks construction. Usually these companies offer turnkey solutions based on EPC[4] contracts – complete with auxiliary equipment and even quay building.

The price of full containment tanks – like Snøhvit - is around 750 Euro per m^3 for the entire tank systems, and about 525 Euro per m^3 for the tank itself. The price of double and single containment is naturally somewhat lower [22,23,24].

The largest above ground LNG storage tank built so far is a 180 000 m^3 tank in Senboku, Japan, for Osaka Gas, finished in 2000. Whereas the larges in-ground tank is a 200 000 m^3 built at the Ohgishima LNG terminal in Yokohama, finished in 2000. This tank was built in concrete having a 36% Ni-steel membrane. Especially in Japan where land prices are high, the bigger tank volumes have had a significant impact on lowering the cost of new plants since they need less ground area to store large amounts of LNG than a multi-tank facility. In Japan approximately 100 storage tanks above the ground have been built at the import terminals, whereas many LNG tanks are built below ground level.

[4] Engineering, procurement, construction

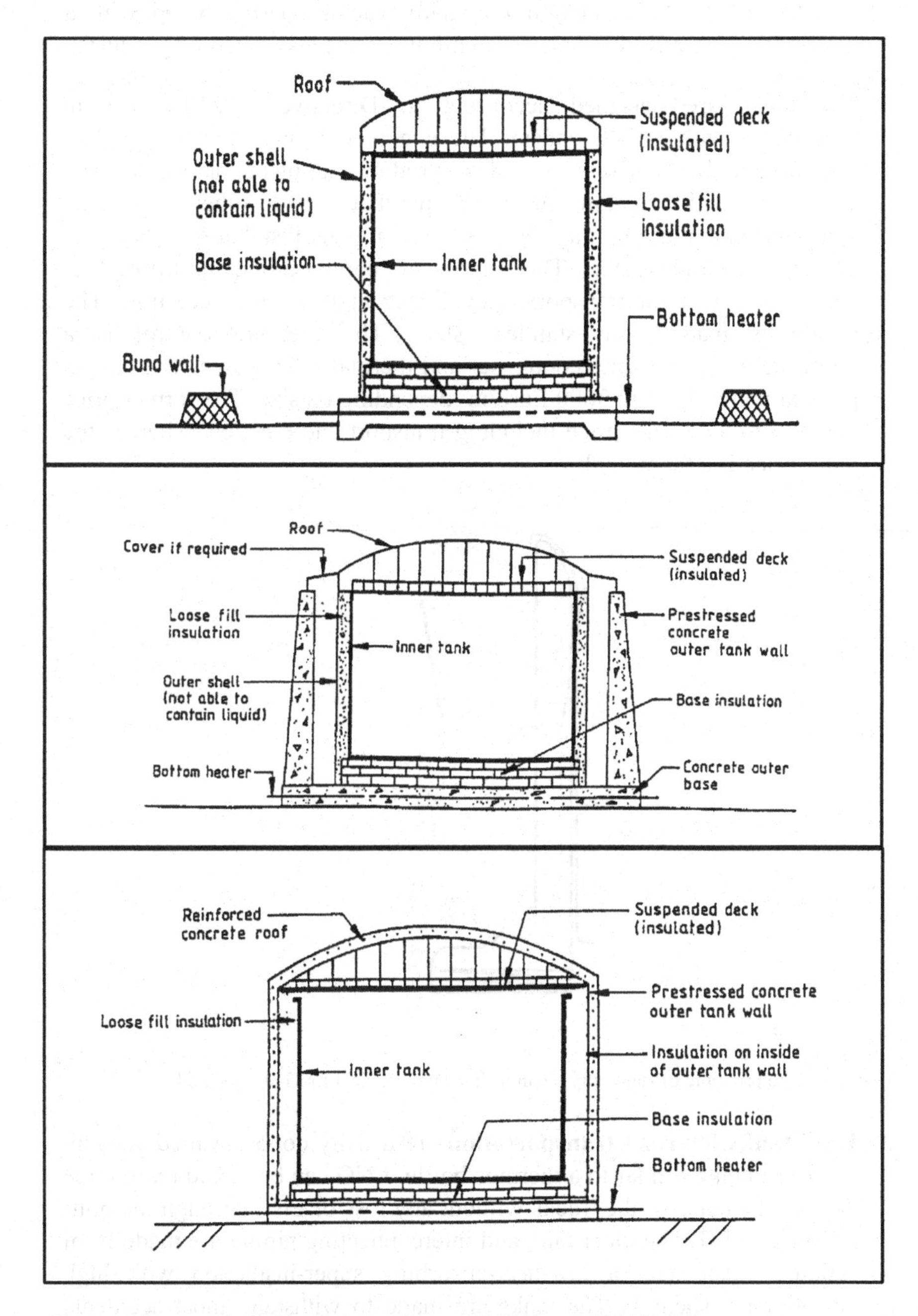

Figure 17 Large-scale LNG tanks according to BS 7777 and API 620 [21]

2. **Engineered vertical and horizontal stationary storage tanks** are typically designed for a capacity in the range from 60 to 500 m^3 and maximum allowable working pressure up to 24 bar for long-term storage

of LNG [12]. The insulation is usually vacuum-perlite based with a molecular sieve adsorber in order to minimise the loss of stored products.

The tanks are designed according to Directive 97/23/EC (or in accordance with ASME codes). The engineered tanks can be supplied with external vaporisers, - either product, ambient or steam heat vaporiser, and vacuum insulated pipelines and other cryogenic components that are required to create a complete installation. The inner vessel is in stainless steel. The outer jacket combines leg and lifting lugs and is designed for transport, easy lifting and low cost erection. The piping is made from stainless steel. The pressure control is a multifunction regulator, economiser and thermal-relief valve. The budget price amounts to 1000-3000 Euro/m^3 depending on size. The former price is a 500 m^3 tank (net price including transport and erection, whereas the latter price is a 50 m^3 tank.

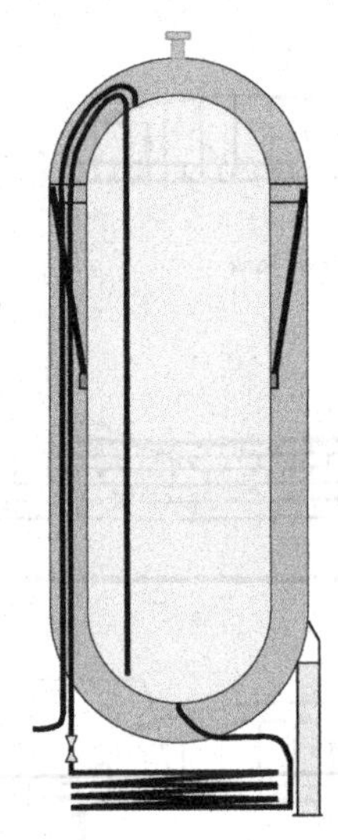

Figure 18 The principle of stationary vacuum-insulated vertical LNG-storage tank

3. **LNG tanks for road transportation** are usually double-walled vacuum insulated tanks similar to a thermos bottle. LNG can be stored up to three days in the tanks of the trucks without any loss of LNG through the boil-off process [7]. The inner tank and interconnecting piping are made from stainless steel with a low-heat-absorbing super-insulation with high performance vacuum. The tanks are made to withstand most accidents that may occur during transportation. Typical capacity of the tank is 50 m^3 net LNG as shown on Figure 19. The transfer of LNG is usually made by pressure - built up by external heater or cryogenic pump. The budget price is 230 000 Euro for the entire system.

Figure 19 LNG tank for road transportation (Chart Ferox, the Czech Republic, type LNG 56/7 with 56.12 m^3, 13.9 m length). The tank is built to withstand rugged road conditions with a three-axle undercarriage with air suspension and protected auxiliary vaporiser. The trailer is designed for easy operation and safety.

4. **LNG tanks for rail transport** are by units of around 120 m^3 net LNG capacity as shown on Figure 20. Vacuum insulated double wall tanks are usually used. Budget price: 500 000 Euro.

Figure 20 LNG transport by rail car (Chart Industries, US, model SR-602, net capacity of 118 m^3 LNG. Design loads: 7g longitudinal and 3g transverse and vertical. Overall length 25 m).

5. **ISO LNG containers** are typically 18 m^3 net LNG capacity as shown on Figure 21. Such tanks are usually vacuum insulated and made for different pressure classes - typically 10 to 22 bar. Holding time is up to more than 100 days. Budget price: 150 000 Euro.

Figure 21 ISO containers for LNG transport (Chart Ferox TVAC Intermodal type for pressure range 6.9 to 22 bar). ISO containers usually include a pressure building vaporiser for pressure transfer [12].

6. **LNG vehicle fuel tanks** are usually double-walled, insulated containers, designed for a maximum working pressure of 16 bar. Usually the car's engine would require a gas pressure varying from 4 to 9 bar. This pressure can be ensured either by a mechanical pump connected to the fuel tank, or by preparing the liquid so that it will maintain the required pressure. Usually the latter option is easier than depending on the reliability of a pump, although the disadvantage is that the density is reduced at the higher pressure which will unable the tank to deliver the full amount of fuel. This arrangement may reduce the vehicle range by as much as 11% [16].

7. **LNG carriers for sea transportation** are either with spherical tanks or with membrane type tanks.

7.1 LNG carrier safety systems

LNG carriers are equipped with gas and fire detection systems. The cargo safety system is extensively instrumented – including a shut down system that is activated on predetermined parameters. Furthermore, the safety equipment of LNG carriers includes sophisticated radar and positioning systems that alert the crew to the surrounding traffic on potential hazards. A number of distress systems and beacons will automatically emit signals if the ship enters difficulty positions [10].

Figure 22 Tank section of a spherical gas tank for ship transportation of LNG [8,24]

8. ECONOMICS FOR SINGLE MIXED LIQUEFACTION PLANTS [20]

8.1 Storage and send-out

Liquefied natural gas plants for peak shaving and vehicle fuel production are generally similar in the pre-treatment and liquefaction areas. However, these plants differ substantially in terms of storage and send-out facilities as shown in Table 2. Indeed, these discrete differences make a huge impact on the capital cost:

- Peak shaving units typically involve large atmospheric LNG storage tanks – equipped with high-pressure LNG pumps and LNG vaporization equipment as required before introducing the LNG in gas phase to a pipeline.
- Fuel plants are capable of storing LNG typically for five to ten days. The LNG will be stored either in low pressure or medium pressure cryogenic tanks (typically 7-10 bar). From these tanks the LNG is loaded on trucks or rail cars, and, eventually, also on smaller ships, to be sent to the market in liquid phase.

Fuel plants are, furthermore, intended to run year round, while the peak shaving units run only in off-peak seasons (usually over the summer from spring to autumn), however, to the extent that storage is full for the high-demand season (winter).

8.2 Project cost

Most facilities in the United States use single containment tanks with a dike for spill control, whereas Japan and eventually Europe are prone to prefer full containment tanks. The trend in larger facilities and those with limited site area is towards full containment. Actual unit installation cost may vary due to site-specific parameters.

Table 4 Comparative project cost [20]

	Peak shaver		**Vehicle fuel**	
	300 ton/day liquefaction		300 ton/day liquefaction	
	100 000 m^3 storage		7 000 m^3 storage	
	4 000 ton/day send-out		600 ton/day send-out	
	200 day/year operation		350 day/year operation	
Drives	**Motor**	**Turbine**	**Motor**	**Turbine**
Power, US¢/kWh	3	5	3	5
Fuel, US$/GJ	3	2	3	2
Capital, million US$	39	43	23	27
Operating cost, US$/ton	23.50	19.50	23.50	19.50
Capital, US$/ton	78.00	86.00	25.50	30.00
LNG to tank, US$/ton	101.50	105.50	49.00	49.50

Figure 23 shows the cost of LNG storage capacity versus tanks size in terms of total investment in single and double containment tanks. The graph also indicates the unit cost (storage capacity per m^3) ranging from roughly 180 to 320 Euro/m^3 for a 100 000 m^3 single containment tank and double containment tank respectively.

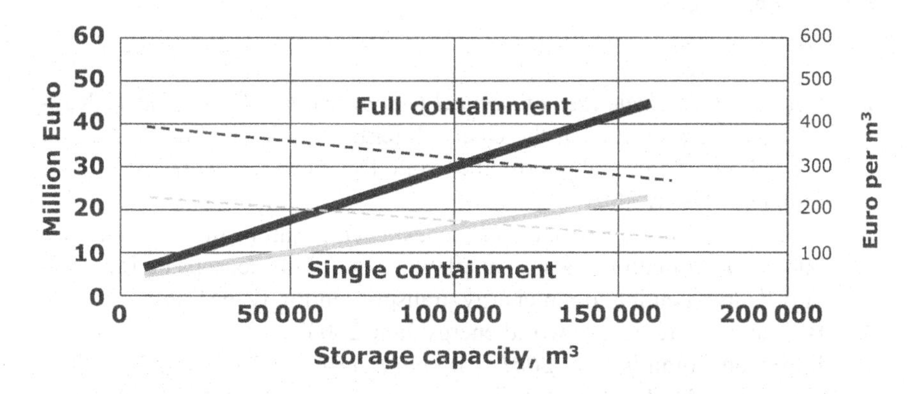

Figure 23 LNG storage capacity cost. The storage capacity cost usually includes some sendout equipment. In the United States most tanks are single containment tanks in contrast to Japan and Europe where full containment are preferred [20]

9. CONCLUSION

LNG is gaining acceptance as an efficient energy carrier for a variety of applications. One advantage of LNG is the high energy density that makes it easy to transport on sea, road and rail – independent on a pipeline infrastructure. Therefore LNG may be used for the provision of gas to regions where natural gas is not available, and eventually where distribution systems for natural gas are lacking. In some cases LNG will also serve as an extension to existing pipelines, and even be used to boost pressure in the distribution system during peak seasons.

Elements that affect the integration of LNG with the prevailing energy and transport systems have been considered. For instance the ability of LNG to be used as fuel for prime movers in marine operations - as in Norway, and as vehicle fuel especially to fleets of busses and trucks operating within a region.

However, any LNG usage includes cryogenic tanks. Therefore, prevalent tank types and storage principles have been drawn up for tanks used for LNG storage and transport.

Furthermore, as LNG offers an option to even-out the fluctuations over the low-demand and the high-demand seasons, LNG should, eventually, be considered as a true alternative to the underground storage of natural gas (UGS).

REFERENCES

1. Hetland, J.: *'Advantages of Natural Gas over other Fossil Fuels'*, Keynote lecture presented at the Opet – International Workshop on Natural Gas; Technologies, Opportunities and Development Aspects, Vaasa, Finland, 31 May 2002
2. *'Energy. Let us overcome our dependence'*, European Communities, 2002. Information also available on http://europa.eu.int and http://europa.eu.int/comm/energy_transport/en/lpi_lv_en1.html
3. BP statistical review of world energy June 2003
4. European Commission: *'2001 – Annual Energy Review'*, January 2002
5. Gochitashvili, T.: *'Transit Perspectives and Utilization Problems of Caspian Gas in Georgian Energy Sector'*, GIOGIE 2002, March 2002, Tbilisi, Georgia
6. Emmer. C.: *'LNG – not just another fuel'*, Vehicle Fuel, LNG Journal, p. 17-19
7. *'Answers LNG FAQS. Frequently Asked LNG Questions'*. Northstar Industries, Methuen, MA, USA. http://northsaring.com/lngfaqs.html, 2003-04-15
8. Aspelund, A.; Brendeng, E.; Einang, P-M.; Meek-Hansen, B.; Mølnvik, M.J.; Neeraas, B.O.; Vist. S.; *'LNG – Technology Evaluation'*, SINTEF Report, 2003, (Restricted)
9. Drewry Annyal LNG Shipping Market Review and Forecast 2002
10. *'LNG frequently asked questions'*, Energy for Wales. http://www.energyforwales.co.uk/faq.html
11. Misawa, R.; Matsumoto, T.; Okamoto, H.; Fukuhara, J.; Kitano, Y.; Handa, K.: 'LNG Satellite Facility', Nippon Kokan Technical Report, Overseas No. 42 (1984), p. 178-182
12. Zeman, J.: *'Liquefied Natural Gas – an attractive alternative for gas-based utility system. Ferox, an Active Participant'*, Podnikatel, November 2002
13. *Price, B.C.: 'Small-scale LNG facility development'*. Hydrocarbon Processing, January 2003
14. Beale, J.: *'"Cold Corner" column'*, First published in the Natural Gas Fuels Magazine, August 1996, available on http://www.ch-iv.com/lng/cc9406.htm – last update September 17, 2002
15. *'Pine Needle LNG Peakshaving Plant'*, http://www.chicago-bridge.com/proj_pine.html
16. Litzke, W-L.; Wegrzyn, J.: *'Natural gas as a Future Fuel for Heavy-Duty Vehicles'*, The Engineering Society For Advancing Mobility Land Sea Air and Space International, Government/Industry Meeting Washington, D.C. May 14-16, 2001, SAE Technical Paper Series 2001-01-2067

17. Beale, J.: *'Developments in Non-Traditional Uses of LNG'*, paper presented at A Global Update & Discussion on Liquefied Natural Gas, September 25-26, 2000, Cyndham Canal Place, New Orleans, Louisiana
18. *'The world's first LNG fuelled ferry receives "Ship of the Year" award in Hamburg'*, Skipsrevyen, http://home.c2i.net/strorvik/glutra02.htm
19. Stokholm, R.; Roaldsøy, J.S.: *'LNG used to Power the Ferry "Glutra" in Norway. – The World first Ferry to run on LNG'*, http://iptnts.ipt.ntnu.no/jsgfag/naturgass/dokumenter/Stokholm2002 Paper.pdf
20. *Company web-site, Eidesvik, http://www.eidesvik.no/english/index.htm*
21. 'Flat-bottomed, vertical, cylindrical storage tanks for low-temperature service', BSi, BS 7777 Part 1. 1993
22. *'Snøhvit Gas Field, Barents Sea, Norway'*, the website for the offshore oil & gas industry, http://www.offshore-technology.com/projects/snohvit/
23. Owren, G.; Moger, J.; Neraas, B.O.: Plant-specific information on LNG plants and storage provided by Statoil, 29 April 2003
24. *'Snøhvit The World's Northernmost LNG project'.* Statoil website. https://www.statoil.com/STATOILCOM/snohvit/svg02699.nsf?OpenDatabase&lang=en

[illegible] the Role and the Emerging LNG [illegible] the Energy Supply Systems [illegible] Industrialized Countries

17. [illegible] "Developments in Non-Traditional Uses of LNG", paper presented at A Global Update & Discussion on Liquefied Natural Gas, September 25–26, 2000, Wyndham Canal Place, New Orleans, Louisiana
18. [illegible] LNG fuelled ferry receives "Ship of the Year" award [illegible]
19. [illegible] used to Power the [illegible] LNG [illegible] International Gas [illegible] Stockholm 2002 Paper [illegible]
20. [illegible] storage tanks for [illegible] 1991
21. [illegible] website at the [illegible]
22. [illegible] Plant-specific information on LNG [illegible] 2003
23. [illegible]

RUSSIAN GAS SUPPLY AND SOME PROSPECTS OF SMALL SCALE LNG UNITS

Estimates In A Nutshell

Vladimir FEYGIN & Yuri RYKOV

ENGO Research Center, Vavilova str., 7, 117997 Moscow Russia

Abstract: In the first part of the paper general present state of Russian gas reserves production and export is discussed. Projections for the period up to 2020 based on various sources are given. It is demonstrated that future gas development in Russia will probably essentially depend on technological innovations. Now Russian gas industry is rather stable, but this situation could change in the future. In the second part some new cryogenic technologies (as an example of possible innovations) for gas industry are reviewed that are based on aerodynamic principles. These technologies are now under implementation in Russia. It is suggested to use supersonic swirled flows for achieving low temperatures and for performing separation process. Such approach in general is new for gas industry and can stimulate innovative process as well as provide certain breakthrough in gas processing.

Key words: gas supply, reserves, production, prices, liquefaction, small scale LNG, aerodynamics, supersonic flow

1. RUSSIAN GAS SUPPLY

In this nutshell survey we describe general state and possible perspectives of Russian gas supply to Europe and FSU countries. Recently Russian Energy Strategy [1] was approved by Russian Government. This fact allows considering more definitely future perspectives. In general the situation can be considered as nowadays rather stable, but it can change in the future. Russian gas industry has huge potential but probably has not enough modern mechanisms for its realization.

J. Hetland and T. Gochitashvili (eds.),
Security of Natural Gas Supply through Transit Countries, 135–156.

1.1 Gas reserves, possibilities of supply

Russia owns approximately one third of world gas reserves. According to International Energy Agency [2] at the beginning of 2001 proven and probable reserves stood at 46,9 tcm. The west Siberian basin has 37 tcm of reserves, or 79% of the country's total. The Continental Shelf (mainly Barents Sea, Sakhalin) has 4 tcm. Gazprom has licenses for the exploitation of 34 tcm of proven and probable reserves, or 73% of the total (60% of company's reserves are concentrated in small number of fields in the Nadym-Pur-Taz region of West Siberia). Gazprom statistics asserts that its share in national reserves is 64% or 29,9 tcm. International Independent Reserves Audit that investigated 84% of Gazprom fields approved 19,4 tcm proven and probable Gazprom reserves. Energy Strategy [1] mentions 127 tcm prognosis gas resources. To illustrate these numbers from the supply point of view, in case of supply numbers at 2000 year level, the supply can last more than 30 years taken into account only Gazprom internationally audited reserves. And with overall proven and probable reserves the supply in 2000 year volumes can last approximately 80 years.

Thus, now major part of reserves is in hands of Gazprom who is lacking finance for investment into industry modernization. Moreover main share of Russian reserves is concentrated in small number of fields in Western Siberia. As Russian Energy Strategy says three of these giant fields are already in the stage of production decline. About 30% of gas reserves are in hands of other companies who are yet not very active in connecting of relevant fields (see below). The increment of the reserves comes mainly from small and medium fields often in hard accessed regions of Russia. So the stability of exploitation of vast Russian gas reserves would be under continuous attention inside and outside this industry and depends on tendencies and reforms in it.

1.2 Tendencies in gas production

Russia will stay major gas supplier to Europe and FSU countries. In coming decades it also will try to enhance the supply market to Asia (China, Japan, Korea, etc.) and to North America (mainly USA). Although Russian Energy Strategy [1] and IEA [3] give different views to the future share of natural gas in domestic energy mix: first source anticipates that this share will shrink, second source – that it will increase; the prognoses changes in natural gas share in domestic energy mix are not greater than 5%. The Russian reserves are still vast, and it can be anticipated that export possibilities will increase and the total supply will have tendencies to increase as well.

In Table 1 the production data are shown according to [2]. 2001 data are taken from author's estimates. Figure 1 illustrates the regional percentage of total production.

Table 1. Russian gas production. Fact, bcm

	1995	**1996**	**1997**	**1998**	**1999**	**2000**	**2001**
Russian Federation	**594,8**	**601,0**	**570,5**	**590,7**	**590,8**	**584,2**	**581,0**
Including:							
OAO Gazprom	**559,9**	**564,7**	**533,8**	**553,7**	**545,6**	**523,2**	**512,0**
- West Siberia (Nadym-Pur-Taz)	519,2	526,9	496,4	515,3	507,1	483,0	473,5
• Urengoy	242,9	242,2	227,2	223,8	209,1	193,3	180,4
• Yamburg	177,8	176,5	169,3	179,6	175,9	168,0	173,0
• Nadym	64,4	65,3	54,0	65,1	72,4	73,6	71,3
• Surgut (Noyabrsk)	34,1	40,3	45,8	46,7	49,7	49,0	48,7
- European part of Russia							
• Orenburg	30,8	28,7	27,0	25,5	24,8	24,1	22,8
Other Companies, including:							
Itera & other independent				2,0	10,6	22,7	29,6
East-Siberian Companies	6,1	6,1	5,7	5,8	6,0	6,0	5,6
Oil Companies	29,0	29,1	29,4	28,9	29,6	32,3	34,7

Note: All Gazprom figures in Table 1 relate to the whole Gazprom enterprises in specified regions.

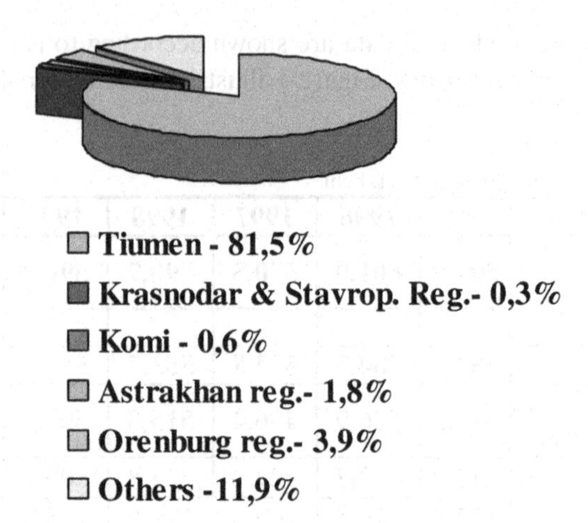

Figure 1. Distribution of Russian gas production, 2001

As it can be seen from Table 1 and Figure 1 main production volumes come from the Gazprom exploited giant fields in Western Siberia. Gazprom production was slightly declining since super-giant fields Medvezh'ye, Yamburg and Urengoi start reducing the output because of the beginning of natural depletion process. But starting from 2002 Gazprom solves this problem by raising production from new Zapolyarnoye field and other fields in the same area. Later on the giant Shtokmanovskoye gas field in Barents Sea and the Yamal peninsula huge gas reserves are planned to put into operation.

In the last years independent producers increased their role in gas production and in the whole production figures. National production was more or less stable. As [1] asserts gas production in 2002 was 595 bcm – an increase in 14 bcm comparing to 2001.

General situation is demonstrated on the map in Figure 2, and Table 2 shows the anticipated production in Russia [2] up to 2020.

Russian Energy Strategy [1] forecasts the increase in gas production from 584 bcm in 2000 (595 bcm in 2002) to 635-665 bcm in 2010 and up to 680-730 bcm in 2020. Thus, it is somehow higher than in Table 2.

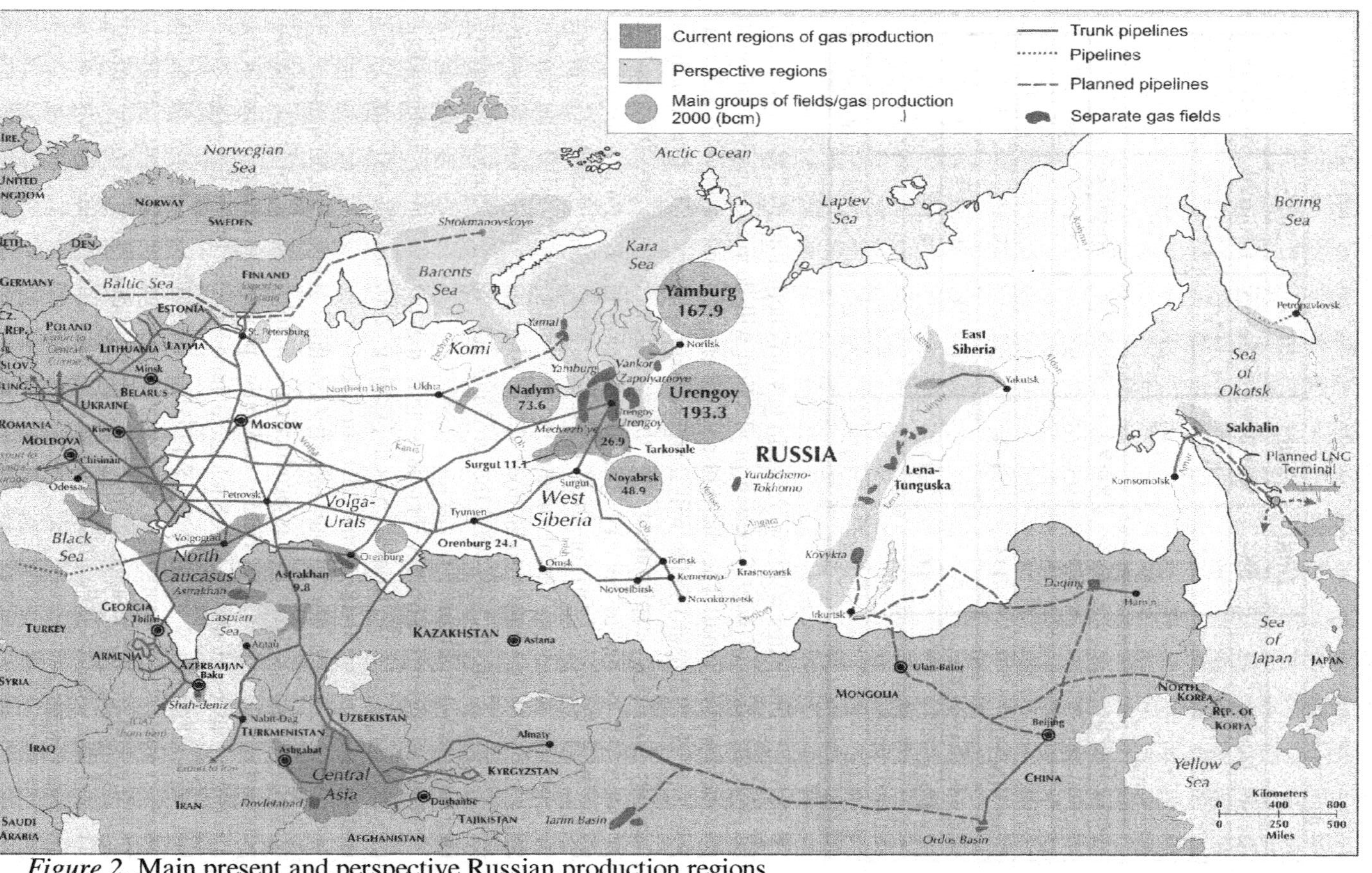

Figure 2. Main present and perspective Russian production regions

Table 2. Russian gas production, Projection, 2000 – 2020, bcm, and shares of the regions in the national production

	2000	2005	2010	2015	2020
Russia, bcm	**584**	**580-600**	**615-655**	**640-690**	**660-700**
European regions	**7%**	**6%**	**12-13%**	**13-14%**	**17-18%**
-Barents sea (Shtokman)	-	-	65%	63%	70%
West Siberia	**91%**	**92-93%**	**80-82%**	**78-80%**	**73-75%**
-Nadym-Pur-Taz	87%	95-96%	95%	94%	80-84%
-Yamal	-	-	-	-	11-16%
East Siberia	**1%**	**1%**	**2-3%**	**3-4%**	**5%**
-Irkutsk (Kovykta)	-	-	60-73%	80-81%	80-81%
Far East	**1%**	**1%**	**3-4%**	**3-4%**	**>4%**
-Saha Republic	47%	50%	<50%	25-45%	31-40%
- Sakhalin	53%	50%	>50%	55-75%	60-69%

Table 2 anticipates stable increase in gas production up to 2020. As it can be seen the share of west Siberian production will decrease though still will provide the main input. The share of east Siberian (mainly Kovykta) and Far East (mainly Sakhalin) fields will increase mainly after 2010. It is worth to mention that the exploitation of these fields belongs to independent producers. It occurs to be very costly to develop giant Yamal reserves and hence IEA foresees start of their operation after 2015 though Gazprom management tells that Yamal production will start before 2010 and before Shtokman comes into production. Thus the IEA projection forecasts increasing gas production with the growing role of independent producers (see illustration in Figure 3).

Russian Energy Strategy [1] says that the production of independent producers will increase from 71,5 bcm (12%) in 2002 to 115-120 bcm (18%) in 2010 and up to 170-180 bcm (25%) in 2020.

But present situation in gas industry contains serious uncertainties. All main future perspectives in gas production – Shtokmanovskoye field, Yamal peninsula fields, Kovykta fields, Sakhalin project – requires significant amount of investments and technological innovations. From the other side, the wear percentage in present supplying capacities is above 50%. In addition since 1998 financial crisis low domestic gas prices were sustained. This invoked the misbalance of fuel for energy production in favor of natural gas and thus secures little amount of money for infrastructure modernization and innovations.

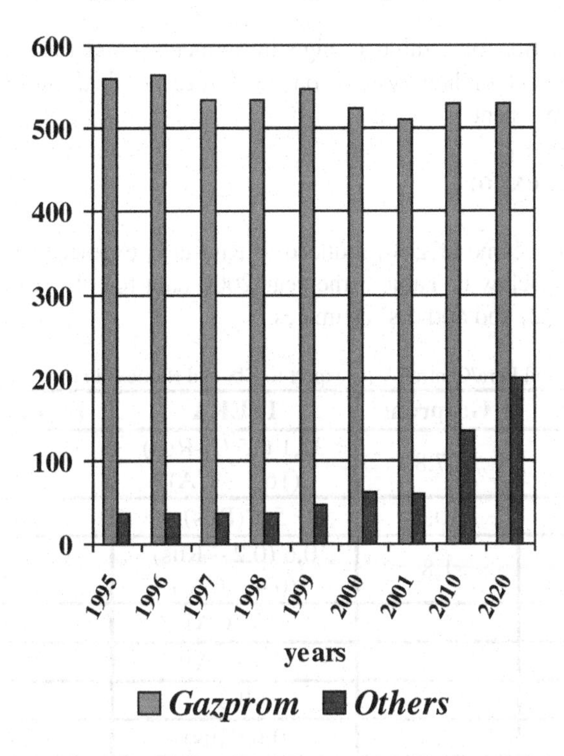

Figure 3. Gazprom & others gas production in Russia

Technologically gas equipment is also in need for improvements. For example, Russian industry only starts production of modern gas turbine facilities for gas transportation and yet has not manufactured sizable amounts of turbines for power generation. As a consequence energy efficiency of operating equipment is significantly (approximately twice) less than by using modern equipment. Because of this and other (mainly structural) reasons overall energy intensity of Russian economy is more than three times higher than in most developed countries. So, it seems that foreign investments should play essential role in future progress, but the deficiencies of law system results now in low level of foreign investments.

So, one can conclude that the situation with the reserves and production capacities is rather stable, has tendencies to grow but nevertheless serious uncertainties exist. This means that there is by no means huge potential in Russian reserves, but most probably the development of these reserves would require attraction of sizable foreign investments. And this relies on the future improvement of law system, on the future political stability and will of Russian government.

1.3 Russian export

Now significant volume of gas produced in Russia is exported to Europe and FSU countries. Below in Table 3 the year 2000 data for FSU export are shown according to [2] and authors' estimates.

Table 3. JSC Gazprom and Itera/Central Asia: export to CIS and Baltic countries, 2000, bcm

	Gazprom	ITERA	Export total
Ukraine	27,8	31,1 (15,0 –Rus) (16,1 – CA)	58,9
Belorussia	10,8	5,8 (Rus)	16,6
Moldova	1,8	0,6 (0,2 – Rus) (0,4 – CA)	2,4
Georgia		0,7 (CA)	0,7
Armenia		1,7 (CA)	1,7
Azerbaijan		0,3	0,3
Lithuania	2,0	0,6 (Rus)	2,6
Latvia	1,0	0,4 (Rus)	1,4
Estonia	0,6	0,2 (Rus)	0,8
Kazakhstan		0,7	0,7
Total	**44,0**	**42,1**	**86,1**

European export is represented [2] in Table 4.

It is anticipated that total volume of Russian export to FSU will not significantly change in the 2020 perspective. So, the main increase in export will come from extended European export. Meanwhile it is worth to mention that Russia has the ambitions to extend its supply and diversify gas export. Sakhalin-2 project plans to export LNG since 2006. Sakhalin-1 project is supposed laying sub-sea pipeline for Japan gas supply. Many sources indicate development pipeline to China from Siberian field of Kovykta. This line would also enable Russia to export gas to Korea and possibly to Japan. A Nation-wide strategy for Russian gas export to Asia and development of regional gas supply system for Russian Eastern Siberia and Far East are now under specification. As World Energy Outlook foresees [3] export to China and Pacific region will exceed 20 bcm by 2030. This is less than 10% of total 2010 export (see forecast below), but marks the desire for supply diversification.

Table 4. Export of Russian natural gas to Europe, bcm

	1997	1998	1999	2000	2001	2002
Former Yugoslavia	3,9	3,7	3,1	3,5	3,7	3,8
Romania	5,1	4,7	3,2	3,2	2,9	3,5
Bulgaria	4,95	3,8	3,2	3,2	3,3	2,8
Hungary	6,54	7,3	7,4	6,5	8,0	9,15
Poland	6,75	6,9	6,1	6,8	7,5	7,1
Czech republic	8,4	8,6	7,8	7,5	7,5	7,4
Slovakia	7,1	7,1	7,5	7,9	7,5	7,7
Central and Eastern Europe sub-total	**42,7**	**42,1**	**38,3**	**38,6**	**40,4**	**41,45**
Greece	0,2	0,9	1,5	1,6	1,5	1,68
Turkey	6,7	6,7	8,9	10,3	11,1	11,8
Finland	3,6	4,2	4,2	4,3	4,6	4,6
Austria	5,6	5,7	5,4	5,1	4,9	5,2
Switzerland	0,4	0,4	0,4	0,4	0,3	0,3
France	10,9	10,9	13,4	12,9	11,1	11,4
Italy	14,2	17,3	19,8	21,8	20,2	19,3
Нидерланды					0,1	1,4
Germany	32,5	32,50	34,9	34,0	32,5	32,3
Western Europe sub-total	**74,1**	**78,6**	**88,5**	**90,4**	**8632**	**87,98**
Europe total	**116,84**	**120,7**	**126,8**	**129,0**	**126,7**	**129,43**

Russian Energy Strategy [1] (new variant) puts the following figures for the forecast of export volumes (do not taking into account re-export from Central Asia), Table 5.

Table 5. Forecast of Russian natural gas total export volume, bcm

	1995	**2000**	**2005**	**2010**	**2015**	**2020**
Export volume	193	194	197-199	217-230	231-242	236-245

The same source anticipates the export to Europe to be in 2020 about 160 – 165 bcm. New variant of strategy significantly decreases export figures. As in previous version export to Europe was estimated as 200 – 210 bcm. This fact also marks certain degree of instability in gas industry conditions and hence in forecasting.

Mr. Yu. Komarov, Deputy President of Gazprom Board, recently indicated (*"Vedomosti"*, June 24, 2003, in Russian) that Gazprom export to Europe will increase from 129 bcm in 2002 to 134 bcm in 2003 and up to 170-180 bcm till 2008.

In [2] it is shown highly illustrative map of Russian export to Europe in 2000. We represent this map below as Figure 4. It can be seen that Europe major consumers are Germany, Italy, France and Turkey. Significant amount is consumed by former SU republics – Belarus and Ukraine, which both import more than three major European consumers together. Now Russian government experiences certain difficulty with Ukraine transit and seeks the possibilities to avoid such way and make bypass through Belarus and North-European pipeline system (see further Figure 5).

In addition Blue Stream supply now is delayed because of disputes with Turkey which overestimated its demand estimates and tries now to reconsider contract terms.

Thus, further development of pipeline system for gas supply to Europe also requires significant investments and investigation from the technical as well as economic points of view.

1.4 Few comments about gas prices and some conclusions

As we have already seen further sustainable development of gas industry requires significant increase of domestic gas prices. For example, in 2002 averaged export price to Europe was above 100 USD/mcm while domestic price was about 27 USD/mcm including excises and VAT. (On the other hand, don't forget that in 1998 export price was at 70 USD/mcm while internal price was at 50 USD/mcm.) Russian Energy Strategy [1] asserts that for the possibilities to make appropriate investments in modernization of gas infrastructure and technological innovations it is required that gas price for industrial consumers, without VAT and distribution costs, will be in 2006 40-41 USD/mcm and in 2010 – 59-64 USD/mcm. In the same time other forecasts based on present state of the art says that the prices for industrial consumers, including VAT and excises, will be in 2006 40 USD/mcm and in 2010 – 50 USD/mcm. Government still debates on the rates of gas price increase which could be met by the population and industry in the coming years. Thus we can see that under financing due to low internal prices most probably will continue.

Such picture causes an additional uncertainty in the possibilities of stable cash flows to support the modernization of gas industry. Moreover the process of liberalization of gas market, which now starts its progress in Russia also, adds another concern in this regard.

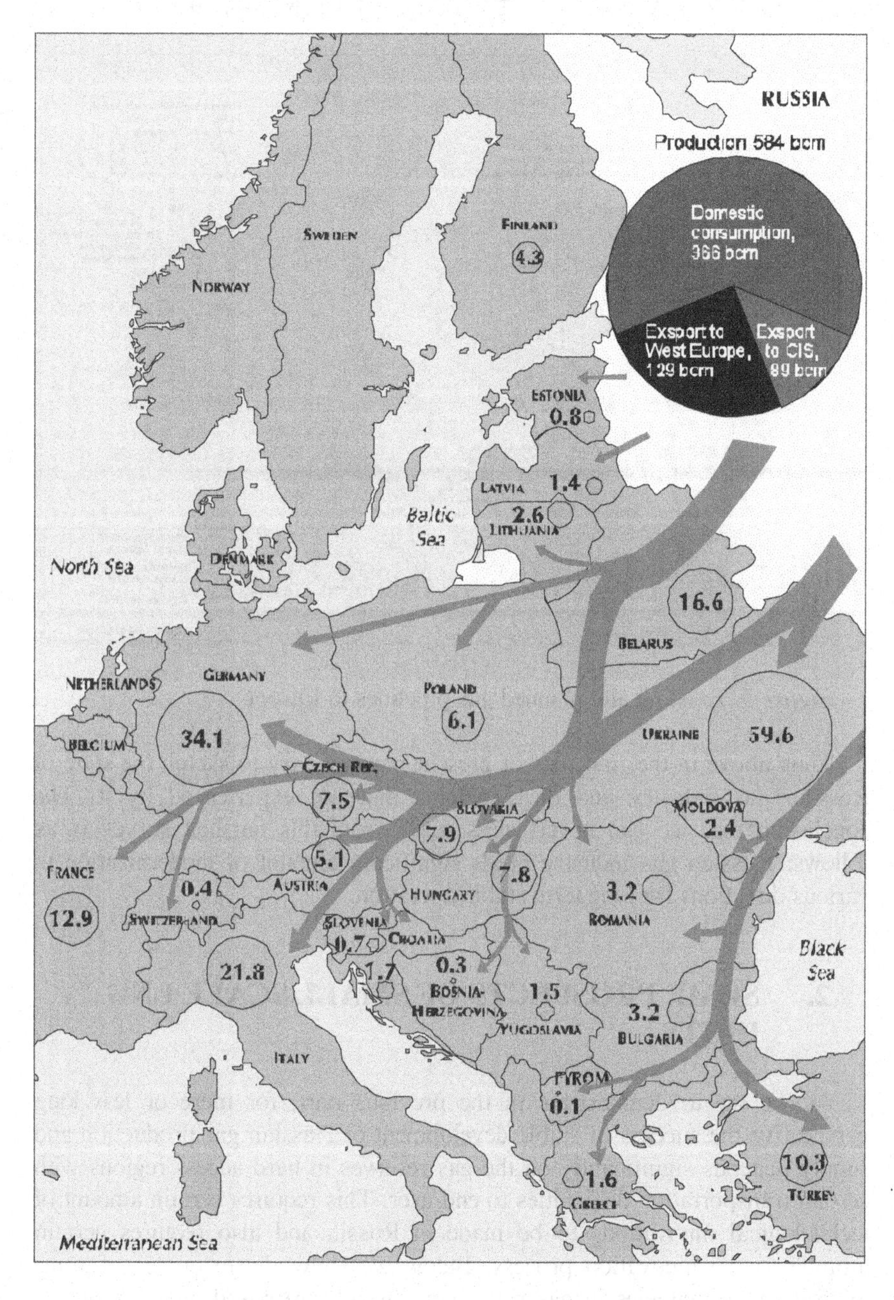

Figure 4. Gas export from Russia in 2000, bcm (includes transit and sales of Central Asia gas to CIS countries). *Note.* Export to other CIS countries, bcm: Armenia – 1,4; Azerbaijan – 0,3; Kazakhstan – 2,7; Georgia – 1,0; Uzbekistan – 0,2.

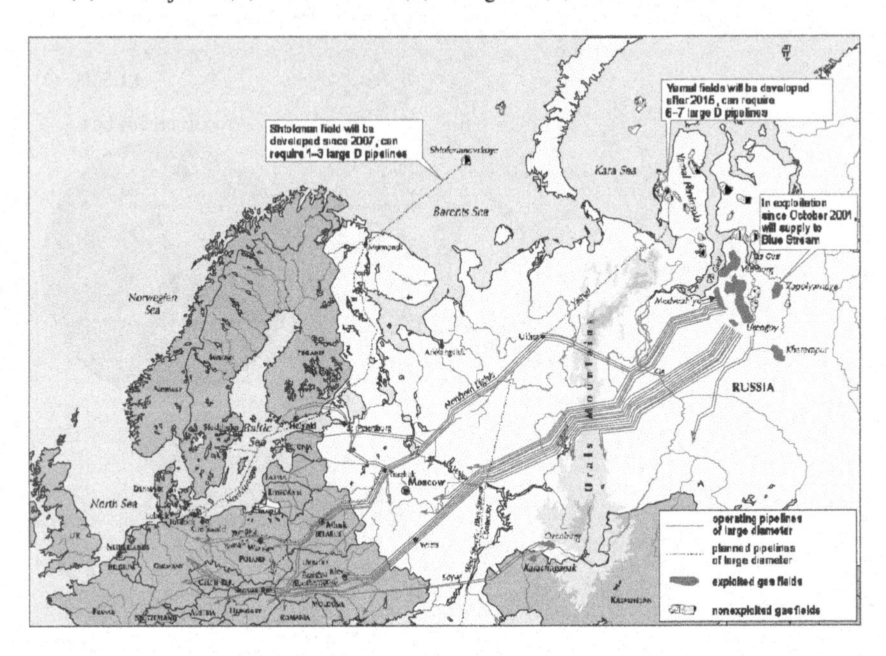

Figure 5. Existing and planned gas pipelines to Europe

Thus above in the first part of present paper we try to sketch the state of Russian gas industry and the problems that are experienced by it. The conclusion that we can nevertheless make from this nutshell survey is as follows: Russian gas industry needs significant amount of modernization in various directions for long term stable operation.

2. SOME PROSPECTS OF SMALL SCALE LNG UNITS

As we try to demonstrate in the previous part, for more or less long perspective the success of stable development of Russian gas production and supply depends significantly on the gas reserves in hard access regions with certain transportation difficulties to end user. This requires certain amount of technological innovations to be made in Russia and also requires certain attitude to the innovation process. Below we show the examples of some innovative technologies in gas processing, which are based on new for gas industry principles of aerodynamics. We believe that these ideas allow at least boosting new ways of thinking in gas industry.

2.1 General position

Last decade LNG trade doubles its volume. Now LNG represents approximately 22% of worldwide cross-border gas trade. And there is a consensus in LNG industry (according to IEA, [4]) that the demand is expected to double to approximately 220 – 270 BCM/y by 2010 and as a consequence LNG trade will experience strong growth. LNG has much higher cargo density than natural gas, so under the development of technology and costs reduction the LNG supply becomes more and more attractive. Figure 6 demonstrates the tendency to equalize the prices for delivered LNG and pipeline NG, see [5].

As BP President Sir J.Braun put it at 2003 International Gas Congress in Tokyo that *the cost of ton of LNG is reduced twice for last two decades and continues to decrease* (*"Vremya Novostei"*, June 3, 2003, in Russian).

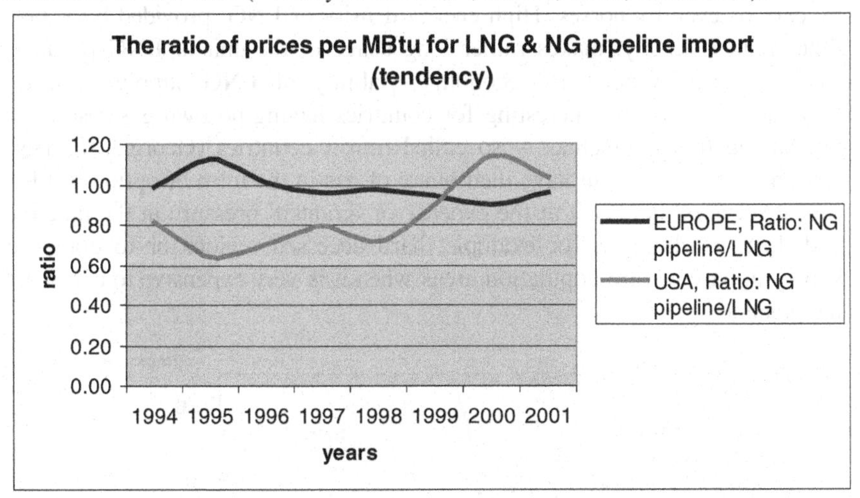

Figure 6. Prices ratio (Source: *IEA, Natural gas information 2002*)

All these trends combining with such LNG advantages as high energy density, safe storage, service as possible future automotive fuel, etc attract our attention to the potential of more wide usage of LNG especially for local needs. Also one has to take into account the possibility of supply to the regions with hard access – there are plenty of such regions, for example, in Russia, Georgia, Armenia, African & South American countries, etc.

In addition long-term tendency for gas market liberalization can be facilitated by inter regional markets trade [6], Chapter 3, through LNG. The growing diversification between consuming and producing countries makes the economies of developed countries to be essentially dependent on import. Meanwhile LNG market is becoming more flexible. Buyers are constantly looking for the options of short-term supply diversification. Spot trade will continue to grow and it grew by 50% in 2001 thus accounting for about 8% of total LNG trade [6], Chapter 3.

So, it is not marginal to consider the possibility of developing small scale LNG facilities for may be worldwide application. Below the Concept Map in Figure 7 is presented where the positions of emerging new technologies are pointed out. Small scale LNG seems to be economically viable for the transportation distances about 500 km and supply volume below 2,5 MMscm/d ~ 900 MMscm/y or 600 – 700 K ton/y.

The small scale LNG supply could be attractive for countries with serious geographic complexities such as mountains, marshes, impassable forests, etc. In this case LNG could be safely delivered by rivers in special tanks or by cars or even by horses. High energy density of LNG, provided by ~ 600 times higher density, makes rather long usage of the unit of LNG product. That creates the point for economic viability of LNG supply projects. Especially it could be interesting for countries having no own gas resources but having transit pipelines – so called transit countries (Georgia, Turkey, Armenia, etc). They can have their share of gas in the form of option gas for transit services; liquefy it at the expense of 'granted' pressure in the pipeline and then distribute to, for example, hard accessed regions or to domestic supply in case of rare population areas when it is very expensive to construct the pipeline.

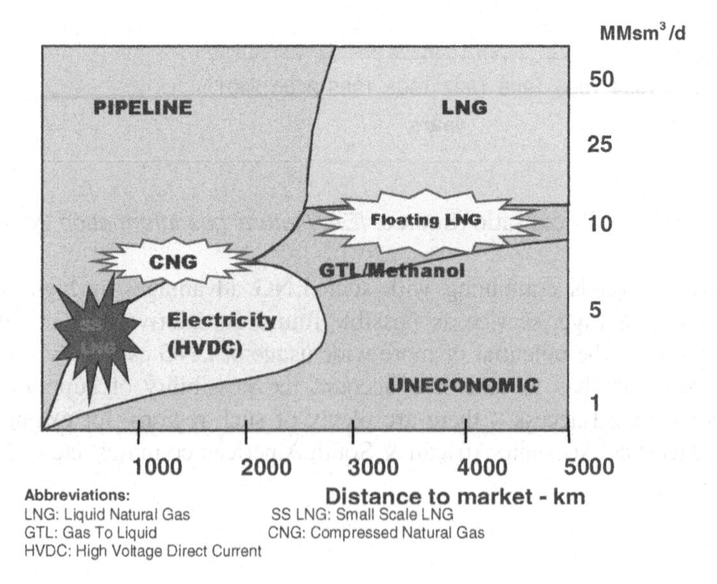

Figure 7. Energy transport. Concept map

Further we describe two technologies that have been now developing in Russia and could be used for the purpose of 'small scale liquefaction'. What these two technologies have in common is the fact that they use aerodynamic, not chemical, principles of work. Taking into account ecological reasons and comparatively simple design of appropriate units, it seems that such technologies can have a perspective and may be not only for small scale LNG but for other scales and generally for solving various problems of gas processing.

2.2 Ranque-Hilsch tube technology

For a number of years a local market for LNG exists in Saint-Petersburg region. The volume of this market is well matched to small scale LNG units. Such units were developed by ZAO "Sigmagaz", which is the affiliate of OAO "Gazprom" through its daughter enterprise OOO "Lentransgaz".

The principle of work of developed unit is based on known methodology of Ranque-Hilsch tube. This effect was discovered in the 1930-es and since that is used for various purposes while ultimate understanding of physical processes is not yet obtained. The principle of work is illustrated in Figure 8.

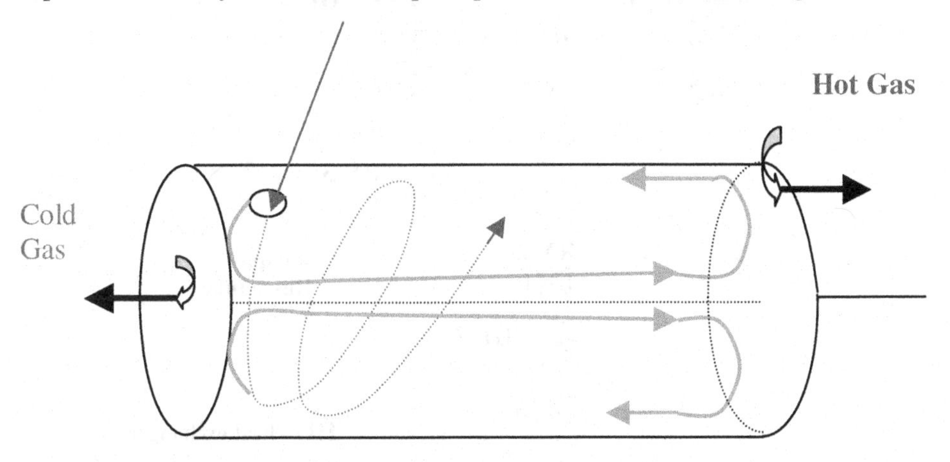

Figure 8. The schematics of Ranque-Hilsch tube

The gas passes tangentially through the hole to the tube which is closed from both sides. Then the gas swirls inside the tube and goes to its opposite side. Eventually gas flow splits into two flows as it is shown in Figure 8: cold and hot flows. Cold flow is used for liquefaction purposes. We are not intended to discuss here physical principles of Ranque-Hilsch tube work. Interested persons can appeal, for example, to recent review [7].

In Saint-Petersburg region several LNG units are now under operation with overall capacity about 10'000 t/y. They are located at Compressor Stations (CS) of various types and therefore have 'for granted' the pressure excess. The inlet pressure of the units is about 75 bars. Under such inlet pressure Rank-Hilsch tube can not be used immediately for liquefaction but is used for efficient gas cooling through several stages. The example of relevant technological scheme from [8] is shown below in Figure 9.

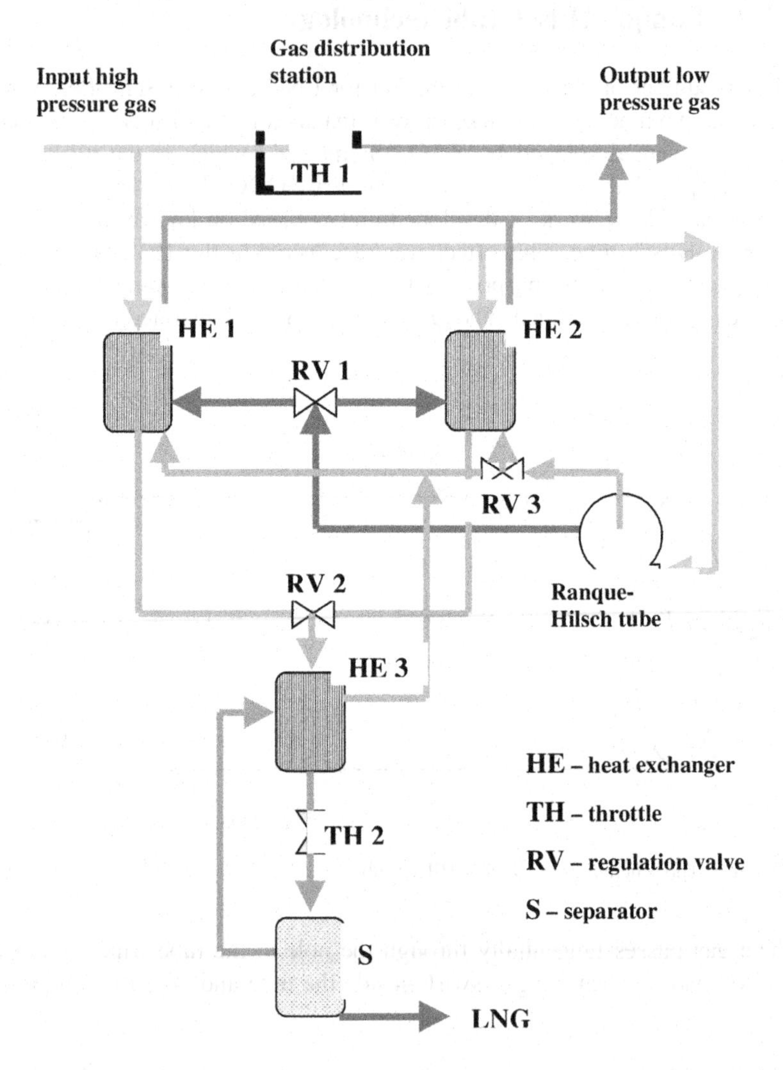

Figure 9. Example of technological scheme using Ranque-Hilsch tube

Input high pressure gas is supplied to heat exchangers HE1, HE2 and to Ranque-Hilsch tube. In the tube the gas flow is separated in two flows with the temperatures of about -60 C (blue line) and of about +20 C (magenta line). Cold flow is supplied to heat exchangers HE1 and HE2 where input gas is cooled up to -50 C. Heat exchangers HE1 and HE2 work in such regime that while HE1 is working HE2 is heated by hot flow from Ranque-Hilsch tube to prevent the formation of undesirable substances, and vise versa. Further the gas flow passes through heat exchanger HE3 where it is cooled up to -80 ÷ -90 C. Then in separator/liquefier S the LNG flow is formed. The pressure of LNG can vary from 2 to 16 bars.

As articles collection [8] reports for local consumers in Russian conditions LNG supply provides cheapest prices for fuel component of heat production.

Table 6. The efficiency of supply for dispersed consumers

Fuel Type	Low heat value (kcal/kg)	Average efficiency value of boilers, %	Approx. cost incl. delivery, USD/ton	Cost of production of 1 Gcal (fuel component), USD
LNG at CS with cooler	11500	92	85,5	8
Coal	4500	67	27,5	9,1
Fuel oil M100	9111	86	72,5	9,2
LPG	11000	92	166,6	16,5
Diesel fuel	10180	89	247	27,3
Natural gas (construction of new pipeline)	11400	92	292,6	27,9

It is also interesting to look through the chart (Figure 10) of gas price formation [8] for consumers in case of LNG supply in Saint-Petersburg region.

Because of low gas prices in Russia and 'granted' pressure excess final price to consumers under the elements from Figure 10 constitutes approximately 60 USD/1mcm.

So, such production of LNG by Ranque-Hilsch tube seems viable [8] to supply dispersed consumers in Russia at the distance ~ 100 km and supply volume ~ 0,1 MMsm^3/d. This surely does not match with Figure 7. But there one has advantages of low gas prices, 'granted' pressure energy and high pipeline cost. So this technology seems to suit Russian conditions and probably can be applied worldwide after further improvement. From the other side, the liquefaction efficiency of such units is rather low ~ 5%. The complexity of physical processes inside the tube makes difficult to essentially increase the efficiency. In view of present gas and energy market liberalization process, which by no means finally will affect Russia, it seems that this technology is able to compete for the consumer.

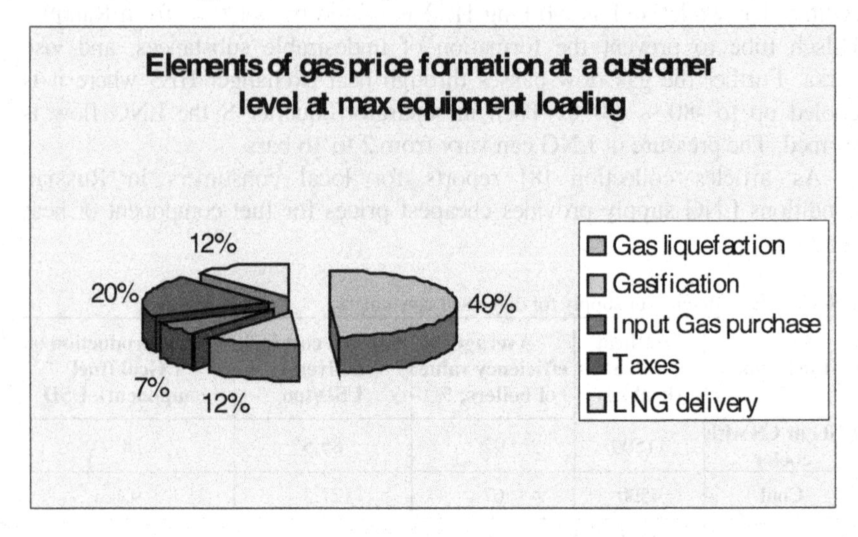

Figure 10. Gas price formation in case of LNG supply

Thus it could be interesting to find another technology of aerodynamic type, which would enjoy the potential of efficiency increase and has more clear understanding of its principles.

2.3 3S™ Technology

Since 1993 ENGO Research Center in Russia in conjunction with its partners has been developing new technology of gas processing – Super Sonic Separation technology (3S).

Some of the advantages of 3S in comparison with conventional technologies for the separation of hydrocarbons from a mixed hydrocarbon gas stream are:

a) small size and therefore reduced space requirements, greater portability and reduced handling and installation costs
b) low capital and operating costs ,

c) no adverse environmental impact,
d) the absence of moving parts,
e) no requirement for routine maintenance,
f) conservation of reservoir energy
g) superior performance capabilities compared to conventional separation equipment and configurations

3S has the potential to be used in the following applications by the gas industry:

A) gas conditioning (dehydration and heavy hydrocarbons separation);
B) propane/butanes separation (LPG);
C) separation of H_2S and CO_2;
D) ethane extraction;
E) methane liquefaction.

In contrast to the technology described in item 2.2 the present one has yet no experience of commercial usage, but only prototype tests at realistic flow rates (~ 500 MMscm/y) for solving problems A), B) from the list above.

Currently a contract with one of leading gas processing companies for installation of pilot units at its facilities is signed.

It is experimentally proved that 3S technology can successfully operate under wide ranges of input pressures and temperatures, namely pressure range is 30 ÷ 80 bars, temperature range is -60 ÷ +20 C.

To understand the potential of the technology regarding small scale LNG let us first consider main principles of its operation. We will constantly refer and compare 3S technology with conventional and widely applied turbo expander technology (TET) in order to elucidate the differences and possible advantages of 3S. As you can see from Figure 11 (see approved *International Patent Application WO 00/23757, filing date 15 October 1999*) the gas flow under suitable pressure and temperature first enters the swirling device 1 and then *twisted flow* passing through sub/supersonic nozzle 2 in approximately isentropic way, so the nozzle acts like an expander in TET. But in TET the pressure energy of gas is transformed to turbine work while in 3S this energy is transformed to the kinetic motion of gas. Because of this, low temperatures arise and the condensation process starts.

In the working section 3 liquid drops of > 1 microns in size form (this fact is confirmed experimentally). Such size of drops is enough for the centrifugal forces are able to act, and the liquid drops start to move to the wall of section 3 simultaneously growing by coagulation process. All this leads to the formation of liquid film (depending on the quality of swirling) and/or enriched gas layer directly near the wall (this fact is also confirmed experimentally). Further through separation channels 4 such in general two-phased mixture is directed to secondary separator of any type, for example, netted one. The operation of this part of 3S unit can be represented as the work of separator of cyclone type.

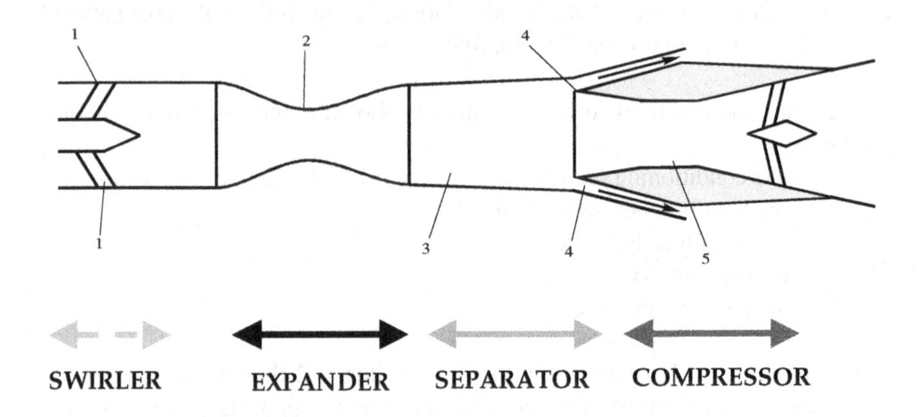

Figure 11. Schematics of 3S installation
Notations: 1 – swirling device, 2 – sub/ supersonic nozzle, 3 – working section, 4 –device for gas-liquid mixture withdrawal, 5 – diffuser

Finally in the diffuser section 5 the pressure is recovered from the kinetic motion of the gas through some kind of shock wave – the fact which is well known in gas dynamics. Certainly the pressure recovers not up to the initial inlet value. The pressure losses (typically 20 – 30% at present) depend on the desired temperature in the working section 3 (as desired temperature is lower the gas movement is faster and pressure recovering is more difficult). This section acts like a compressor in TET.

Let us also mention that the temperature drop depends only on pressure differential inlet – outlet and does not depend on the inlet pressure value. So, in principle the unit can work with reasonably low initial pressure to achieve low temperatures. Here we do not bother ourselves with the dependence of condensation process intensity on pressure value.

Thus the whole configuration from Figure 11 works as the combination of several devices of TET (of course in TET there is no any swirler).

3S unit is working quite well under sub and transonic regimes to substitute conventional JT valve. Experiments show that 3S is much more effective than JT valve, especially under positive inlet temperatures when JT valve generally has no effect.

Moreover the comparison of 3S and TET installations under transonic and supersonic regimes clearly shows 3S advantages, providing that gas flow is not essentially lean.

One can see that while flow is sonic 3S installation is practically equivalent, as far as energy requirements are concerned, with TET installation for recovery rates below 85% (the parameters of work of 3S in this case are approved experimentally). With the increase of Mach number 3S efficiency grows and for essentially supersonic conditions the compressor capacity which is required for reaching the same recovery rate is twice less than TET energy needs. To have the same advantages for high recovery rates and moderate M it is necessary to design more sophisticated technological schemes, but it is also fairly possible.

Moreover 3S is practically insensitive in contrast with TET to the presence of liquids in the flow. And as Shell, who now is developing the dehydration technology which is also based on high speed flows, has reported [9], [10], under such rapid flow conditions the hydrates formation is hardly possible.

Finally, let us show the *ideal* thermodynamic cycle description that elucidates the advantages of 3S compared to TET in achieving low temperatures.

For 3S and TET cycles we start from the same pressure and temperature and end with another outlet pressure (the same for two processes) but with different temperatures (see below the reasons for this).

The operation of TET. First, the gas flow is cooled in heat exchanger under approximately constant pressure, then gas flow works in the turbine and is cooled further. (Note that this process is not isentropic, i.e. even in ideal case the losses are inevitable). Thus it founds itself under the condensation curve and the condensation process takes place. Further, gas again passes through heat exchanger and finally is additionally compressed by compressor.

The operation of 3S also starts with heat exchanger. Then *by isentropic expansion* gas is cooled down. This process allow to enter deeper under condensation curve and therefore to have more favorable conditions for condensation. Further in the diffuser the pressure recovery takes place. This process is isenthalpic because such recovering is achieved through shock waves of different type. And finally gas flow again passes through heat exchanger. End points of both cycles lie on the same isobar, but due to the use of compressor outlet temperature for TET is higher.

Thus these two installations are principally different. 3S allows reaching lower temperatures because 'early' *starts isentropic process* and has significant pressure drop within the device. But this pressure drop can be recovered in the diffuser. Moreover it is insensitive to the presence of liquids (which in turbo expander can damage the turbine) and seems not to experience the hydrates formation.

It also differs essentially from Shell device, since from the beginning of the process 3S unit operates with the *swirled flow.* Swirling flow also facilitates achieving the temperatures for the central part of the flow that are lower than ones achieved through isentropic expansion.

Combining with its small size 3S installation can be used for cooling purposes in LNG technological schemes and has at least the perspective for the problem of small scale natural gas separation and liquefaction.

REFERENCES

1. Russian Energy Strategy up to 2020, www.mte.gov.ru
2. Russian Energy Policy, Review 2002, *International Energy Agency.*
3. World Energy Outlook 2002, Chapter 8, *International Energy Agency.*
4. Natural gas Information 2002, Part I, *International Energy Agency.*
5. Natural gas Information 2002, Part III, *International Energy Agency.*
6. World energy outlook 2002, *International Energy Agency.*
7. Gutsol A.F. Ranque Effect. Methodical notes, *Uspehi Fiz. Nauk*, **167**, No. 6, 665 – 687 (1997) (in Russian).
8. The perspectives and experience of application of LNG at OAO "Gazprom" objects. *The materials of scientific and technological council of OAO "Gazprom", SPb December 2001*, Moscow 2002.
9. D. Page, M. Lander, S. de Kruiff, Twister – a revolution in gas separation. *Exploration and Production Newsletter – November 1999 / SIEP 99-7011*, 29 – 31.
10. Twist in the tale, *Offshore Engineer July 2000*, 2 – 3.

PART III: STRATEGIC RESERVES AND GEOLOGICAL SURVEY

UP-TO-DATE RESEARCHES AND FUTURE TRENDS IN UNDERGROUND GAS STORAGE FACILITIES: A STATE OF THE ART REVIEW

Dr Fabien FAVRET
SOFREGAZ, France

Abstract: The Underground Gas Storage (UGS) facilities are mainly built in hydrocarbon reservoirs (depleted oil & gas fields), in aquifers and also in salt caverns using proven techniques based on the best use of geology and geophysics. Sustained efforts from industrial R&D are devoted to these widely used conventional techniques with the aim to both improve their performances (working gas volume and daily output) and reduce their costs (development, operation and maintenance). Storage facilities in porous and permeable media, oil and gas depleted reservoirs and aquifers account for more than 95 % of the working gas volume capacity world-wide. The technique of storage in salt caverns is also very often used, in particular in the US where the gas market is deregulated, but also in some western European countries where the market on the impulsion of EU is being liberalised. Costs have been divided by two with the development of larger caverns, due to the advances in knowledge of the mechanical behaviour of salt rock and also to the increase in the specific working capacity with the use of higher maximum operating pressures. Techniques aimed to accelerate the availability of salt cavern facilities have been developed, in particular in the US. New techniques, still at the R&D stage, and construction of demonstration units, will offer alternative solutions for underground storage in regions where geology does not permit the development of conventional storage. These are caverns in thin salt layers and lined rock caverns. The expansion of world gas demand will lead to a considerable increase in storage activity in the decades to come, and many new facilities will have to be constructed to meet future needs. This paper also highlights technological innovations in the UGS field.

Key words: underground gas storage, state of art, new technologies

J. Hetland and T. Gochitashvili (eds.),
Security of Natural Gas Supply through Transit Countries, 159–193.

1. INTRODUCTION

Natural gas is assuming a preponderant role in world energy supplies. Gas penetration in the energy sector is very large, reaching about 23 % world-wide in 2001 and with a forecast increase up to 28 % by 2025[1]. Due to heating requirements, this sector has to deal with wide seasonal variations in consumption.

In addition, the growth of international trade has profited increasingly distant sources. The technical and economic constraints of production facilities and long-distance transport infrastructures mean that generally supplies flow remain more or less constant throughout the year, with a load factor close to the maximum (to maximise the profitability of Exploration & Production gas projects).

As identified by EU, the use of "less secure" supplies lead the gas companies to be worried about any supply shortages, be it technical or political.

Theses developments require the creation of an infrastructure capable of dealing with the problem of fluctuating demand versus rather constant supply flows while still guaranteeing supply security. Among the solutions which can be considered, underground gas storage (UGS) facilities seem to be one of the most reliable and cheapest additional facilities.

1.1 Use of Underground Gas Storage (UGS)

1.1.1 Modulation

The initial function of a UGS facility is to balance gas consumption and resources at all times:

- seasonally,
- monthly,
- daily,
- and intra-daily,

mainly for the residential and commercial sectors, where demand is especially sensitive to changes in temperature (use of gas heating systems).

In addition, storage allows the operators to balance the peak winter demand. The relative peak demand on the coldest day of the year is a very important parameter for the gas industry, because it conditions the size of the gas infrastructures. For example, in France, during the maximum daily send-

[1] Source IEA International Energy Outlook 2003

out, UGS facilities have been sized to supply more than 60 % of the demand during several days. The same applies for summer peak demand of power in country where air-conditioning is preponderant.

In fact, storage facilities provide many more services:

- they can play an important role in reducing or eliminating the risk of supply shortages (interruption of a supply source, technical problems in the pipeline system, etc.),
- they enable gas transportation and distribution companies to operate their networks more efficiently and with more reliability/flexibility throughout the year by allowing a high load factor, thus reducing the final cost of gas distribution,
- they allow multi-annual supply adjustment,
- but also, they allow storage during low price periods for selling during better favorable market price periods, optimising the global exploitation costs for oil & gas companies.

1.1.2 New services

In addition to their modulation function, UGS facilities offer commercial possibilities. The use of underground storage to hedge against seasonal and/or monthly differentials in gas prices is already developed in the US, and being developed in western European countries (especially in UK). The profits to be made due to price fluctuations can be considerable. As a result both the supply and the demand side are leading storage for more speculative purposes. In these countries, storage also influences prices directly, and it is why a lot of trading or shipping companies:

- either have their own UGS facilities,
- or rent available capacities on the UGS open market offering Third Parties Access (TPA) to storage facilities.

1.2 Different types of UGS

Natural gas can be stored in UGS facilities, or in Liquid Natural Gas (LNG) receiving terminals and peak-shaving units which can supply gas at a high rate in the cold season and over a short interval.

UGS represents the most effective means. In 2000, there were 602 underground storage sites world-wide, with a working gas volume capacity of 310 bcm[2] and a maximum daily output around 4.46 bcm/d. The two other alternatives offer limited capacities:

[2] 1 bcm = 1 billion cubic meter = 10^9 m^3

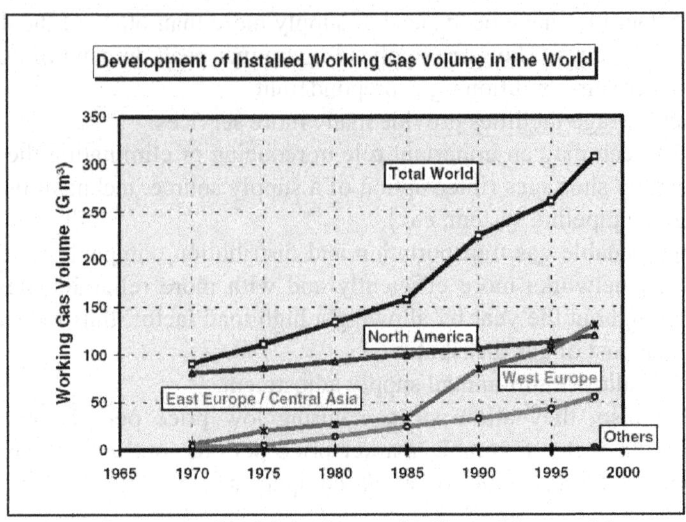

Fig. 1 - Development of UGS in the world (from 21st world gas conf.).

- there are less than 80 peak-shaving units in service, with a storage capacity of 2-3 bcm,
- and 30-35 LNG receiving terminals with a storage capacity of 7-8 bcm.

So, the three main storage techniques used are underground and depend largely on the existing geological formations:

- depleted oil & gas fields,
- aquifers,
- salt caverns.

Storage in porous reservoir is the most convenient way for storing gas in order to meet winter demand (seasonal modulation). The cheaper means of storage is often in depleted fields. In the absence of such structures, however, natural gas can be stored in aquifers or in salt caverns. Examples of natural gas storage in abandoned mines can also be found. Other alternatives have been investigated, such as lined rock caverns.

1.2.1 Depleted oil and gas fields

Gas storage in depleted oil and gas fields is the most world-wide used method and often the cheapest one. Most of these are depleted gas reservoirs, although a few depleted oil reservoirs are also operated for the purpose. The first gas storage experiment (injection) was made in a gas field in Welland County, Ontario (Canada) in 1915. The first gas storage facility in a depleted reservoir was built in 1916.

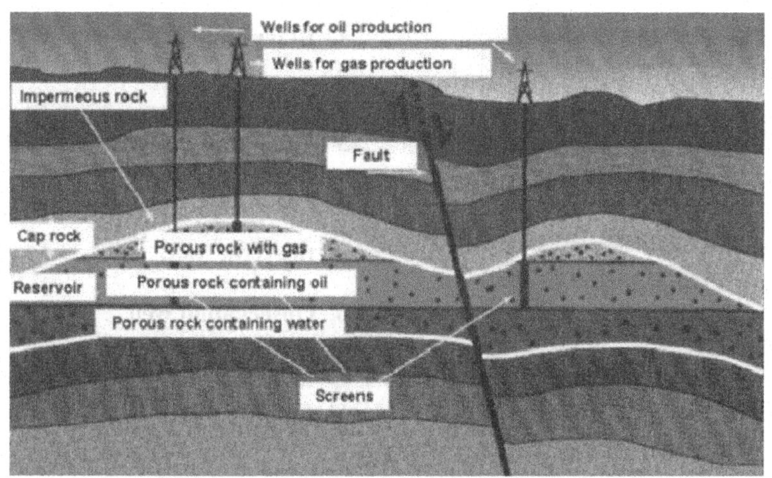

Fig. 2 - Schematic cross section of a depleted oil & gas field

It was a gas field in Zoar near Buffalo, New York (United States). This UGS is considered to be the oldest in operation in the world.

Today, there are more than 460 storage facilities located in depleted reservoirs world-wide. The principle of a storage facility in a depleted reservoir is simple, because the reservoir formerly contained gas or oil. Hence it satisfies the permeability and porosity conditions required for storage. However, before developing gas storage in a depleted field, it is indispensable to check whether it corresponds to the required production goals (high throughputs over short periods[3] from 500 to 5000 Mm^3 [1 Mm^3 = 1 million cubic meter = 10^6 m^3], see table 2) and the tightness of the cap rock (impervious formation upper to the storage structure).

1.2.2 Aquifers

Aquifers for gas storage were first used in 1946 in Kentucky (United States). They are around 80 storage facilities in aquifers in the world today, most of them in the United States, the former Soviet Union and Western Europe (France, Germany and Italy). The principle of aquifer storage is to create an artificial gas field by injecting gas into the voids of an aquifer formation (involved gas volumes are similar to depleted fields). For this reason, the following geological conditions are necessary:

- an anticline with sufficient closure,
- a porous and permeable reservoir,
- and, an excellent quality and tight cap rock.

[3] withdrawal periods = cold periods = winter time = between 90 and 120 days

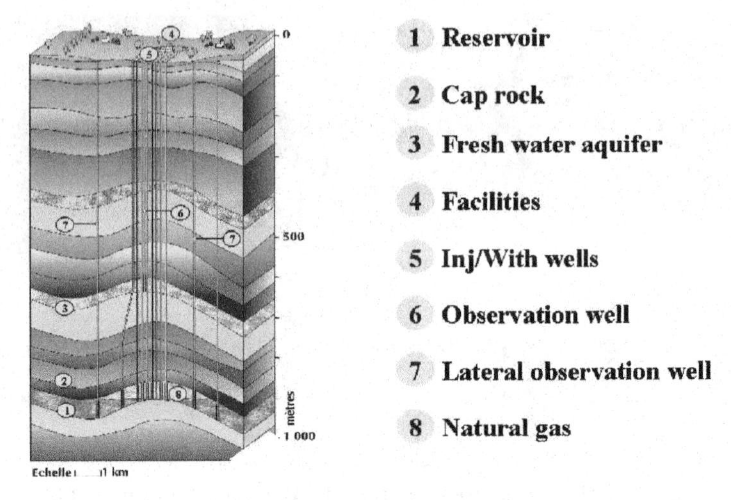

Fig. 3 - Schematic cross section of an aquifer converted into UGS

1.2.3 Salt caverns

Salt caverns have been used to store Liquefied Petroleum Gas (LPG) for a long time, but the technique is relatively recent for natural gas. It was introduced in the United States in 1961, in Saint-Clair County, Michigan. Today, there are around 60 storage facilities of this type world-wide, 27 are located in the US. The number of this type of storage is increasing rapidly (a lot of new projects especially in Western Europe).

The principle consists in dissolving the salt by leaching with fresh water (sometimes sea water) and removing the brine via a single well, which then serves for gas injection and withdrawal. These reservoirs serve to store relatively smaller quantities of gas (ranging from 50 to 500 Mm^3, see table 2) than in aquifers or depleted fields. The storage capacity for a given cavity volume (several hundreds of thousands cubic metres) is proportional to the maximum operating pressure, which mainly depends on the depth.

Salt caverns which are also a useful complement to the large porous reservoirs, offer other advantages:

- high deliverability,
- high degree of availability,
- short filling period (using large compression units),
- low percentage of cushion gas,
- total recovery of cushion gas when decommissioned.

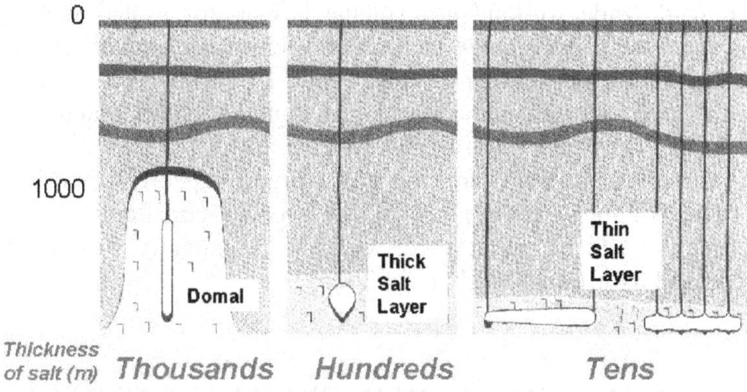

Fig. 4 - Different types of UGS in salt caverns

Thus the combination of the two types of storage, in porous reservoirs which are generally used to guarantee basic demand to meet seasonal modulation, and storage in salt caverns, which are generally operated to cover peak demand, makes for high withdrawal rates even at the end of the withdrawal period.

1.3 UGS facilities world-wide and prospects for 2025

Storage needs vary enormously from country to country, depending on the energy situation, and gas supplies:

- the distance between production centres and consumption areas,
- the rate of penetration of gas in the different consumer sectors,
- the domestic gas production, ...

Table 1 – UGS statistics world-wide (derived from 21st world gas conf.)

Area	Nb Porous Storage	Nb Cavern Storage	Total Nb UGS	Working Gas Volume (bcm/y)	Withdrawal capacity (bcm/d)
Western Europe	56	22	78	55	1,09
Eastern Europe & Central Asia	64	3	67	131	1,00
Canada	31	7	38	14	0,20
USA	383	27	410	106	2,15
Australia & Japan	7	0	7	2	0,01
Other Countries	2	0	2	2	0,01
Total	543	59	602	310	4,46

The world-wide working gas volume has almost tripled since 1970. During the 1980's, gas storage facilities developed very rapidly in Europe, where the volume of working gas rose from 11 bcm in 1980 to 53 bcm in 1997. In 2025, world-wide natural gas demand could reach 3,900 – 4,300 bcm[4]. By then, natural gas will account for 28 %[5] of world energy supplies. Current working gas storage capacity is equivalent to a little bit more than 10 % of the world's consumption, but for this percentage to be maintained, world-wide capacity would have to be increased by around 100 bcm in 2025[6].

2. COST REDUCTION IN EXISTING STORAGE TECHNIQUES

2.1 Existing costs

From the projects already implemented, a comparison between UGS and LNG storage costs can be established as follows:

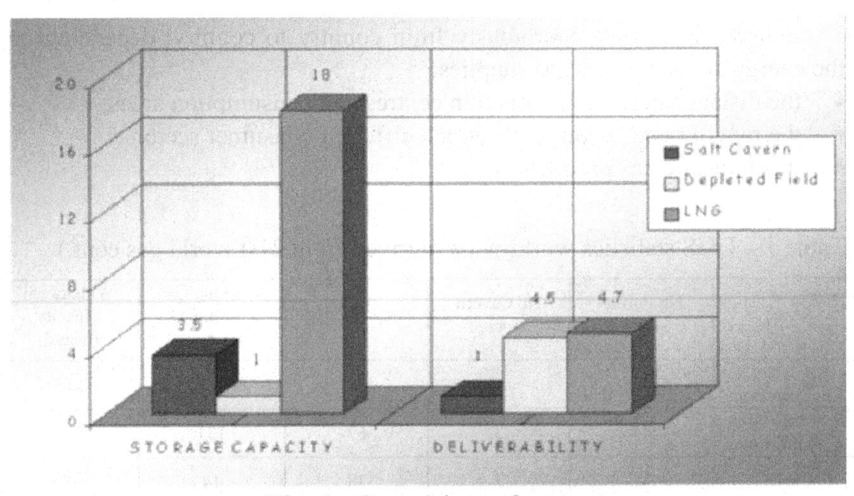

Fig. 5 - Comparison of costs

[4] Consumption range from CEDIGAZ statistics (May 2003) for years 2010 and 2020 (note: the IEA has predicted in 2003 a forecast of 5280 bcm by 2025!!!).

[5] Source IEA International Energy Outlook 2003.

[6] World-wide total working gas volume in 2025 can be estimated between 390 and 430 bcm, to be compared to the 310 bcm working gas volume in 2000 (see Table 1).

Table 2 – Average capacities and costs (in US$) for UGS facilities

Costs	Depleted Fields	Aquifers	Salt Caverns
Range of possible Working Gas Volume (Mm^3)	300 – 5000	200 – 3000	50 – 500
CAPEX[7] (US$/$m^3$)	0.05 – 0.25	0.3 – 0.5	0.4 – 0.7
OPEX[8] (US$/$m^3$)	0.01 – 0.03	0.01 – 0.03	0.01 – 0.1

Having a look on Fig. 5, it can be easily assumed that UGS facilities are always cheaper than LNG storage. From the past experience in UGS and available studies [1], it is also possible to give the average costs (investments and operation) for UGS facilities (see table 2). Investment costs average ratios can also be assumed and are given in Table 3.

Table 3 – Investment costs average capacities and costs (in US$) for UGS facilities

Costs	Ratios
Exploration	10 %
Wells	20 %
Gaz treatment and compression facilities	30 %
Cushion gas for porous reservoir[9]	40 %

[7] CAPEX = CAPital EXpenditure
[8] OPEX = OPeration EXpenditure
[9] mainly for aquifers and oil fields, but it has to be adjusted depending on the depletion level for gas fields; and for salt caverns this percentage covers roughly cushion gas and leaching facilities.

2.2 Market and service price of UGS in EU countries

From the storage tariffs published on Internet, it can be estimated that the storage tariffs in Europe in 2003 are between 3.5 and 7 $cts/m^3, with an average price close to 6 $cts/m^3.

Another way to estimate the storage services price is to compare with alternative solutions. Without UGS access, a gas shipper can use the following other solutions:

- modulation of supplies,
- customer interruptible contracts,
- short term market trading,

or can be exposed to the following risks:

- Take-or-Pay clauses,
- automatic balancing cost (which are very dissuasive in all EU countries).

All these services or risks have a cost which can be estimated in EU in 2003 between 3.2 and 10 $cts/m^3, with an average cost close to 8 $cts/m^3.

2.3 Storage in porous reservoirs

Storage in porous and permeable formations, in hydrocarbon reservoirs and aquifers, represents more than 95 % of the working capacity of all the storage facilities in the world.

2.3.1 Improving the use of porous structures

2.3.1.1 Geological description, knowledge and seismic monitoring

The techniques used for exploring storage structures and reservoirs are those developed by the oil industry for the research and production of hydrocarbons. The storage industry benefits from R&D spin-offs from the oil industry whose annual turnover is around 600 BUSD[10], i.e. more than 100 times that of the storage industry which is estimated at 4 BUSD.

With drilling techniques dating back to the 19th century and differed logging and seismic monitoring dating to the first half of the 20th century, the techniques used are antique but have taken benefits from the progress in computer technology, above all in the 1980's for seismic monitoring (3D seismic since 1990's and 4D seismic in the 2000's).

[10] 1 BUSD = 1 billion US$ = 10^9 US$

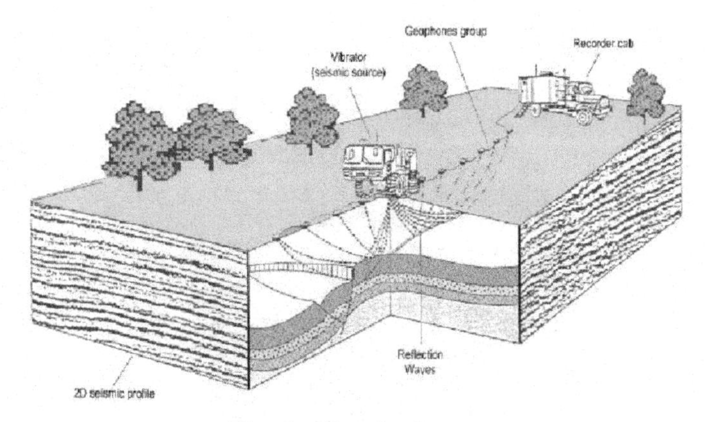

Fig. 6 - 2D seismic

Today, detailed seismic monitoring allows small structures to be spotted and small-scale discrepancies (accidents, faults with little slip) and even gas-liquid interfaces (bright spots) and lateral variations in facies (stratigraphy) to be detected.

The search for structures able to be used as storage facilities differs from, and in a way is more complex than, the search for hydrocarbon deposits. In fact, just because a well has indicated a good, suitably covered reservoir does not mean that the structure is appropriate for storage purposes. It is necessary to prove that the gas to be stored there will not escape and therefore ensure the caprock's continuity and the closure of the structure.

Certifying whether a structure can be used for storage purposes is a complex and costly operation, especially for aquifers, involving at the very least one (3D) seismic survey, drilling and many measures and tests (interference measurements between wells). In the field of certifying structures for storage use, the precision of 3D seismic exploration allows uncertainties to be minimised. It reduces the number of wells to be drilled for the certification and permits storage wells to be better located within the reservoir, and so, lowering the number of development wells required. The geological and geometrical characteristics of underground structures play a major role in determining the principal performances of an underground storage facility, and therefore, help optimising the development and the operation of facilities.

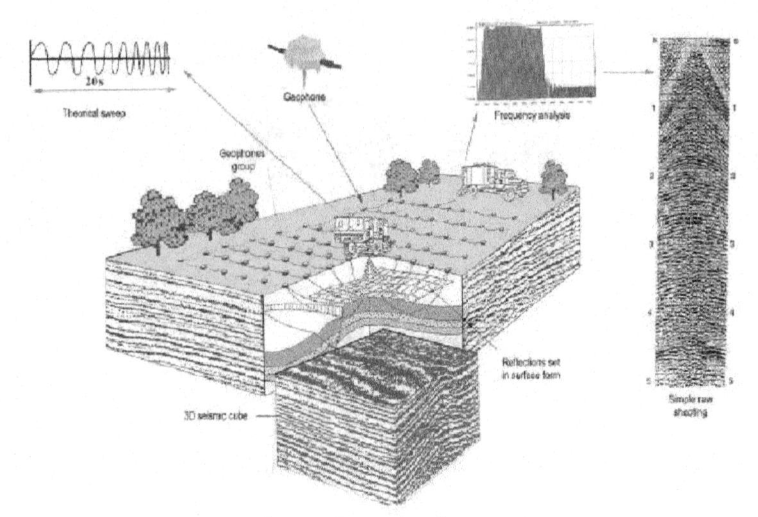

Fig. 7 - 3D seismic

Geological description has profited from advances in the processing of logging information and from seismic exploration and geostatistics, a science developed since the end of the 1970's in relationship with the increase in computational power [2]. However, it was the progress of computer graphics and imaging techniques in the 1980's that really revolutionised geological description methods. This computerised description is fully needed to create the huge amount of data (geometric characteristics and petrophysical properties of reservoirs) which have to be used for fluid flow simulation.

Monitoring by 3D repeatable seismic measures [3], also commonly named 4D seismic, is still, more or less, at the research stage (even in E&P field) but early tests seem promising. This system is based on techniques such as the use of seismic sensors placed at regular intervals on the surface or in wells (permanently) [4]. Advanced multi covering seismic such as AVO (Amplitude Versus Offset) can help in better investigating petrophysical properties of reservoirs.

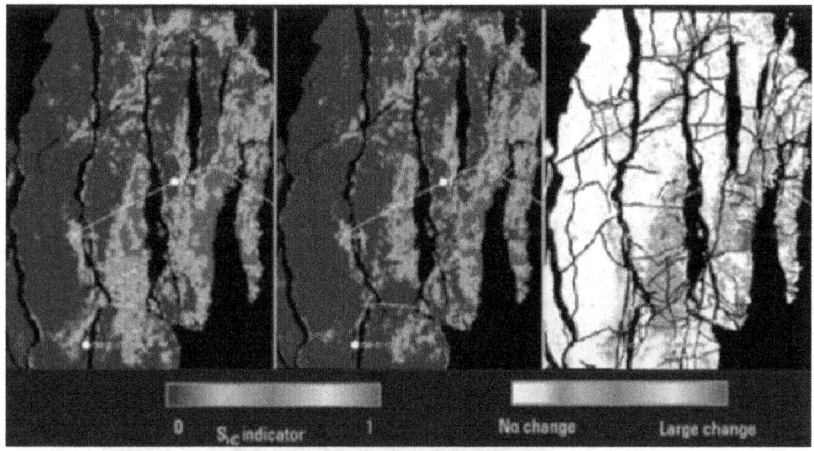

Fig. 8 - 4D Seismic (from KBB)

The seismic measurements help to determine the limits of the gas bubble in the reservoir and its development between two measurements. The results from the seismic monitoring are also used to refine the reservoir simulation model and to improve production predictions. The primary advantage in the application of seismic monitoring to storage facilities is, on one hand, the ability to control the progression of the gas bubble towards the critical spill points in several directions in order to maximise the filling of the reservoir, and on the other hand, to identify the areas with large accumulations of gas so that better production wells may be put there. Rather than just filling the reservoir to its optimum capacity, the number of storage wells can also be cut back and the number of observation wells can be substantially reduced. Intensive fluid flow numerical simulation coupled with geostatistical spatial models of the reservoir properties (porosity, permeability, saturation ...) can also help in better understanding reservoir behaviour during exploitation, especially to quantify uncertainties. These studies are helpful for optimising recovery and well implementation.

2.3.1.2 Knowledge and description of fluid flows

Advances in knowledge and models of fluid flows in UGS have increased simultaneously. As a result of the increasing performance and falling costs of computer systems, calculations have been able to progress in terms of their accuracy, rapidity and cost. Fluid flow simulations give a clearer picture of how gas is distributed in the reservoir at any moment and any place, provided rock properties are known. In this way they allow the assessment of the working volume, the peak withdrawal rate, the number and location of new wells required and finally the minimum gas cushion to be injected to guarantee performances [5].

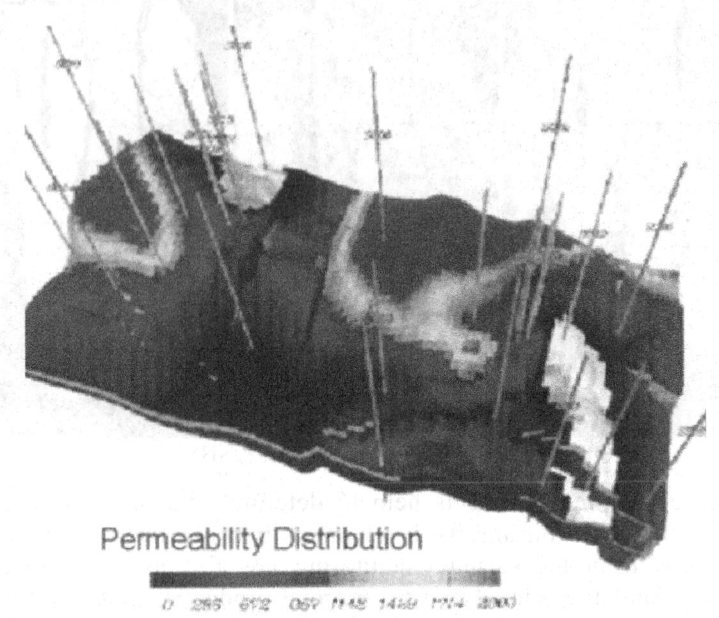

Fig. 9 - Example of 3D modelling

Fluid flow simulations are also very useful to predict water or liquid production but also to refine the operational strategies (where and which quantities to be injected or withdrawn on which wells and/or in which layers).

Growing calculation speed, due to progress made in computer technology and also in simulation software allows an increase in the number of cases studied and their complexity at the same time. It is now possible to study in details tens of possible site development situations at the time of the pilot studies, so adapting the service offer to the client's requirements. Instead of establishing an optimum functioning based on constraints mainly fixed through by experience, we can now fix many optimum plans for each need anticipated by the customer. The result is a more global optimisation of the storage system.

Increasing calculation accuracy has led to many improvements such as reports by the simulation of data measured on-site, or the "tuning" of models, which now relate to longer time periods, are more accurate and above all are obtained quicker. It has been noted that, in practically all cases, the most detailed description of the reservoir had been accompanied by a better tuning from the first simulation. Fewer tuning "tricks" and greater accuracy leads the engineer to have more confidence in his result, because the uncertainty of the simulations can be decreased.

Furthermore, an increase on the complexity of phenomena taken into account has been made possible. For example, nowadays the description of faults is very detailed, several gas constituents (compositional simulation) can be monitored within the reservoir and the reservoir - wells - wellhead equipment system automatically taken into account [6].

These improvements permit a quicker exploration of a larger area amongst a site's possible developments, so enlarging the possibilities. Finally, lower calculation costs have made the detailed study of the majority of sites possible and have multiplied the number of studies which it is profitable to undertake. Although it may be difficult to evaluate the growth in profitability resulting from modelling advances because this concerns the entire system, it is clear that the performance of storage facilities is estimated with an increasing level of accuracy.

2.3.2 Investment reduction of the cushion gas

When natural gas is stored in porous formations (especially in aquifers), cushion gas accounts for the largest part of the investment, representing about 30 to 40 % of the development cost of aquifer gas storage facilities. It is possible to reduce significantly this cost by replacing the natural gas with an alternative cushion gas. The gain depends on the price of natural gas, the rate of asset tax and the cost of alternative cushion gas. This technique requires specific know-how, and success depends on the following scientific and technological aspects being monitored:

- hydrodynamic analysis and conditions of inert injection related to storage operation or development,
- geochemical studies,
- simulation studies for the optimisation of inert injection and storage operation,
- location of inert gas wells,
- gas mixing monitoring,
- inert gas manufacturing technology.

Being one of the leaders in this technique since 1974, Gaz de France has experimented gas mixing operations in 7 UGS facilities and has a wide experience in inert injection and more generally in mixing gases [7] [8]. These operations take place under very different practical conditions and each of them was designed so that the withdrawn gas can be sent out without any specific operating constraints (calorific value) over several decades.

Monitoring such operations implies specific measures and modelling tools which correctly handle gas mixing phenomena. When such conditions are obtained the total saving is estimated at 20 % of the cushion gas cost.

2.3.3 Mono-bore and large diameter wells

In order to increase the deliverability of UGS, more and more wells using large diameter completion are drilled every year. This technique is very good to increase gas flow when there is no production of liquid (oil, condensate or simply water) because increasing diameter will decrease gas velocity and so can lead to some troubles for the production of liquids[11]. In addition, and for minimising the pressure drop along the production tubing, nowadays more wells are designed with a completion without diameter reduction in order to limit gas flow perturbation. These wells are called "mono-bore" wells [9].

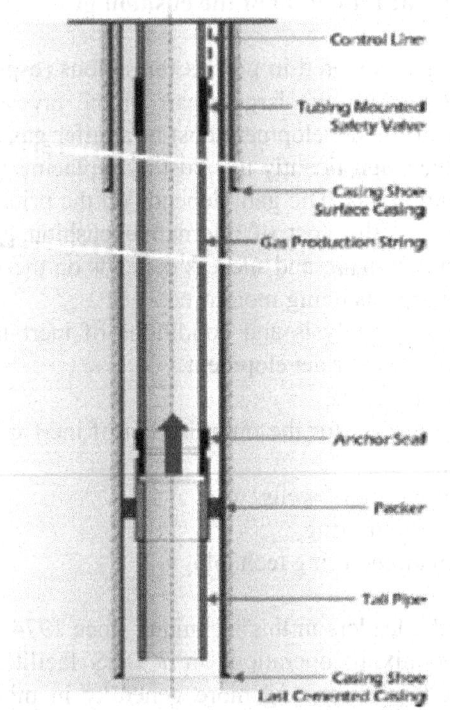

Fig. 10 - Mono-bore completion (from KBB)

[11] Even in aquifers, it is sometimes better to produce gas and water than nothing (water blocking) even with huge pressure drops involved by water perturbation of gas flow.

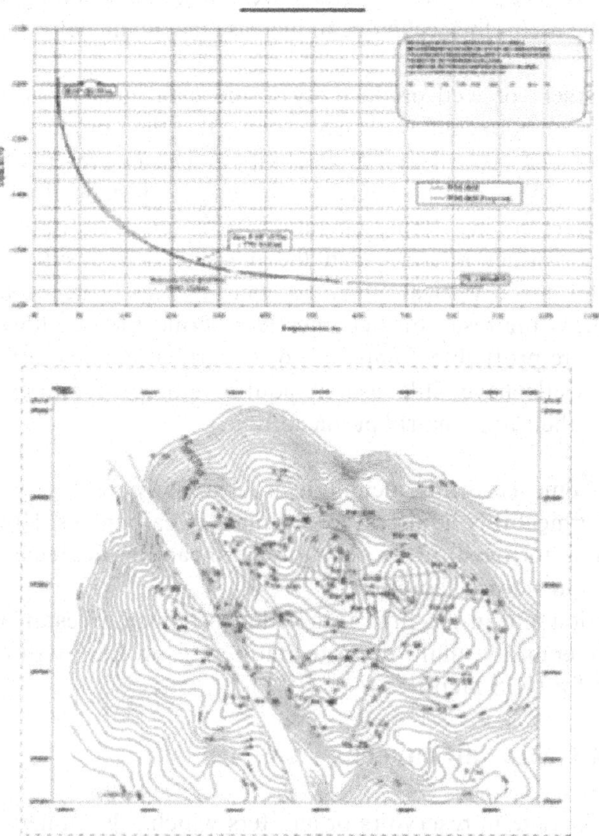

Fig. 11b - Example of horizontal wells design

2.3.4 Horizontal wells

Horizontal drilling [10] [11] appeared in the 1950's in some areas of oil production, with commercial success. This technique is now generally used all over the world in oil and gas fields (E&P).

The application of this technique to UGS is starting slowly with only a few successful horizontal gas storage wells having come on stream. Indeed, in 2000, out of some 10,000 storage wells in the 600 UGS facilities in the world, only less than 100 were horizontal wells. At the same time, at least more than 100 new horizontal wells were planned, pointing out the fast growth in the use of this technique to increase maximum deliverability[12].

Enhancement of well productivity is the main goal of horizontal well drilling. Indeed, horizontal wells have shown that their productivity is 1.5 - 6 times higher than vertical well productivity in the same reservoir, depending on reservoir quality and horizontal drain length. Horizontal wells can also minimise water coning during operation, if the drain trajectory is always above the gas-water interface and the pressure drop inside the drains is smaller than in a vertical well, so causing less water coning during withdrawal. As the reservoir becomes less permeable, the horizontal well becomes more profitable compared to its vertical counterpart. Therefore, horizontal wells have little use in aquifer storage facilities which are selected for their high matrix permeability.

But in some case this technique is absolutely necessary to develop very low permeability aquifers, for example in carbonates. However, they are profitable for most of the very low permeability depleted fields to be converted into storage facilities.

In addition and in the future, in order to reduce investment cost ratio versus deliverability, multi-drain technology from E&P will be probably introduced for UGS facilities.

2.4 UGS in salt caverns

Unlike storage in reservoirs or aquifers which rely on natural voids in porous and permeable rocks, with storage in salt caverns, the gas is stored in man-made, solution-mined caverns. Geology is only the starting point, and engineers design and construct the project.

Because it is quite a recent technology (1961), these types of storage gas have also taken benefits from geotechnical advances (rock mechanics, mining techniques, etc.) made possible by the considerable development of computer technology since the 1970's. This explains why reduction in investments is particularly noticeable in this area where development costs have been divided by two between the 1960's and the 1990's.

[12] The best gas production well in the world is not a horizontal well but a simple deviated well located in Norge UGS (close to Groningen in Netherlands) with a maximum daily deliverability over 12 MMcm/d.

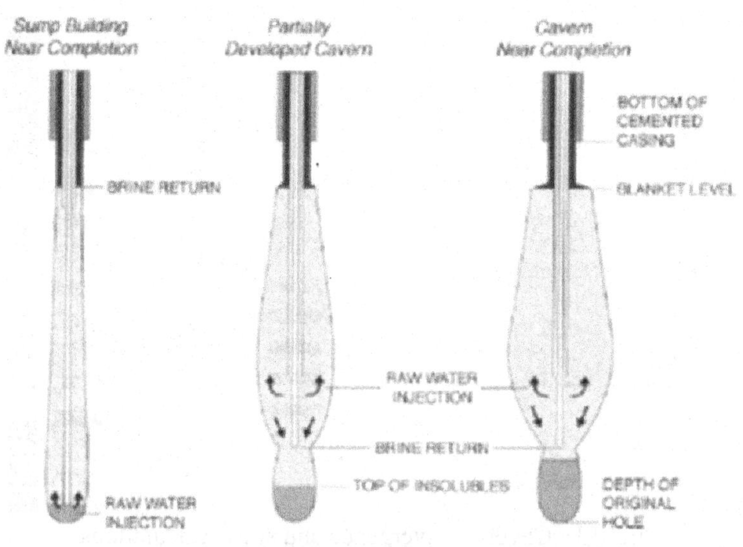

Fig. 12 - Solution mining in salt dome

2.4.1 Geotechnical advances

2.4.1.1 From knowledge of salt rock to modern cavity sizing tools

The first caverns constructed during the 1960's and 1970's used as a starting point the methods employed for sizing salt rock mines (gallery sizes, pillars and chambers). Elastoplasticity provided information about the final suite of the galleries, pillars and chambers (over the total service life).

In the 1970's, various observations carried out at the caverns at the Eminence salt dome site (Mississippi, USA) and in Tersanne (France) led to alarming results, reporting geometric volume losses of several tens of percentage points which, moreover, did not stabilise. These results drove operators into launching large R&D programmes with research institutes, universities and mining schools, programmes which included theory, laboratory and on-site tests and measurements.

At the end of the 1970's and the early 1980's this programme showed the highly non linear elastoviscous behaviour of salt rock. From the 1980's, 2D software allowed the sizing of isolated or network caverns and from the 1990's, 3D software permitted the sizing of any cavity as well as the prediction of changes in cavity geometry during operation.

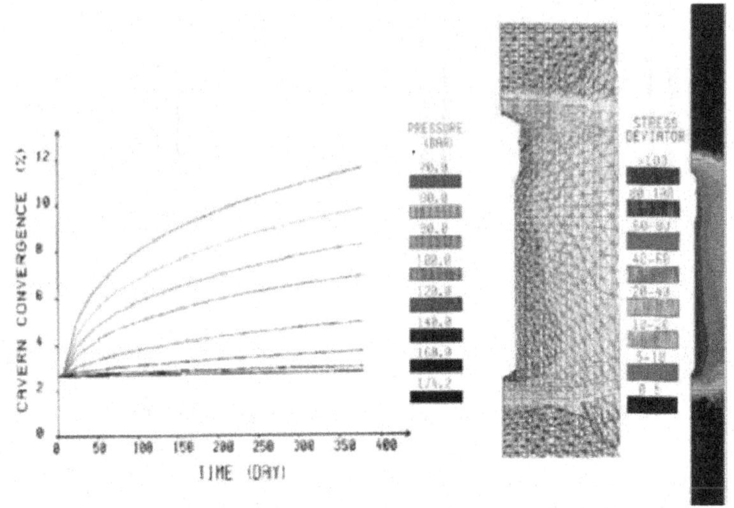

Fig. 13 - Cavern convergence and stress calculations

2.4.1.2 Increasing geometric volume

Investment in storage in salt caverns is the sum of:

- the investment proportional to the working quantity of gas (solution mining, compression, treatment),
- the investment proportional to the number of caverns (platforms, wells, storage mains and manifolds).

Increasing cavity geometric volume is therefore a key factor to reduce investment ratios. For this reason, increasing geometric volume has been a constant preoccupation since the 1960's. At this time, experience of salt mines and liquid hydrocarbon storage caverns was of benefit but there was no knowledge of the behaviour of salt rock under strong and periodic stress conditions due to natural gas operation by compression and expansion. Moreover, the first caverns developed in the USA (Marysville, Michigan, 1961), in Canada (Melville, Saskatchewan, 1964), in Armenia (Abovian, 1964), in Germany (Kiel, 1969), and in France (Tersanne, 1968) were purposely of a small dimension and volume (30,000 - 100,000 m^3) because of the classical engineer's prudence.

In the 1970's, engineers, whose knowledge was growing, constructed always larger projects, from 200,000 m^3 to 400,000 m^3 by the end of the decade. The results of the R&D work previously mentioned led to larger caverns being built, ranging from 300,000 - 600,000 m^3 in France and Germany from the 1980's and reaching 1,000,000 m^3 in the US in the 1990's.

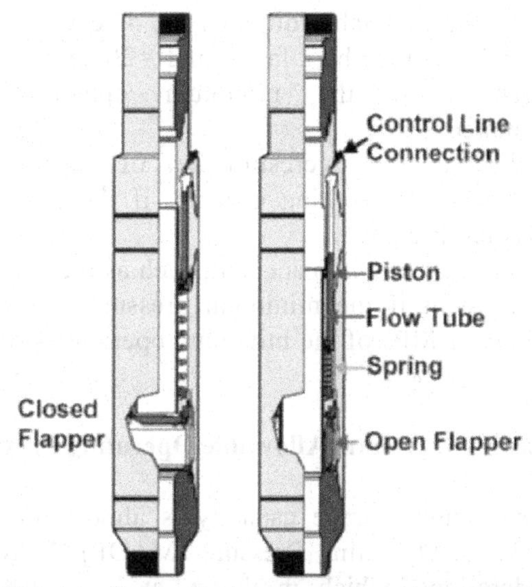

Fig. 14 - Subsurface safety valve (from KBB)

Safety considerations (increased risks resulting from concentrating storage capacity in a small number of large facilities), as well as performance reasons (the flow is limited by pressure loss in the wells) made it appear that the maximum possible volumes (400,000 - 600,000 m^3) had been reached in Europe where well diameter is limited by the installation of a buried blow-out preventer (SSV for Subsurface Safety Valve).

2.4.1.3 Lowering the minimum operating pressure

Lowering the minimum operating pressure has two consequences:

- increasing the working gas capacity,
- lowering the investment in cushion gas.

In the 1960's and 1970's minimum operating pressure was fixed by trial and error.

It was either too low in certain sites, as it was the case in Eminence, Mississippi (USA) and Tersanne (France), which led to important volume losses through creep, and consequently to losses in storage capacity, or it was too high, so causing over-investment in cushion gas and losses of potential working storage capacity.

From the beginning of the 1980's, knowledge of salt rock behaviour and particularly of its creep under high stress allowed the calculation of an optimum minimum operating pressure. This value results from a compromise between:

- the cost of the inevitable diminishing of cavity volume through creep (which can vary by a factor of 1 - 20 depending on salt rock characteristics) if the minimum operating pressure is underestimated,
- costs linked to the overestimation of cushion gas and the underestimation of working capacity if the minimum operating pressure is too high.

Even taking other factors into account, such as the need for additional compression equipment if the minimum pressure is too low (shallow sites), a decrease by 1 MPa of the minimum operating pressure results in savings around10 – 15 %.

2.4.2 Increasing Maximum Allowable Operating Pressure (MAOP)

A cavity's maximum storage capacity is almost proportional to its Maximum Allowable Operating Pressure (MAOP). MAOP is fixed by tightness considerations, which in theory at least, are based on the impermeability (and plasticity limit) of the salt rock to hydrocarbons when pressure remains lower than stresses equal to the weight of the overburden (approximately 0.023 MPa per meter depth).

Therefore, the first way to maximise MAOP is to have the deepest possible caverns, which is unfortunately not always possible for geological reasons. It is also not always desirable because of loss in geometric volume through creep which increases quickly with depth above all below a depth of 1,000 m and which is very perceptible within certain time steps (see before).

The other way is to approach as closely as possible the point above which problems with gas leakage begin to occur while still leaving a reasonable safety margin.

At the end of the 1980's, the Solution Mining Research Institute (SMRI) has launched an R&D programme over several years [12] which aimed at clarifying the confining conditions of gas and many other products. This programme included theory and on-site tests which took place at the Etrez site (France). It was possible to evaluate critical values such as the salt permeability versus pressure, fracture strength, self-healing, behaviour of the well cementing, etc.

As a result it was discovered that MAOP was fixed at a conservative level. The R&D led to the formulation of scientific methods for calculating maximum operating pressure and value of 0,019 - 0,021 MPa per metre in depth, even more for certain sites. Before, pressure was usually between 0.016 MPa and 0.019 MPa per metre depth.

This approximately 10% increase in the MAOP permits an increase in working gas capacity of approximately 15%. Considering the extra investment linked to MAOP increases (e.g. increases compression ratio and bursting resistance), savings are between 12 and 14%.

2.4.3 Solution mining techniques

Solution mining the caverns represents about 25 – 35 % of the investments. Taking one to several years to complete, it is a long process which requires large water resources (7 - 9 m^3 per m^3 mined) and which produces just as much brine with a salt concentration of 260 - 310 kg/m^3, which is used by chlorine and sodium chemistry or re-injected into the sub-soil or even pumped into the estuaries or the sea.

2.4.3.1 Fewer pulling jobs and echometric measurements

Solution mining requires two concentric tubings though which the water is injected and the brine removed. Various measures must be taken so that the cavity stays within the planned theoretical boundaries (geotechnical sizing studies). The tube shoe depths must be modified in stages, echometric measurements must be performed and a blanket such as LPG, oil nitrogen or even natural gas must be used in order to protect the upper part (roof).

Solution mining software (e.g. SALTxx from SMRI or INVDIR from Gaz de France) and control formulae for blanket movements (SURMOVINER from Gaz de France) were continually developed between 1960 and 1990 as a result of progress in computer technology.

The optimisation studies for solution mining techniques which were performed using these computation tools made possible a considerable reduction in the number of steps and therefore of round trips and echographic surveys. This explains why in the 1960's, the first 100,000 m^3 salt caverns created by Gaz de France had required 10 steps whereas now a 500,000 m^3 would only need 5 steps.

A more than 10% reduction (taking account of time gains and avoided pulling and survey costs) in the cost of mining is another consequence of the technical advances made.

For salt caverns in salt domes, where mining is easier because of the larger available height, the results are not as easily perceptible.

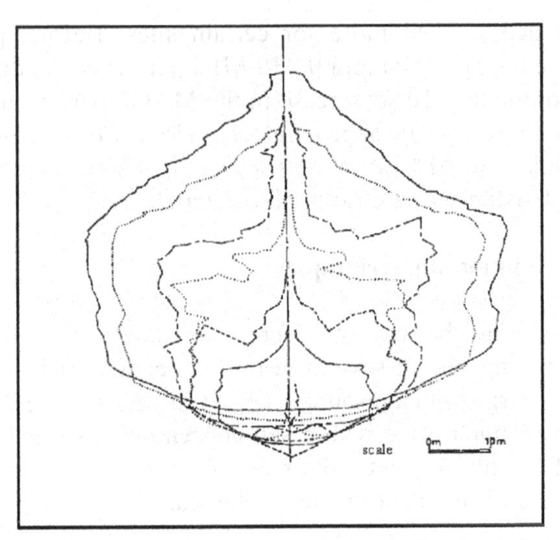

Fig. 15 - Managing cavern shape during leaching

2.4.3.2 Solution mining under gas (SMUG)

Gas (natural or inert) under pressure can be used as blanket material at the cavern roof during conventional solution mining. However, it is generally more convenient to use diesel oil or LPG at that stage (better for cavern shape management).

Once the cavern is completed, the well is re-completed for dewatering (first gas filling) and gas storage operation. Gas is then injected through the annulus between the dewatering string and the production casing (or tubing), and brine is produced through the dewatering tubing. When the cavern is full of gas, the dewatering string can be left in place, or cut, or snubbed out of the well. The overall process of solution mining and dewatering can take 18 months to several years.

One way to put a cavern into gas storage operation sooner is to perform solution mining under gas (SMUG) [13]. The cavern is initially developed by using conventional solution mining. However, its upper section is, on purpose, developed up to its designed final diameter while its lower section is not. The design also calls for a few modifications of the wellhead and solution-mining strings in order to allow for gas storage in the cavern upper section while continuing solution mining the lover section.

The upper section of the cavern is then dewatered and used for gas storage operation while solution mining is resumed in the lower. Gas store, in the upper section acts as a blanket for continued solution mining of the lower section. The gas-brine interface is closely controlled during the process and maintained at mid-cavern.

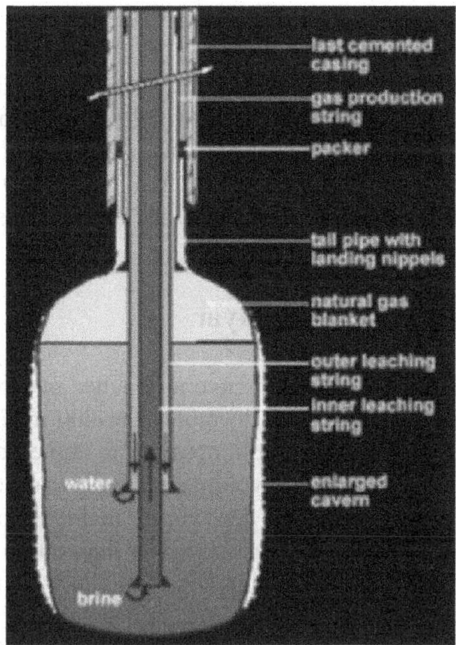

Fig. 16 - Solution Mining Under Gas operation (from KBB)

Once the lower section has reached approximately the same diameter as the upper, the same well completion and wellhead configuration can be further utilised to optimise storage operation and continue to develop the cavern capacity. Indeed, water can be injected through those two strings when gas is produced out of the outer annulus, and, conversely, brine can be produced out of the cavern through those two strings when gas is injected back into the cavern. The gas-brine interface is thus no longer maintained at this same level, but is constantly moving up and down. It needs to be controlled even more closely in order to keep roof protection and avoid gas overfilling. SMUG can be also resumed at any time, if needed.

There are numerous advantages to this improved process. First, most of the gas stored in the cavern can be produced and the cushion gas requirement is minimal. Gas deliverability can also be maintained longer at the same level, since water injection keeps gas pressure in the cavern higher. Moreover, every time the cavern is refilled with water, more capacity is created for storage. Finally, the cavern can be further developed by resuming SMUG.

This process has been successfully used in the US to develop salt caverns for storage at Moss Bluff (Texas) and Egan (Louisiana).

3. EMERGING TECHNNOLOGIES

New storage techniques still being researched will allow underground storage facilities to be developed in areas which geology does not permit the use of traditional storage facilities [14]. This is the case for caverns in thin salt layers and lined rock caverns. In addition a lot of R&D works have been made also to improve operation, maintenance and optimisation of existing UGS facilities.

3.1 Caverns in thin salt layer

Conventional large volume salt caverns leached out from a single well require a thick salt layer of 150 m to more than 400 m. Even with thinner salt layer, of 60 - 100 m, it is still possible to mine caverns using this technique although their geometric volume will be lower, between 50,000 m^3 and 100,000 m^3. Many areas of the world, mainly those bordering on sedimentary basins, have thin salt layers of less than 60 m. Gaz de France is examining the technical and economic conditions in which these layers can be used for gas storage.

Relying on geomechanical studies conducted using the GEO3D modelling tool, tunnel-shaped caverns of 1,000 - 3,000 m^2 cross-sectional area, stretching almost horizontally over several hundred metres and with a volume between 100,000 m^3 and 1.000,000 m^3 have been declared stable. The main difficulties (drilling techniques while controlling the well's trajectory - at least one well is horizontal - well equipment techniques and above all solution mining) lay in controlling the solution mining of tunnels in thin salt layers which are almost always interspersed with insoluble layers. Unlike with mining traditional caverns, gravity no longer plays an essential role.

Two test caverns (in Alsace in the north east of France) have validated a solution mining computational tool incorporating the complex laws of solution mining and fluid mechanics (convection, advection and diffusion) developed by Gaz de France. Even with investment costs 15 - 20% higher than those relating to conventional salt caverns, this technique has every chance of becoming essential at the beginning of the 21st century in Europe and North America. Gazprom also plans to use this technique to develop new storage facilities near Kaliningrad (Russia).

3.2 Lined rock caverns (LRC)

The main goal of lined rock cavern (LRC) technology is to provide storage capacities for countries where the lack of suitable geological formations does not allow any other form of underground storage facility [15].

The main principles are to store gas at high pressures (15 - 25 MPa) in lined rock caverns at relatively shallow depth (100 - 200 m). Gas-tightness is provided by a steel liner. The liner is supported by a concrete layer, the purpose of which is to transmit the pressure forces from the liner to the rock. The role of the rock is to resist to the gas pressure.

The geological requirements for the localisation of LRC storage facilities are related to the rock mass quality. The LRC concept has been successfully tested between 1988 and 1993 in a 130 m^3 chamber at a depth of 50 m in Grangesberg, Sweden. Gaz de France and Sydkraft joined together in 1997 to invest in a demonstration plant in Southern Sweden (Skallen). The geometrical volume is 40,000 m^3 which is the half size of commercial cavern. The working gas capacity of a commercial facility will range between 20.10^6 and 200.10^6 m^3 in several caverns. This pilot UGS is under commissioning and final tests will be made next summer (2003) in order to be in operation during winter 2003/2004.

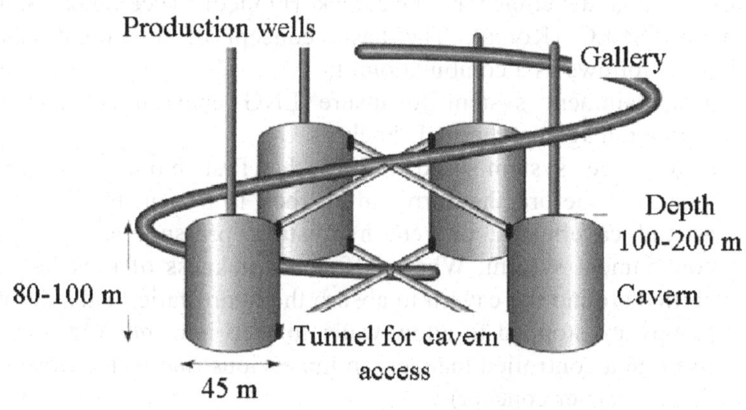

Fig. 17 - Schematic scheme of LRC

The construction cost, 2 - 4 times higher than for conventional facilities, will remain far less than the construction costs of LNG peak shaving facilities. Due to high deliverability, LRC storage facilities can be cycled several times per year. In this case, the service cost will be comparable to that of a conventional facility.

3.3 LNG caverns

Many attempts have been made to store LNG in underground storage facilities but with very limited success. A lot of failures have occurred due to thermal stresses generating cracks in the host soil leading to gas leaks and to increase the heat flux between LNG and ground. These facilities have been decommissioned due to their excess boil-off rate.

New R&D works are still underway in order to solve the past problems. One pilot, which is not strictly a LNG tank, is under study (founded by Department Of Energy in USA) to develop the concept of salt caverns used as off-loading tanks in LNG re-gazeification plants. The underground part is classical (see above) but the main new developments will be in the re-gazeification process which will use the "Bishop Process" technique to vaporise at highly rates the gas. The principle which consists of a pipe-in-pipe heat exchanger seems simple but has to be developed in order to deal with extremes temperatures (from -160°C to 20/30°C).

Another pilot plant testing a new concept for LNG underground storage is also under construction in South Korea at Daejong. This facility is very similar to Lined Rock Cavern UGS, but it will contain directly LNG using more or less the same technique used in above surface LNG tanks.

This pilot is developed by Geostock (France), Technigaz (Saipem group) and SKEC (Korea). The basic concept of this plant can be described as follows as a combination of:

- a containment system to insure LNG containment and rock protection against thermal shock,
- a drainage system used during the first months of storage operation, before the surrounding rock is frozen, to drain water around cavern and prevent hydrostatic pressure acting against containment system. When sufficient thickness of rock has been chilled around the cavern to absorb the hydrostatic loads, drainage pumps are stopped to allow water to seep into the cold rock and form, in a controlled manner, an impervious ring of ice (forming a double barrier concept).

The containment system is more or less the same than developed by Technigaz in the 1960's for LNG carriers. LNG in these types of tanks will be injected and withdrawn using pipes.

The pilot being developed has the following size:

- a 3,5 m x 3,5 m section,
- 10 m long,
- 20 m below soil level,
- and will contain 110 m^3 of LNG.

Tests are expected during autumn 2003. If the tests are conclusive, industrial plants will follow using 20 m x 30 m section gallery (or 50 m x 50 m) of hundreds meters long, for a global available volume of more than 100,000 m^3 of LNG.

3.4 UGS operation and maintenance

A lot of new tools have been also developed to help operators in daily management of existing UGS facilities. The list is quite long and it's not possible to be exhaustive but it seems important to mention the followings major new technologies:

- gravel-pack to limit sand production in unconsolidated porous reservoir,
- polymers both to limit water production and also to reinforce unconsolidated porous reservoir [16],
- reliability and flexibility assessment mathematical and software tools to check, and so, to secure the design, and so, to improve the availability of UGS facilities,
- use of permanent sensors (pressure/temperature gauges) or of optic fibers inside the well to have real-time temperature and/or pressure logging measurement especially in salt caverns to better determine the position of gas leaks or gas-brine-interface (during leaching), and also, to better estimate the bottom hole pressure during injection periods,
- development of new logging tools such as Nuclear Magnetic Resonance in order to improve the measurement and the determination of the fluids in place in porous reservoirs,
- multi-fluid flows measurements inside the well and especially in horizontal wells,
- sand detectors to better know the condition of sand production [17],
- software to determine wells' damages and to recommend maintenance works (from Gas Research Institute),
- software to provide 3D images of the reservoirs geometry and of the rock properties,
- development of new advanced 3D seismic sensors in order to better investigate the geomechanical behaviour of the reservoirs,
- accurate well and global UGS performances calculations especially in porous reservoirs,
- ...

3.5 UGS optimization

3.5.1 Generalities

In addition to the above technical improvement and new technologies to emerge, some UGS owners operating several UGS facilities (like Gaz de France operating 13 UGS facilities in France, 6 in Germany and involve in 3 others UGS facilities in Slovak Republic) have done R&D works in order to solve the technical and economical optimisation problem due to the multiplicity of available sites [18] [19] [20]:

- in which site to inject and which volume,
- where to withdraw,
- where to increase total stored volume,
- where to decrease,
- best optimal profile (when) to maximise the performances and minimised the costs,
- ...

3.5.2 Problem to be solved

Primarily and therefore at short term (excluding the effect of inflation), updating, etc.) the optimized economical use of this type of industrial tool can be characterised by the search for a minimum complementary investment (here rather a charge) for a maximum (respectively minimum) potential gain (respectively a loss).

The revenue for a gas industry is easily related to sales and in particular with regards to underground storage by the winter sales whether given as a volume (refer to the adequacy of requirements versus supply contracts and contracted volumes) or a peak period (refer to high marginal cost of the peak period cubic meter of a supply contract).

The investment (or more precisely the charge) itself is characterised by a volume of gas to purchase and to store during the summer period and also by the storage profile needed to reach the stock level required to cover requirements (refer to supra).

Thus, in this type of technical-economical optimisation approach it is possible to express the storage requirements in two complementary manners:

- a volume to be withdrawn during the winter period (meet seasonal needs),
- a peak period supply to provide during heavy demands (cold periods).

To satisfy these two goals under the best technical-economical conditions and taking into account the different behaviour of facilities (porous reservoir active to varying degrees, variable volumes and rates available, salt caverns, gas treatment facilities, connection to the network, distance to the different points of supply and consumption, etc.) it is necessary to define the best overall operating strategy for these storage facilities. Accounting for the above considerations, this strategy is mainly based on the control of the two following main parameters:

- the level of gas stored at the beginning of winter (globally and also for each facility),
- the breakdown between the volumes to store and to withdraw over time (monthly profiles) and also between facilities (variable breakdown keys).

Naturally, this strategy must be compatible with the multiple operating constraints of the facilities (minimum and maximum volumes, minimum and maximum reservoir pressures, minimum and maximum stored volumes, gas treatment and compression facilities saturation, etc.), of the transmission network (transportation situation) and of the supply contracts (contracted volumes and rates).

3.5.3 Tools

To determine these optimal conditions, a tool is necessary which, for each facility, can characterise performance under various and varied conditions but at the same time simple enough to be able to perform simulation and optimisation at facility level.

Even though it may be relatively easy to characterize the short-term performance of a salt cavern facility by a relation directly relating the maximum possible flow rate to the in-situ gas volume and the number of cavities available the same is not true for porous reservoirs where the dynamic aspect (effect of injection/withdrawal profiles) is important.

For the latter mentioned, the tools normally available in oil and gas industries generally incorporate the following characteristics:

- tools for modelling the subsurface using a simplified physical method: These approaches are therefore precise and really powerful but need training and dedicated computers and personnel because they are often complex to implement (Multi: Gaz de France in-house tool, ECLIPSE, GOCAD, etc.),
- others tools for modelling the gas treatment installations (process engineering type model: HYSIM, etc.).

Consequently, and even though today these various tools can be interfaced, they rarely make it possible to easily couple the subsurface and the surface simulators efficiently unless long and costly computer development has been carried out, or models require a prohibitive calculation time, especially when multi-site "burst" type simulations are required to solve an "optimisation" type problem.

Nonetheless, this coupling is a mandatory requirement to assess and simulate correctly the overall performance of a facility and must be associated with satisfactory performance levels to carry out multi-facility optimisation. Thus, and to meet the targeted problem, R&D works have been done last years in order to develop dedicated new tools (ex: in Gaz de France) using, at the beginning, simplified models both for the subsurface and gas treatment facilities modelling.

After years of experiments and the nonetheless effective use of these tools making it possible to refine and systematically optimise control of facilities, the tools due to increased performance and possibilities made available by software and other computer equipment, has been improved by increasing the calculation horizon, beyond two gas years by using adapted models (closer to classic reservoir models by taking into account a simplified physical representation). At the present stage with R&D works still continuing, they now include more and more complexity (risk management, stochastic simulations, etc), and works are focused on:

- integration of transmission constraints by coupling the storage models to a simplified gas network model,
- introduction, if possible, of daily management of each facility (day by day control),

but also take into account true economical data such as:

- the marginal cost and the variable of the stored and drawn off cubic meter per facility,
- the transmission costs,
- the real costs for imported supply.

These improvements are fully indispensable to be competitive in a fully deregulated gas market.

4. CONCLUSION

UGS, which is a relatively recent technique, has been developing rapidly and is now an essential part of the gas chain. Stimulated by the changes taking place on the gas market, underground gas storage will continue to be developed in the next years.

The expected growth in gas demand, increased dependence on imports and the growing need for flexibility will undoubtedly strongly influence future storage requirements. In addition, the storage infrastructure is gradually offering new opportunities (new services, storage for third parties, storage for transit, etc.), enabling the owners of the facilities to strengthen their position on the market.

So we will probably see a growing UGS market structured as follows:

- more flexible UGS facilities in the mature gas markets (mainly US and western Europe) in order to match with the needs (trading) [21],
- high capacity UGS all along the gas supply chain in order to secure the gas supplies which will come from increasingly distant countries (Middle-East, Siberia, ...).

The investments required to develop and operate storage facilities represent a major share of the cost of gas supply. Sustained efforts from industrial R&D are devoted to the widely used conventional techniques (storage in oil and gas fields, storage in aquifers, and storage in solution mined salt caverns) with the aim to both improve their performances and reduce their costs. Beside the potential of improvement in the performance on the existing sites, the development of new technologies continues to guarantee the consumer security of supply at a reduced cost.

Thus, by 2025, savings in terms of development costs could be similar to those observed in the past. These research efforts to reduce storage costs contribute to reinforcing the competitiveness of gas supply and consequently to satisfy the end user.

REFERENCES

1. "UN/ECE Study on Underground Gas Storage in Europe and Central Asia", Y. Guérini, UN/ECE, The Global Gas Village; The case for Underground Gas Storage in Europe - EnergyWise – Paris, France, 17 & 18 May 2001.
2. "Integrating geomechanics and geophysics for reservoir seismic monitoring feasibility studies", S. Vidal, Gaz de France, P. Longuemare, Institut Français du Pétrole, F. Huguet, Gaz de France, SPE European Petroleum Conference held in Paris, France, 24-25 October 2000.
3. "Seismic Monitoring for Optimising the Operation of Underground Storage Facilities", F. Verdier & F. Huguet, Gaz de France. 20th World Gas Conference. Copenhagen, 10 - 13 June 1997.

4. "Performances survey of aquifer UGS facilities", F. Favret & F. Huguet Gaz de France, ARTEP (French Research Association on Oil Operating Technics), France, September 1997.
5. "Development and Application of Underground Gas Storage Simulator", Masanori Kurihara, Jialing Liang, Japan Oil Engineering Co., Ltd., Fujio Fujimoto, Japan National Oil Corporation, Long Nghiem, Peter Sammon, Computer Modelling Group, Ltd., SPE Asia Pacific Conference on Integrated Modelling for Asset Management to be held in Yokohoma, Japan, 25-26 April 2000.
6. "Case Study of Gas BTU Control For Operational Strategy and Inert Gas Injection Planning", Lebon, M., Nabil, R., Lehuen, P., Gaz de France, SPE Annual Technical Conference and Exhibition held in New Orleans, Louisiana, 27-30 September 1998
7. "Cushion Gas Alternatives", IGU Committee A Report. Study leader: V. Onderka, PhD, Geogas a.s., 20th,World Gas Conference. Copenhagen, 10 - 13 June 1997.
8. "Using inert Gas: Almost Twenty Years of Experience", G. Meunier & F. Labaune Gaz de France. 20th World Gas Conference. Copenhagen, 10 - 13 June 1997.
9. "Effect of the Wellbore Conditions on the Performance of Underground Gas-Storage Reservoirs", Omer Inanc Tureyen, Hulya Karaalioglu, Abdurrahman Satman, Istanbul Technical , SPE/CERI Gas Technology Symposium held in Calgary, Alberta Canada, 3-5 April 2000.
10. "Horizontal Wells in Underground Storage", IGU Committee A Report. Study leader: D. Bourjas, Gaz de France. 20th World Gas Conference. Copenhagen, 10 - 13 June 1997.
11. "Peak Gas Production from Deep Aquifer Gas Storage by Horizontal Well" K. Homann, VEW Energie AG, H.J. Kretzschmar, DBI CTUT Gmbh, 20th World Gas Conference. Copenhagen, 10 - 13 June 1997.
12. "Increasing Maximum pressure in Salt Caverns", P. Desgrée & M. Fauveau, Gaz de France, 20th World Gas Conference. Copenhagen, 10 - 13 June 1997.
13. "Solution Mining and Storing Natural Gas Simultaneously – Operational Experience", Jack W. Gatewood, Market hub Partners. Michel Dussaud, Charles Chabannes, Jérôme Jacquemont, Sofregaz US Inc., L. Cherouvrier, Sofregaz, SMRI Spring 1997 Meeting, May 11-14, Cracow, Poland.
14. "Underground Gas Storage, Technological Innovations for an Increased Efficiently", Marie Françoise Chabrelie, Michel Dussaud

Sofregaz US, Daniel Bourjas and Bruno Hugout Gaz de France, 17th Congress of the World Energy Council 1998.

15. "The Potential Market for. Lined Rock Cavern Storage of Natural Gas", Per Tengborg, Sydkraft, Muriel Rosé, 20th World Gas Conference. Copenhagen, 10 - 13 June 1997.
16. "A Successful Polymer Treatment For Water Coning Abatement in Gas Storage Reservoir", A. Zaitoun, Institut Français du Pétrole, T. Pichery, Gaz de France, SPE Annual Technical Conference and Exhibition held in New Orleans, Louisiana, 30 September-3 October 2001.
17. "Monitoring The Solids In Well Streams of Underground Gas Storage Facilities", M. Megyery, Geoinform Ltd., T. Miklós, MOL Hungarian Oil and Gas Co. J. Segesdi, Geoinfrom Ltd., Z. Tóth, Elcom Ltd., SPE International Symposium on Formation Damage Control held in Lafayette, Louisiana, 23-24 February 2000.
18. "Optimising Your Natural Gas Assets" F. Favret Sofregaz, EnergyWise UGS workshop, Amsterdam, Netherlands, April 2003.
19. "Optimal use of the GDF's set of underground gas storage" F. Favret, E. Rouyer, D. Bayen & B Corgier, Gaz de France, 21st World Gas Congress, Nice, France, June 2000.
20. "Increase of working gas capacity in existing Gaz de France storage facilities", F. Favret, E. Rouyer & Y. Muller, Gaz de France, French Gas Conference, Toulouse, France, July 1997.
21. "Operational management of Gaz de France underground gas storage with flexibility and for optimal purposes", F. Favret, Gaz de France, Nafta Gas congress, Malacky, Slovakia, May 1998.

UNDERGROUND GAS STORAGE IN GEORGIA
Geological, technical and economic aspects

Mads CHRISTENSEN, David PAPAVA and Michael SIDAMONIDZE
Christensen: Team Leader and Senior Energy Consultant at RAMBOLL Consulting (Denmark). Papava: Petroleum Geologist. Sidamonidze: Reservoir Engineer
EU TACIS Project on Rehabilitation of Gas Transmission System in Georgia, June 2002-June 2003

Abstract: Underground gas storage (UGS) plays a key role in gas supply both under normal operational circumstances as well as in case of supply emergencies. Most underground formations in Georgia with UGS potential have previously been identified and investigated. Two areas have been screened: Depleted or partly depleted oil fields east of Tbilisi and a large aquifer belt on the Black Sea coast. The EU TACIS project has demonstrated the technical and economic feasibility of a UGS facility at the Ninotsminda oil field, 60km east of Tbilisi. Estimated unit costs for storage services from that facility are comparable and in some cases more favorable than fee rates charged by EU UGS operators for similar services.

Establishing UGS will double Georgia's security of supply-index and gas supply from Azerbaijan will quadruple the index compared with today's situation. But foreign investors are reluctant and awaiting major improvements in the energy sector's commercial performance and abolishment of political interference in the sector's regulation.

Key words: Underground gas storage, gas supply, depleted oil fields, technical and economic feasibility, foreign investors, commercial performance, political interference, energy sector regulation.

1. INTRODUCTION

Underground gas storage (UGS) plays a key role in EU gas supply both under normal operational circumstances as well as in case of supply

J. Hetland and T. Gochitashvili (eds.),
Security of Natural Gas Supply through Transit Countries, 195–226.

emergencies and there are economic and strategic reasons why gas storage should be located close to the market. Gas companies therefore seek, as far as geology and economy allows, to spread storage facilities as well as possible and to locate them as near to large demand centres as possible i.e. preferably not too far from large cities.

Underground storage serves several functions including:

- Strategic reserve for security of supply in case of disruption (particularly used in Member States with high dependence on non-EU gas imports);
- Seasonal load balancing to match peak demand (gas is pumped into the storage during the spring and summer and typically withdrawn from October/November to February/March);
- Achieving daily balance;
- Arbitrage of gas prices i.e. commercial optimisation of variations in gas prices e.g. around periods of recalculation of gas prices (e.g. beginning of quarters) and more generally as a commercial tool in liberalised markets (notably in the UK). As a gas price in a competitive gas market is expected to increasingly reflect demand and supply for gas, new patterns of price variations and volatility may be expected. Under such circumstances, it should be expected that gas from storage would be released in case of high prices hence limiting volatility.
- Overall system optimisation including facilitating swaps;
- Transmission-support such as mitigating localised capacity constraints or critical pressure thresholds.

While there may be short-term adjustments with regard to requirements for storage and the wish of market players to carry the costs of gas storage, it is generally expected that availability of storage facilities will become increasingly important over time due to growing EU gas demand and import dependency and thus the need for additional storage for security of supply reasons. Furthermore, additional need for storage will exist for load balancing and due to increasing import dependency and relative declining flexibility from domestic production.

The availability of storage and equivalent alternative flexibility mechanisms as an integrated part of the overall gas supply system is crucial for an efficient operation of the gas system.

The relative importance of underground gas storage differs among the Member States and some Member States are even without storage capacity on their own territory. On average, the EU gas storage capacity is equivalent to approximately 50 days of gas consumption (or 14 % of total energy consumption). Austria has the highest storage capacity, equivalent to 115 days of average gas demand, France has storage for 95 days, Germany and Italy for around 80 days and Denmark for around 65 days. The UK, Greece, Belgium, Spain and The Netherlands have storage equivalent to in the order of 10-20 days of average gas consumption while the remaining Member States have no gas storage capacity.

The Commission of the European Union has recently prepared a draft Directive that stipulates a target of 60 days of gas consumption. At present it is, however, uncertain when this target will be adopted and thus also when it can be fulfilled.

A number of Member States and large gas companies have in various ways specified conditions, which need to be fulfilled in relation to security of supply or availability of storage for existing and new entrants into the market:

- In Italy, for example, new entrants importing non-EU gas into the Italian market are required to hold gas stocks equivalent to 10% of the annual supply.
- In Spain, overall gas supply dependency upon any single external supply source must not exceed 60% and there is an obligation on gas suppliers to keep gas reserves for at least 35 days of supply.
- In the UK, security of supply standards are defined to meet "1 in 20 years" peak day demand and "1 in 50 years" winter duration. Similar standards are applied in The Netherlands and France and other Member States.
- The French gas system has also been designed in order to be able to withstand (notably through strategic gas stocks) disruption of the largest source of supply for up to one year.
- In Denmark, the integrated gas company, has designed its back-up and storage capacity to be able to continue gas supplies to the non-interruptible market with no alternative fuel switching capacity in case of a disruption of one of the two offshore gas pipelines supplying gas to Denmark.

Countries like Sweden and Finland that relies on gas imports from only one supplier - Denmark in the case of Sweden and Russia in the case of Finland - has not established their own gas storage capacity but relies on storage capacity in Denmark and Latvia respectively.

2. GAS STORAGE POTENTIAL IN GEORGIA

Underground gas storage facilities are normally established by injection of natural gas into three types of sub-surface structures:

- Caverns that are leached out in salt structures,
- Porous dome-style structures that are sealed to the top of an impermeable layer of e.g. clay stone (an aquifer gas storage), and
- Oil/gas fields that are fully or partly depleted.

Most of the underground formations in Georgia with potential for development as gas storage facilities have been already identified. Extensive work was done to characterize the country's geology during the Soviet period. Much the relevant information is available in Georgia from various research and academic institutions.

In corporation with Georgian geological experts that has been assigned to the project team and with Saknavtobi[1], an initial screening of possible sites for underground gas storage was carried out[2]. Initially it was decided to rule out the potential for gas storage in natural caverns and artificial caverns created in salt formations. Likewise, it was decided to exclude the possibility of establishing gas storage in abandoned or partly exhausted mineral formations. Both options are believed to be non-feasible because it will be very expensive to develop storage facilities with capacities that are insufficient in relation to the requirements.

Thus underground reservoir storage is the most realistic approach to be used in Georgia. Two geological areas have been identified a good options for long term storage: the depleted or partly depleted oil fields in Eastern Georgia and a large aquifer belt on the Black Sea coast. In the screening phase the fields were selected on the basis of the depth to the top seal and a rough estimate of the potential gas storage volume.

The current situation as regards licensing of blocks for petroleum exploration in Georgia is presented in Figure 2.1.

[1] The Georgian state owned oil company

[2] In accordance with the contractual requirements, the EU Contractor prepared a 'Report on Potential Underground Gas Storage' that was submitted in September 2002.

Figure 2-1 License blocks for petroleum exploration in Georgia

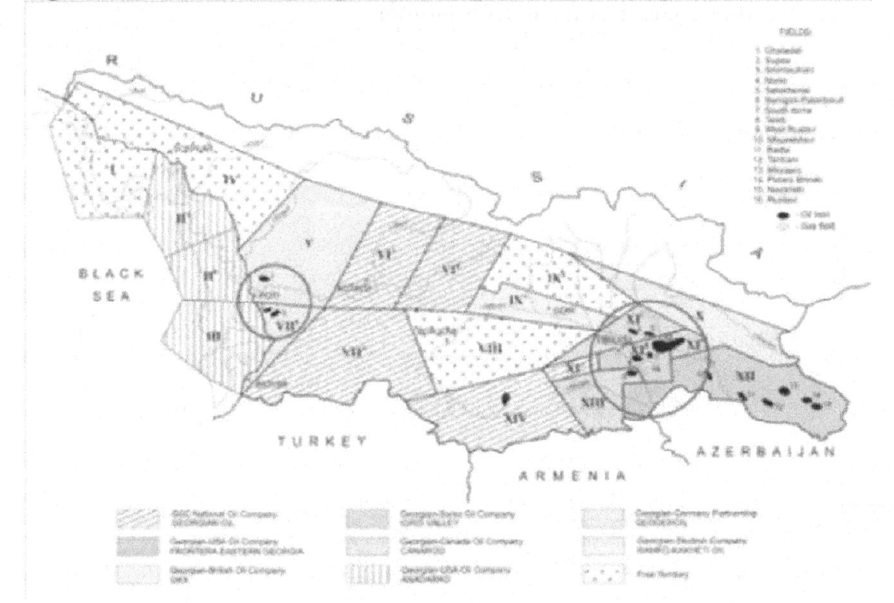

8 possible sites for underground gas storage (UGS) were selected:

- Six oil and gas fields in the area east of Tbilisi (Ninotsminda, Samgori, Patardzeuli, Samgori South Dome, West Rustavi and Rustavi.)
- Two oil fields in the western area, near the town of Poti: Chaladidi and Shromisubani.

The 8 UGS candidates were chosen among other oil and gas fields because these sites (with some variation) are well documented (e.g. 2D seismic surveys and a number of wells) and the reservoirs has a proven seal, which has been able to withhold a gas overpressure for million of years. This is important both for safety reasons and for operating the gas storage without risks of gas leaking away in the substrata.

Main characteristics of the 8 fields are presented in Table 2.1.

Table 2-1 Main characteristics of UGS locations

Name	License holder	Type	Top	Estimated work volume [Mill. m^3]	Depth SSL [m]	Pressure [bar]
Ninotsminda	CanArgo	Oil field/gas cap	Middle Eocene	750	-1750	225
West Rustavi	CanArgo	Oil field	Middle Eocene	850	-2300	235
Samgori	Ioris Valley	Oil field	Middle Eocene	107	-1750	225
Patardzeuli	Ioris Valley	Oil field	Middle Eocene	142	-1750	225
South Dome	Ioris Valley	Oil field	Middle Eocene	230	-1650	199,5
Rustavi	Ioris Valley	Gas condensate	Middle Eocene	295	-3200	334
Chaladidi	GEOGEROIL	Oil field	Upper Cretaceous	800	-3000	243
Shromisubani	GEOGEROIL	Oil field	Meotian	170	-3795	527

The Ninotsminda, Samgori and Patardzeuli reservoirs may not be considered as separate reservoirs but rather as a series of connected anticlines with Ninotsminda at the top followed by the deeper situated Patardzeuli reservoir and finally the lowermost Samgori reservoir.

Most of the selected fields are exhausted or partly exhausted oil and gas condensate reservoirs located in the Tbilisi area while two of the fields, located in the western area near the town Poti, are oil fields under exploration. The depth to the top seal of most of the fields is less than 2 km. The structures are anticlines except for the Shomisubani field, which is tectonically screened.

A number of geological features of some of the fields are presented in Figures 2.2 to 2.5.

Figure 2-2 Iso-line Map and Cross section of the Chaladidi Oil Field, West Georgia

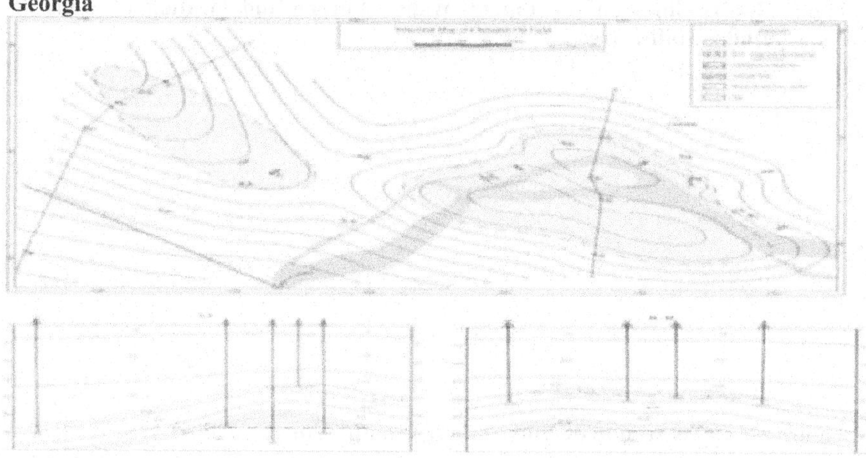

Figure 2-3 Iso-line Map and Cross section of Middle Eocene, Tbilisi Area

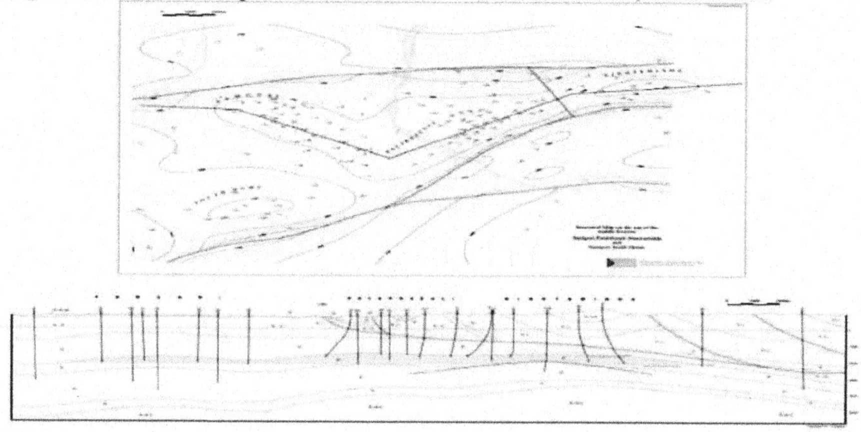

In the Tbilisi area, the Middle Eocene sequence is a volcanic erupted structure with a very high density of interconnected micro fractures and a relatively smaller number of macro fractures with apertures in the mm range. The matrix rock is of low porosity and low permeability and is not considered to contain any significant reservoir volume. The fields have been producing since 1974 by open-hole wells of which a large number have experienced water break through.

Figure 2-4 Iso-lines on the top of Middle Eocene and location of wells at Ninotsminda, Tbilisi Area

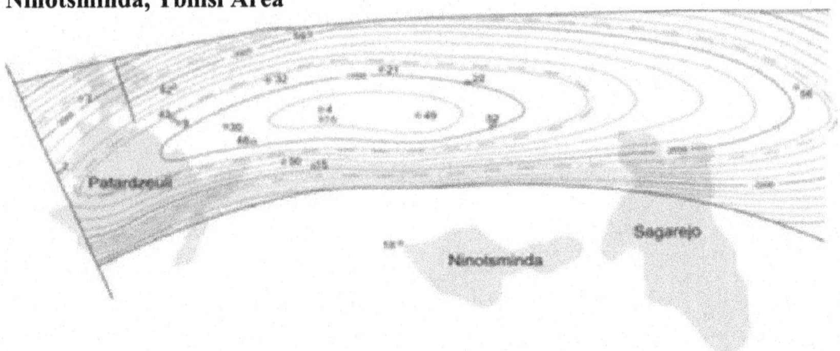

Figure 2-5 Cross-section of Ninotsminda (South – North), Tbilisi Area

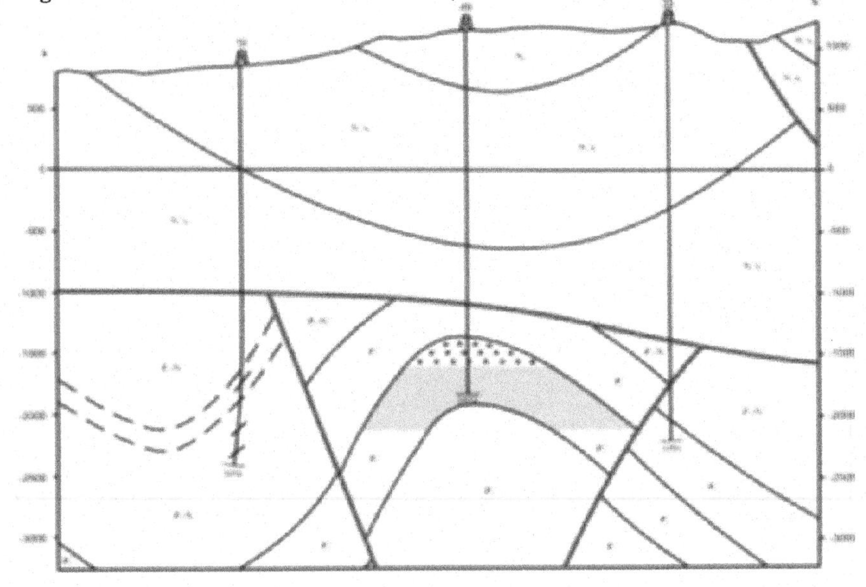

The fold is complicated by trust fault; as a result Neogene sediments impend from the north onto Paleological sediments of the Ninotsminda dome. That dome is believed to have a 2-storey construction. The tectonic of upper storey, which is mostly represented by Neogene sediments, does not correspond to Eocene sediments construction of lower storey. Wells under the thrust prove up cast debris, which is spread along north and south limbs of the Ninotsminda fold that configured horst construction of the whole fold.

Toward North East (near the village of Patardzeuli), the Patardzeuli dome experiences bending (to –1,650 meters) and is prolonged to almost the same latitude of the Ninotsminda dome with its highest part at –1,350 m. The Ninotsminda dome gradually immerses in the direction of the Giorgitsminda village.

The screening report included an initial ranking of the 8 fields by an overall assessment of the following parameters: The estimated work volumes, the buffer or cushion gas volumes, the reservoir depths and pressures, the reservoir type and status and the distances to pipelines and consumers. Also the average permeability and the amount of available data have been taken into consideration.

Table 2-2 Initial ranking of the selected 8 potential UGS sites

Name	Type	Estimated work volume [Mill. m3]	Initial ranking
Ninotsminda	Oil field/gas cap	750	1-2
West Rustavi	Oil field	850	1-2
Samgori	Oil field	107	5-6
Patardzeuli	Oil field	142	5-6
South Dome	Oil field	230	4-5
Rustavi	Gas condensate	295	7-8
Chaladidi	Oil field	800	3-4
Shromisubani	Oil field	170	7-8

Thus the initial ranking put Ninotsminda and West Rustavi on top of the list. Both fields are located near Tbilisi. Both fields were expected to have a high working volume and by including nearby fields the working volumes may be expanded further. The fields are located near the gas customers. The estimated capacity of the West Rustavi fields is larger than the capacity of the Ninotsminda field but the need for buffer or cushion gas on the Ninotsminda field is much smaller than what is needed for West Rustavi field. This assumes that the gas cap at Ninotsminda is actually intact.

As the TACIS project has developed, the initial ranking of Ninotsminda as the most promising site has been confirmed through an extensive dialogue with representatives of the project beneficiaries in particular Georgian Gas International Corporation (GGIC)[3] and Saknavtobi. Thus the EU Contractor has focussed his activities, including the development of a reservoir simulation model, towards the Ninotsminda oil field that is party depleted and has a gas cap.

Figure 2-6 illustrates the grid- and cubic structure and the location of existing and new wells in the 3D-model that was developed for Ninotsminda.

[3] GGIC is the Georgian state owned gas transmission company

Figure 2-6 The 3D-model of Ninotsminda

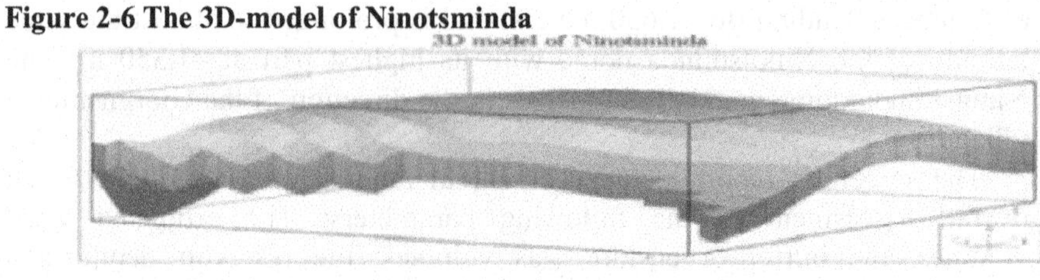

The purpose of the simulation was to investigate the technical feasibility of using the oil field as underground gas storage. Hence, the focus has been on matching the general reservoir behaviour and field injection/withdrawal performance. The reservoir model has been calibrated and operated in order to investigate if the Ninotsminda oil field could perform in accordance with a number of scenarios defined for the common purposes of reservoir simulation and gas supply-demand balance. Figure 2-7 illustrates this[4].

Based on the reservoir simulation the overall conclusion is that the Ninotsminda oil field can technically be used as an underground gas storage facility. The developed reservoir simulation model is able to match reasonably well with the historically observed oil and gas production, and with average reservoir pressure for the Ninotsminda oil field.

In all scenarios it was possible to inject the required volumes of working gas within a period of 1- 5 months. The subsequent withdrawal of identical volumes of gas from the storage was possible in all scenarios[5], except Scenario IV, without using additional cushion gas. In Scenario IV, it was possible to withdraw only 465 and not the required 480 million m^3 gas, if no additional cushion gas was to be injected. It was demonstrated, however, that lowering the minimum well bottom hole pressure by 10 bar to 80 bar and optimising the individual well rates, would enable the withdrawal of the required 480 million m^3 gas.

[4] Bullets are the observed data and full curves are reservoir simulation results.

[5] Please refer to Chapter 4 of this article

Figure 2-7 Match of observed oil & gas production and reservoir pressure

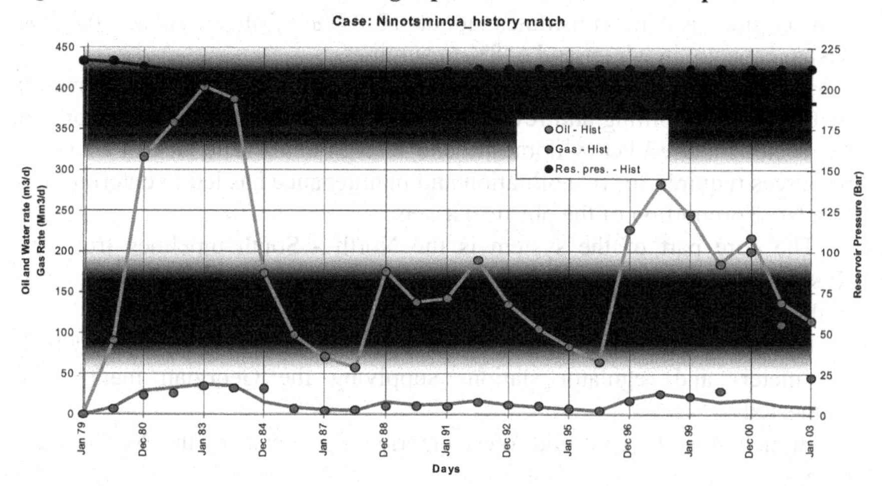

3. BASICS ABOUT GEORGIA'S GAS SECTOR

Natural gas in Georgia was introduced in 1959 with commissioning of the Karadagh – Tbilisi gas pipeline from Azerbaijan.

In 1965 the 700 mm Vladikavkaz – Tbilisi gas pipeline was put into operation. From 1965 to 1970 natural gas was delivered to Georgia from two independent sources, Azerbaijan and Russia. During 1970 – 1978 gas was delivered from Iran to Georgia. After that period Georgia is supplied with gas from Russia only while the South Eastern pipeline connection from Azerbaijan (and Iran) is not in operation.

During the Soviet period, Georgia developed its gas transmission network intensely. Gas became the main fuel in Georgia with consumption peaking in 1990 at 6 billion m³, and covering about 60% of all energy consumption. At that time 48 towns, 230 villages, 587,000 apartments and 3,100 industrial, agricultural and municipal enterprises were supplied with gas in Georgia. The penetration of natural gas covered almost 45% of the potential users.

The independence of Georgia after the collapse of the Soviet Union changed many aspects. However, gas maintained its role as an important source of energy although at a much lower level. Today, the demand is below 1 billion m³ and represents around 30 % of all energy consumption. In addition to the gas consumption in Georgia, the transmission system transits up to 1.5 Bcm per year to Armenia. The actual throughput of the gas transmission system constitutes about 12–15% of the system's original design capacity in winter and 6–8% in summer.

Faced with massive increases in the price of imported gas in combination with ineffective billing and revenue collection during the 90-s the Georgian gas sector incurred heavy domestic and foreign debts. The lack of financial resources required for rehabilitation and maintenance has led to deterioration of a large proportion of the physical assets.

The core part of the system is the North - South pipelines from the Russian border to the Armenian border. The largest volumes are transported in the DN 1200 and DN 1000 pipeline constructed during 1980-1994.

Parallel to this line an older DN 700 and DN 500 line is located to which the meter- and regulator stations supplying the Georgian market are connected.

Branch lines to East and West Georgia are linked to the North-South main system:

- A North West branch runs from Saguramo to Sukhumi with connections to Java, Bakuriani, Ambrolauri, Kobuleti and Poti.
- A South West branch runs from Tsiteli to Alastani.
- An eastward branch loops from Rustavi to Jinvali.

When gas consumption in Georgia was peaking in the early 1990'es a compressor station in Kvesheti boosted the capacity of the system. This station is now mothballed and out of operation. Construction of a second compressor station in Saguramo was never completed.

Georgian gas main pipelines total length is 1,940 km (by July 1, 2002). The total geometric capacity (= gas pipelines accumulation possibility or 'line packing' at 1 bar) is slightly exceeding 550 000 m^3. Figure 3-1 presents the existing gas transmission system in Georgia.

Today's gas transmission company, GGIC, has developed from being a State Department in the period prior to 1996, to becoming a joint stock company in 1997. GGIC owns all the high-pressure transmission trunk lines, the 2 compressor stations and a number of meter-/regulation stations as well as gas processing and other facilities. As of December 1999 GGIC has created a separate daughter company Georgian Transportation Company (GTC) responsible for maintenance and operation of the entire gas transmission system.

During the past few years, gas has been supplied by the Russian company Itera, via the local Georgian import/trading company Sakgazi (which is 100% Itera owned), using the transmission network owned by GGIC and operated by GTC (100% owned by GGIC), to the end-users, including approx. 20 local distribution companies (LDC's), main industries, the Gardabani power plants and transit to Armenia.

Figure 3-1 Georgia's gas transmission system

Georgia's gas consumption in 2002 was 699 mcm or 18% lower than in 2001. The total consumption is now 30% lower than the 1 Bcm-level in 1999 and 2000. The largest consumer segment in 2002 was the industry sector, which consumed 297 mcm or roughly the same as in 1999 and 2000 after a drop to 200 mcm in 2001. The residential sector consumed 221 mcm or 9% less than in 2001. The biggest change was in the power sector that consumed only 165 mcm or 58% less than in 2001.

Table 3-1 illustrates the sharp decline in the gas consumption and the major shifts in the structure of gas consumption between 1989 and 1999-2002. Figure 3.2 compares the historic gas consumption with the different gas demand projections, which have been prepared under the EU TACIS project.

Estimates of Georgia's gas storage requirements have to be based upon a combination of information about the actual monthly gas consumption and projections for the future annual gas demand. In accordance with the contractual obligations, the EU Contractor has made projections of the future gas demand[6].

[6] A Report on Gas Demand Forecast for 2005-2020 was submitted in October 2002.

Table 3-1 Changes in gas consumption level and structure

	1989		1999-2002	
	%	Bcm	%	Bcm
Residential & Commercial	13	0.8	26	0.2
Industry	48	2.9	31	0.3
Power generation	38	2.3	43	0.4
Total	100	6.0	100	0.9

Figure 3-2 Past gas consumption and future gas demand projections

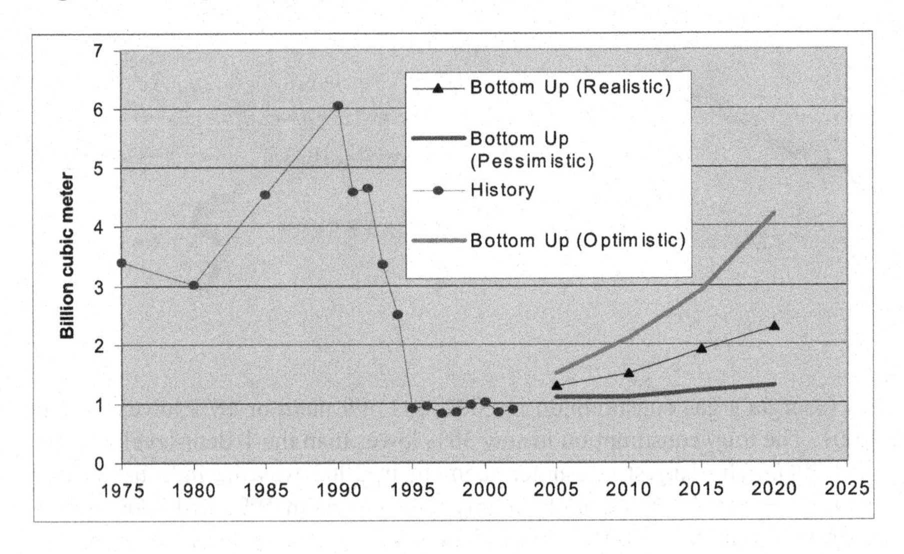

The gas demand was projected in three different economic scenarios: A pessimistic, a realistic and an optimistic with annual GDP growth rates of 3, 7 and 11 percent respectively for the period 2005-2020. The forecast uses two different methodologies: 1) A macro-economic Top-Down method based on the development in GDP per capita, which compares the economic growth in Georgia to that of other countries, and its resulting impact on the energy demand; and 2) a more detailed sector-wise Bottom-Up analysis, focusing on the expected growth in each of the main gas consuming segments.

It is obvious that the future gas demand develops very differently in the three scenarios but the previous peak level of 6 Bcm is unlikely to be achieved within the next 20 years. This implies that it is important for the gas transportation company, GGIC to have a flexible strategy for developing its infrastructure to match requirements that are very variable. One of the ways to reduce the level of uncertainty is to urge for better and more precise economic forecasting from the agencies responsible for this kind of planning tools.

It should be emphasised that the present gas demand forecast is (much) less optimistic than other forecasts prepared in recent years by the Ministry of Fuel and Energy and GGIC respectively.

4. GEORGIA'S NEED FOR GAS STORAGE

Improving the reliability of gas supply has been a goal for Georgia (and Azerbaijan) since the early 1990's. Prior to the break up of the USSR, the Caucasus republics operated as a part of a unified gas supply grid that was dispatched from Moscow. The grid supplied substantial volumes of gas from Russia during the autumn and winter to the gas fired power plants in the Caucasus. The value of the gas energy was returned during the spring and summer months, when there was surplus hydropower capacity in the Caucasus.

Today there is some gas storage capacity in the Caucasus. Azerbaijan has two UGS facilities, in Kalmaz and Karadag that both are located SW of Baku[7]. At the opening of Kalmaz in 1976 the working storage capacity was 1.3 Bcm but the reliable capacity has declined to 0.3 Bcm, largely due to the deterioration of the compressors and other mechanical equipment. The Karadag facility was built in 1986 with an original working volume of 1.0 Bcm. The current working capacity is 0.2 Bcm. The Armenian UGS facility in Abovian is partially owned by Gazprom, which is able to control its operations. It has a working capacity of approx. 0.2 Bcm.[8]

Storage for security of gas supplies to Georgia i.e. a strategic gas reserve is mainly associated with the difficult political situation between Georgia and Russia, not least in connection with the situation in Chechnya. The existing 2 import pipelines are located close to Chechnya and may be exposed to sabotage and other incidents that may jeopardise the reliability of the gas supply to Georgia and the transit to Armenia.

From a commercial point of view, Georgia's negotiation position towards Russian gas suppliers is weak due to the lack of alternative suppliers and due

[7] Brooks Howell & Co. LLC: Definitional Mission Report on UGS project in Georgia. Prepared for U.S. Trade & Development Agency (USTDA), 3 May 2002

[8] According to a report on 'Caucasus natural gas resources, perspectives of Caspian gas transportation to the European market', prepared by OPET Caucasus (an EU supported organisation) in Tbilisi, 2001

to the historic record of weak payment discipline including the accumulated debts. This situation will change when the plans for a gas pipeline, the South Caucasus Pipeline (SCP) from Baku via Tbilisi to Erzurum in Turkey, based upon gas from the Shah Deniz field in the Caspian Sea, are implemented.

The agreements with Georgia concerning the SCP pipeline system entitles Georgia to receive a transit fee or option gas in the form of 5% of the gas volumes transported. In addition, Georgia can buy supplementary volumes of gas at a special price during a 20-year period. These supplementary volumes range from 200 million cubic metres (mcm) in year 1 to 500 mcm during years 6-20 of the SCP-operation.

The determination of the necessary gas storage capacity is based upon the approach of primarily using the storage as security of supply in case of disruption of the gas imports from Russia. Secondarily, the gas storage could potentially be used for both security of supply and for seasonal load balancing to match peak gas demand in extreme situations caused by several days of extremely low temperatures. But there is less need for seasonal storage because there is surplus capacity in the import pipelines and the transmission system to handle today's gas consumption as well as the gas demand projected in 2020 and in fact beyond that point.

The capacity of gas storage is essential for many purposes: It defines the size of the working gas volume, the number of wells for injection and withdrawal and other equipment of the facility. These parameters determine the requirements to the reservoir simulation model and the cost estimates for the required investments in equipment (compressors, pipelines, etc.) at the UGS site. Thus the overall feasibility of the storage is very dependent on the capacity of the gas storage in relation to the gas demand.

Any underground gas storage facility must have capacities that fit to the country's specific gas supply security strategy. Two aspects must be included in the analyses: Firstly, the volume requirement, which depends on a political decision (as defined by the President or the Government) or a commercial decision (as defined by GGIC) on how many days or months the storage should be able to continue supplying the non-interruptible[9] consumers in case the gas supply is disrupted for technical or commercial reasons. Secondly, the required capacities for injection of gas into and withdrawal of gas from the storage are important from the point of view of

[9] One emergency response measure is the use of interruptible contracts with consumers that have alternatives to gas supply. This means that in case of serious gas supply or distribution problems, the consumers with interruptible contracts will not receive gas. This strategy will allow continuing deliveries to the non-interruptible consumers. Interruptible consumers are usually power stations and other large consumers. The non-interruptible consumers are typically the residential, commercial and most industry consumers.

estimating the technical layout, the related investment costs of the underground gas storage facility. All in all, this will define the feasibility of the underground gas storage in relation to other alternatives.

It is a common strategy for countries that rely on a large share of gas in their primary energy balances to have gas stored in volumes sufficient to cover gas demand in the coming one or two months. Thus during summer there are relatively small requirements to gas storage. During winter the need for gas storage is much larger. Figure 4-1 presents this varying monthly need for security of supply storage-volumes. The presently required storage volume based upon a one-month security of gas supply-strategy has a maximum in December, where around 70 mcm is needed (based upon 2002 data). From April to October the storage required for one month of security of supply is only around 25-35 mcm.

Figure 4-1 Monthly gas consumption during 2002

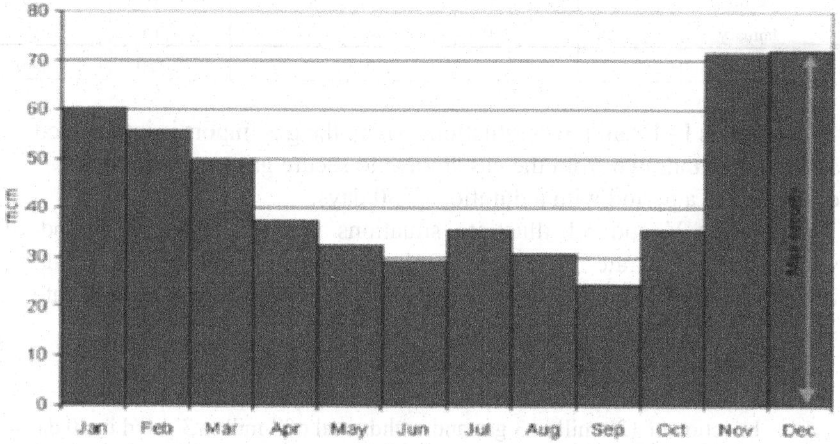

In order to accommodate with the need for consistent parameters in the different contexts, a number of scenarios have been defined. These scenarios are presented in Table 4-1.

Table 4-1 Brief description of UGS scenarios

	Security of Supply only				Security of supply and seasonal peak shaving	
Scenario	I	II	III	IV	V	VI
Withdrawal gas rate (mcm/day)	4.0	8.0	8.0	16.0	4	8
Gas Injection rate (mcm/day)	1.5	3.0	3.0	6.0	1.5	3.0

No. days withdrawal for SoS	30	30	60	30	30	30
Volume withdrawn for SoS (mcm)	120	240	480	480	120	240
Volume withdrawn for Seasonal variations (mcm)					30	60
Volumes, total (mcm)	120	240	480	480	150	300
Monthly gas withdrawal rate (mcm/month)						
December					15	30
January	120	240	240	480	15	30
February			240		120	240

Scenarios I - IV all cover situations where the gas import is interrupted and gas is withdrawn from the gas storage to secure gas supply to the customers in a period with a duration of 30 days.

Scenarios V and VI illustrate situations where a 2-month period of withdrawal at low rate for seasonal peak shaving (in December and January) is followed by a 1-month period (in February) of withdrawal at high rate to secure gas supply when gas import is interrupted.

The different scenarios can be summarized as follows:

- I Injection of 120 mill. m3 gas and withdrawal of 4 mill. m3 gas/d in 30 days
- II Injection of 240 mill. m3 gas and withdrawal of 8 mill. m3 gas/d in 30 days
- III Injection of 480 mill. m3 gas and withdrawal of 8 mill. m3 gas/d in 60 days
- IV Injection of 480 mill. m3 gas and withdrawal of 16 mill. m3 gas/d in 30 days
- V Injection of 150 mill. m3 gas and withdrawal of 0.5 mill. m3 gas/d in 60 days followed by withdrawal of 4 mill. m3 gas/d in 30 days
- VI Injection of 300 mill. m3 gas and withdrawal of 1 mill. m3 gas/d in 60 days followed by withdrawal of 8 mill. m3 gas/d in 30 days

5. SECURITY OF SUPPLY

The gas supply chain is a complex lifeline, which is exposed to serious risk and significant consequences should something go wrong. Some main characteristics in the case of Georgia are:

- o Natural gas is produced under difficult conditions in remote terrain
- o The gas is then transported under high pressure through very long pipeline networks
- o Passing equally troublesome terrain, and crossing borders in interconnected systems, with rapidly changing ownership and jurisdiction
- o Ending in fine meshed grid vitally integrated in the economies and social welfare of the countries in which it is consumed

If something goes wrong, it may impact on the entire chain, and the consequences can therefore be severe. It could for example be a major increase in price, or a complete disruption of supply, with the risk of ending in social and economic unrest. Disruptions and accidents will happen, but can be reduced in impact and frequency by careful planning. This short text gives an insight to the factors, which have an impact on security of supply and which instruments the individual countries can use to improve their own security of supply.

The Danish consulting firm, RAMBOLL, has developed a model that assesses security of gas supply on a national scale. The model is an attempt to quantify and incorporate a large number of determinative factors into a single easily understandable figure expressing relative security of supply. The model cannot forecast supply disruptions or the probability of such. The model leads to an indexation or ranking of countries. However, just as important is it that one can see which factors are important for security of supply for the different countries. The model reflects that there are several different instruments a country can use to reach a higher security of supply.

RAMBOLL has identified 12 important factors, which determine security of supply. These factors can logically be grouped into two categories: Political/Economic and Technical:

Political/Economic factors

- Export share of production: The model will penalise countries that have a high export share of production, as this will decrease the countries flexibility.
  - o This would be e.g. Norway, Russia, Netherlands and Denmark.
- External suppliers: It is good for security of supply to have several external suppliers, as this means that a country will be less affected if something dramatic happens to/in a single country.
  - o Germany has 4-5 suppliers, where as e.g. Georgia, Belarus, Poland, Estonia, Latvia, Lithuania, Finland and Sweden have only one supplying country.
- Gas share of Primary Energy Consumption: If the share of gas, out of total primary energy consumption, is very high will the country also be

more vulnerable to disruptions. It is therefore associated with a certain negative consequence to security of supply, when gas has a high share of the total primary energy.

- Border crossings: How many borders does the imported gas need to pass from production to consumption. The more borders the gas needs to go through, the higher is the risk that problems will arise.
    - A pipeline that goes from say Denmark to Germany, gives a higher security of supply measure than a pipeline from Russia, through: Belarus, Ukraine, Slovakia, the Czech Republic and which then ends in Germany.
- Country risk: Different countries are perceived to have different risks.
    - Security of supply for Germany is therefore higher if gas is imported from for example Norway than from Russia; this is especially the case when including the risk of doing business in the transit countries (listed above).
- Replaceable share of consumption: Gas supply to certain consumers, such as electricity producers, is interruptible whereas it is almost 'impossible' to interrupt supply to households because they have no alternatives to gas supply. It gives flexibility to be able to cut off certain consumers, if necessary. Thus security of supply is higher if it is possible to replace a certain share of the gas consumption by different energy sources.
- Production share of consumption: If a country can produce its own gas, it can avoid many risks. Security of supply is therefore increased if a country produces parts of its own gas consumption.
    - Countries with a high production share compared to own consumption are for example: Norway, Russia, Denmark and Poland. Countries without gas production are: Estonia, Latvia, Lithuania, Finland and Sweden.

Technical factors

- Reserves: Large reserves will ensure that a country can be self reliant on gas in the future. This will add to security of supply.
    - This is especially the case for Norway, Russia and to a lesser degree for the Netherlands, Denmark, Romania and Poland.
- Months supply in national storage: When a country has its own gas storage it becomes less vulnerable to short time disruptions in the import pipelines. It will therefore increase security of supply to have a national storage.
    - Countries like Lithuania, Estonia, Finland and Sweden have no national storage whereas Latvia can supply the whole country for more than two years with its own national gas storage.

- Import pipelines: The higher the number of import pipelines, the less impact will a problem with a single pipeline have. Additional pipelines will therefore increase security of supply.
- LNG terminals: Are adding to security of supply, just as if it was another pipeline.
- Offshore pipeline risk: It is more difficult to repair an offshore pipeline than an onshore pipeline. It is therefore associated with an increased risk, or lower security of supply, to receive your gas from an offshore pipeline compared to an onshore pipeline.

The RAMBOLL Security of Supply (SoS)-model has been used to calculate SoS-indices for Georgia in order compare its current situation with other European and CIS countries, see figure 5-1.

Due to Georgia's specific combination of a high share of gas consumption, no diversification in technical supply with only one supplier and no domestic storages, the SoS-index for Georgia in today's situation comes out as the third lowest.

By establishing a UGS facility with a capacity of 1 month's gas consumption, the SoS-index will increase by a factor of 2 but Georgia will still be ranked below the average index, see figure 5-2.

The effects on Georgia from the implementation of the Shah Deniz and SCP projects are two-fold: Firstly the SCP project will increase focus on Georgia as a 'beneficiary' and host country because a large part of the investment in the SCP pipeline will be put on Georgia's territory. Secondly, the gas from Shah Deniz represents a new opportunity for Georgia to diversify its gas supply in the way that Russia will no longer remain the sole supplier of gas to Georgia.

This represents an important step towards increasing the technical security of gas supply and it also holds the potential of Georgia gaining commercial benefits from the fact that Georgia is entitled to purchase supplies of 'supplemental gas' in addition to the transit gas payments in the forms of 5% of the transported volumes in the SCP.

Establishing a UGS facility has the potential of optimising the benefits of those new supplies of gas in addition to the gas that is imported from Russia and the transit fees that are received from transiting gas through Georgia to Armenia.

By adding Azerbaijan as the second gas supplying country through the SCP pipeline the SoS-index will increase again by a factor of 2, and it brings Georgia's score close to the median value of the SoS-index for all the involved countries, see figure 5-3. This gradual improvement in the security of gas supply is illustrated in the 3 figures overleaf.

Figure 5-1 Georgia's security of supply index – today's situation

Security of supply index – today's situation: One supplier, one import line

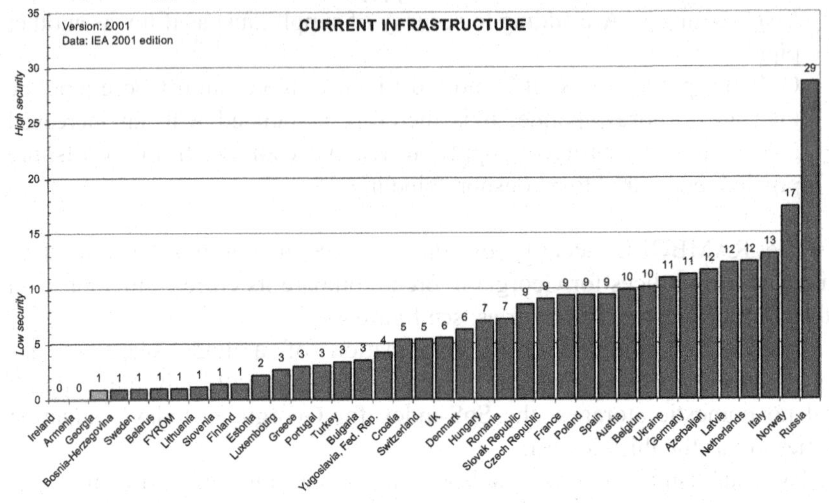

Figure 5-2 Georgia's security of supply index – with 1 month of gas demand in storage

Security of supply index – One supplier, one import line + Gas Storage

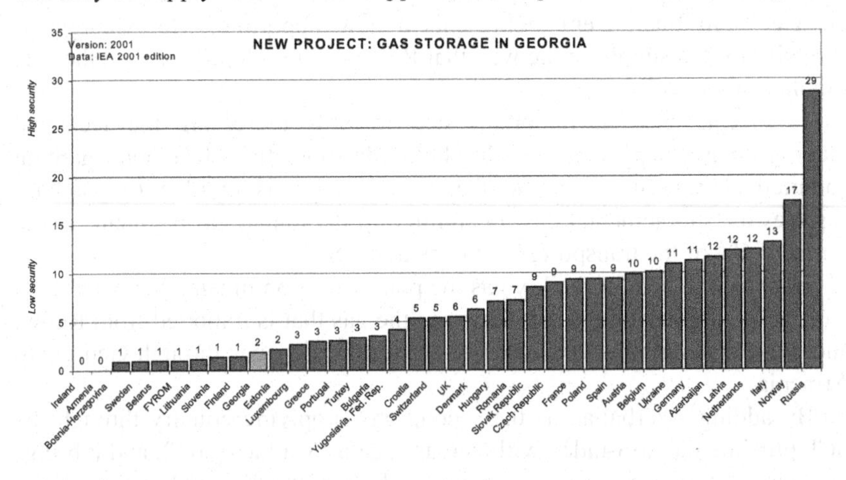

Figure 5-3 Georgia's security of supply index – 1 month of gas in storage and South Caucasus Pipeline

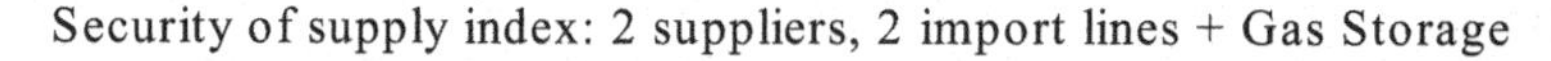

Security of supply index: 2 suppliers, 2 import lines + Gas Storage

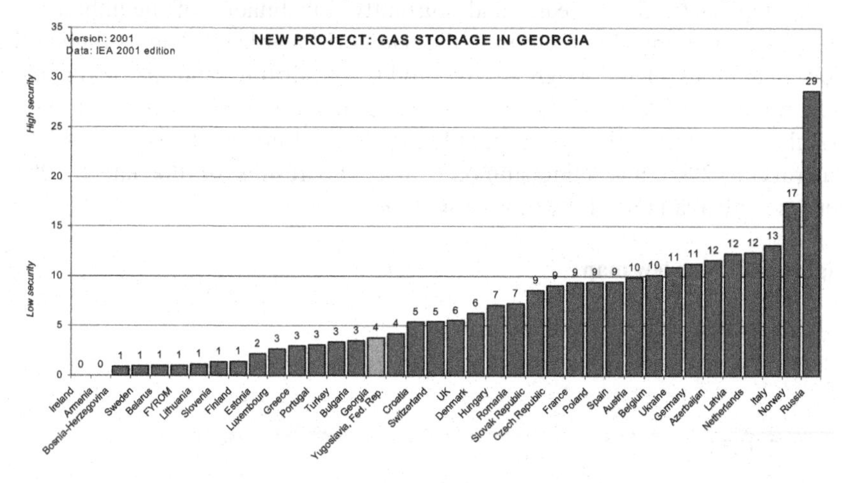

6. THE PROPOSED UGS FACILITY AT NINOTSMINDA

The geological site that has been selected as the most promising location for establishing an underground gas storage facility, the Ninotsminda oil and gas field, is located 40 km east of Tbilisi in the valley of the river Iori. There are 3 villages Patardzeuli, Ninotsminda and Sagarejo located to the west, south and southeast of the Ninotsminda oil field respectively. 2-2,5 km south of the top of the field (the dome part) there is central trunk road and 5 km south there is the west-east ongoing railway. The most western part of the Ninotsminda oil field is practically adjacent to the outskirts of the village Patardzeuli; the distance between the village Ninotsminda and the top of the structure is 2 km. The distance from the field to the village of Sagarejo is 3 km.

Main cattle-breeding farms outside the villages in the surrounding area are not shown on the topographic map (figure 6-1). The surrounding area, which is not forested, is used for pastures and cultivation but only outside the border of the licensed areas allocated for oil and gas recovery in Ninotsminda.

During last 15-20 years and currently inhabitants of neighbouring villages are employed in relation to the Samgori-Patardzeuli-Ninotsminda oil field operations including crews for production, drilling and major repairs. According to the last official statistics for population (based upon the 1989 population census) the local population is as follows: Sagarejo 14439, Patardzeuli 2903 and Ninotsminda 1969. The results of the latest 2001 census were available at the time of writing.

Figure 6-1 Overview map

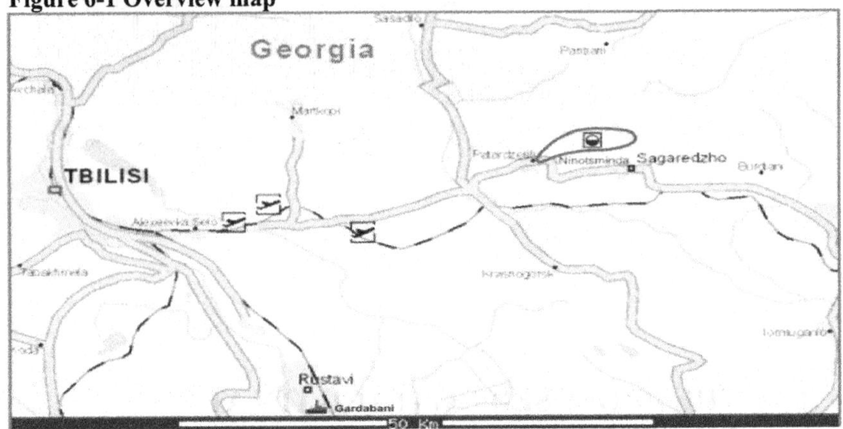

Ninotsminda is situated in a mountainous terrain, mostly covered with wood, crossed with secondary roads and paths. Here are a lot of small ravines. The river Iori flows 6 km south of the top of the field, from west towards east. The altitudes of the terrain vary from 800 to 1200 meters above sea level. The altitude increases from south to north. Relief locates on 1100 m above sea level in the dome part of Ninotsminda.

The Ninotsminda oil and gas field is located in the XIE license block area. The current license owner and operator is the British Canadian oil company CanArgo that was awarded the license in 1996 for a period of 25 years. The license area is 70km^2.

Commercial oil production was initiated in 1979 from middle Eocene volcano-genic deposits (Well # 2Nin.), and commercial free gas production rates were confirmed in 1982 from the gas cap (Well # 9 and 16). The accumulated oil production until 2002 is approx. 1.5 million tons and the accumulated gas production in the period 1997-2002 is approx. 150 mcm.

Figure 6-2 The existing high-pressure transmission pipeline (from Russia) and the proposed pipeline to connect with Ninotsminda UGS

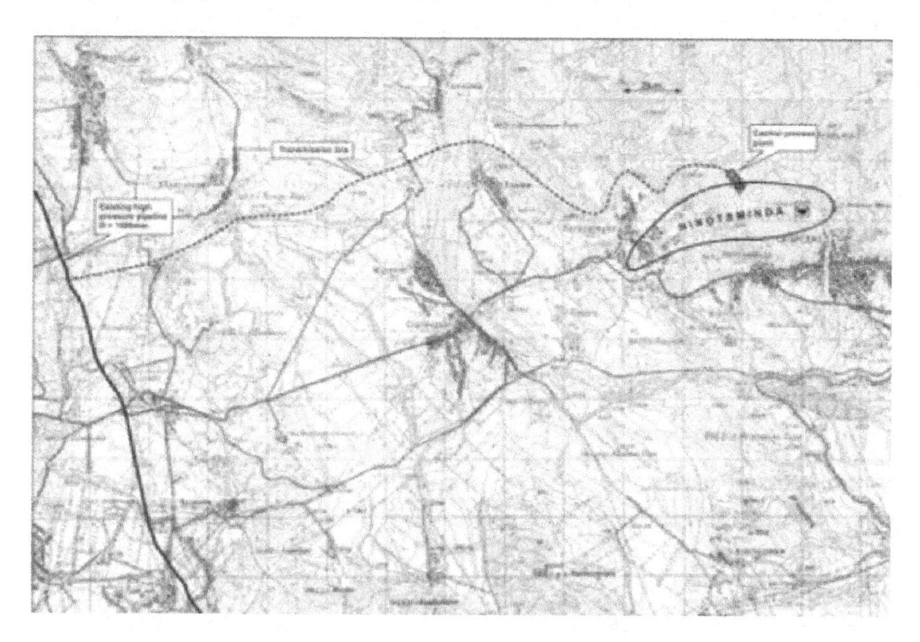

The straight distance between the nearest point of the existing DN 1000 mm gas transmission pipeline and Ninotsminda field is 22 km and the length of the proposed route for the new connecting pipeline is 25 km. See figure 6-2.

A cost estimate has been prepared for the Ninotsminda UGS facility (see figure 6-3) – based on in-house RAMBØLL data. An adjustment has been introduced to take into account an estimated price difference for work carried out in Europe and Georgia. Particularly, construction and civil engineering works are presumed relatively cheaper in Georgia (a factor of 0.5 to 0.6 has been applied in this case) [10]. The price level is year 2003.

[10] Although the cost estimate has been based on European data, the figures have been corrected for the different geographical location and the general cost level in Georgia.

Figure 6-3 Layout of above ground facilities: well sites and central processing plant

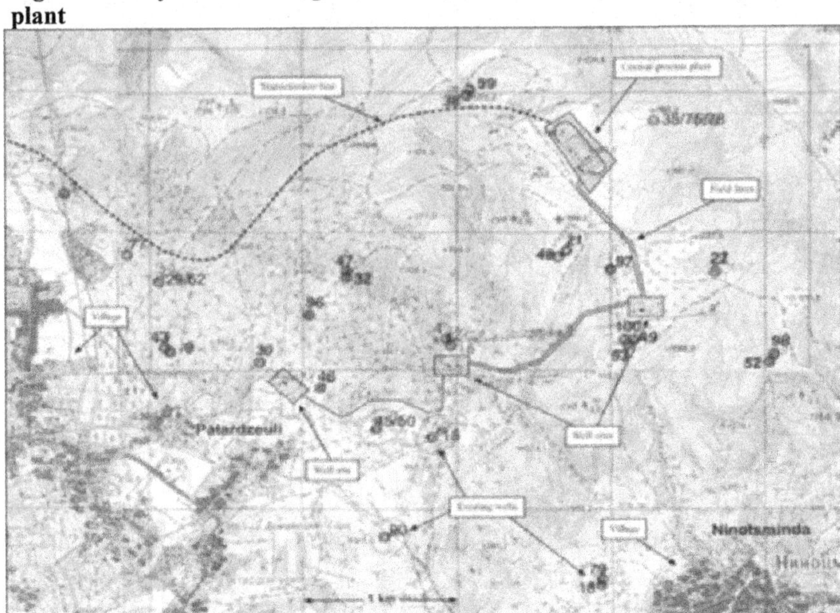

The investment cost for the main project components are presented in Table 6-1 along with the main data on capacities for working volume, injection and withdrawal of gas.

Table 6-1 Cost estimates (price level June 2003) – Ninotsminda UGS

Case	# Wells		*Working volume*	With-drawal capacity	Injection capacity	Total capital costs 2003 Million EUR			
			mill m^3	mill. m^3 per day	mill. m^3 per day				
	New	Work-over				Above Ground Facilities	Wells	Con-necting pipeline	**Grand Total UGS**
I	2	1	120	4.0	1.5	45	9	5	**59**
II	3	2	240	8.0	3.0	67	14	7	**88**
III	3	4	480	8.0	3.0	70	16	7	**93**
IV	4	4	480	16.0	6.0	114	20	10	**144**
V	3	1	150	4.0	1.5	46	13	5	**64**
VI	3	4	300	8.0	3.0	70	16	7	**93**

(Source: RAMBØLL)

7. ECONOMIC ANALYSIS OF UGS PROJECT

The cost of gas storage varies mainly with the size of the working volume and the capacities for withdrawal and injection of gas. In general, a large facility is more costly than a smaller one, but it does also have 'economics of scale', which gives lower unit costs. The cost of gas storage in Georgia varies with size of working volume and capacities for withdrawal and injection of gas. A large installation is more costly than a smaller one, but does also have 'economy of scale', which gives lower unit costs. Table 7-1 and Figure 7-1 illustrate this.

Table 7-1 Key figures for the Georgia UGS cases

Case	Working volume	Withdrawal capacity		Injection capacity		EUR/ 1000m^3
	mcm	mcm/day	%*	mcm/day	%*	
I	120	4.0	3.3	1.5	1.3	125
II	240	8.0	3.3	3.0	1.3	88
III	480	8.0	1.6	3.0	0.6	50
IV	480	16.0	3.3	6.0	1.3	69
V	150	4.0	2.7	1.5	1.0	108
VI	300	8.0	2.7	3.0	1.0	75

*) Withdrawal /injection capacities as daily capacities in relation to the yearly volumes. Thus, in Cases I, II, and IV the stored gas volumes can be withdrawn in 30 days.

Case I, with the smallest working volume of 120 mcm and a withdrawal capacity of 4 mcm/day, comes out as the most expensive with a unit costs of 125 EUR/1000m^3. At the other end of the scale, Case III has the largest working volume of 480 mcm and withdrawal capacity of 8 mcm/day, and it comes out with the lowest unit costs of 50 EUR/1000m^3 yearly.

The calculated unit costs of operating a gas storage facility in Georgia have been compared with the rates for renting gas storage capacity charged by various gas transmission / storage companies in EU countries. The companies' storage rates are divided into 2 groups, low and high flexibility respectively in order make comparisons more realistic. A high flexibility requires larger withdrawal and injection capacities and is thus more costly than renting or building storage with lower flexibility. Figure 7-2 presents this comparison.

Cases III and VI may be compared with Group 1 (low flexibility). It turns out that the most feasible of the Georgia UGS cases; Case III and VI have unit costs (43 and 64 EUR/1000m3 respectively) that are within the range of the fee rates for gas storage services in Group 1. Furthermore, Case III offers a slightly higher and thus more favorable flexibility (the withdrawal capacity is 1.67%) than the cheapest EU gas storage service provider, Dynergy (1.5%).

Figure 7-1 Economics of Scale

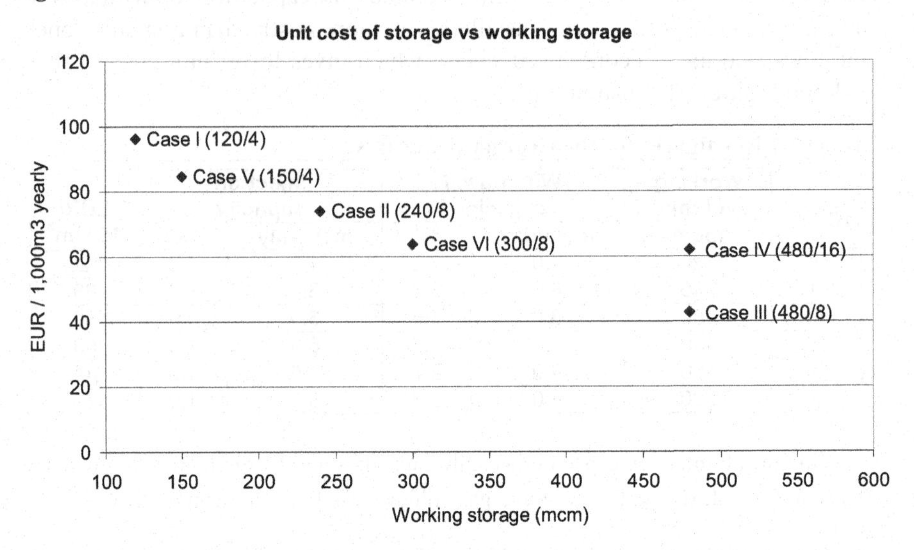

Figure 7-2 Withdrawal capacities versus unit costs

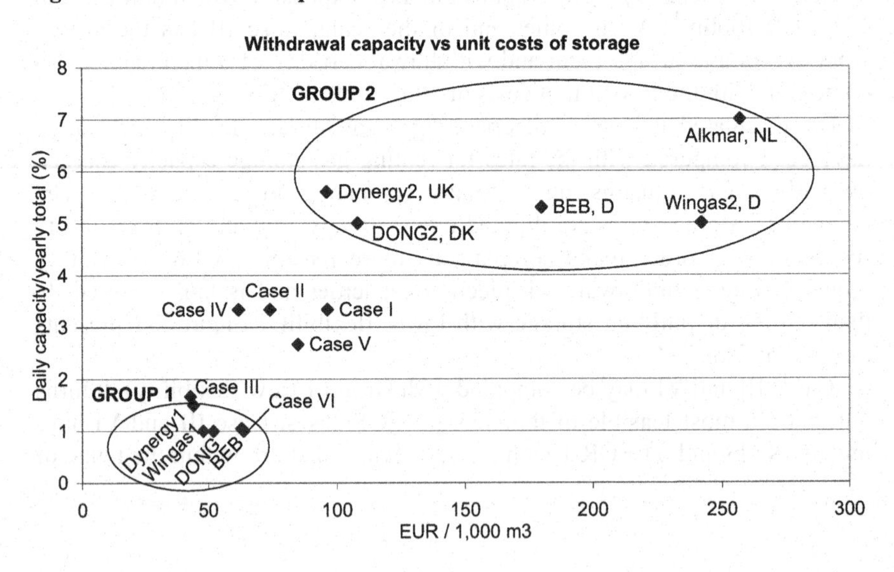

The other Georgia UGS cases (IV, II, V and I) can better be compared with Group 2 (high flexibility). It turns out that these cases offer withdrawal capacities that are lower i.e. less favorable from a technical point of view. However, the Georgia UGS unit costs for are either cheaper (Cases IV, II and V) than or comparable (Case I) with the fee rates in Group 2.

Introducing gas storage in the chain of gas supply from the producer to the gas consumer can be compared with taking out insurance on top of paying for the gas delivery. The investment costs will have to be recovered from the consumers by adding a storage cost element on top of the gas tariffs charged today. The size of this storage cost element varies with the size and thus feasibility of the gas storage facility.

A storage cost element, ranging from 8 EUR/1,000 m^3 (Case I) to 13 EUR/1,000 m^3 should be added to the 120 mcm or 480 mcm of working gas that is to be injected into the storage in Cases I and III respectively. This 'insurance premium' is to be added to the transmission tariff and then recovered from the transmission of the total volume of imported gas.

Today's transmission tariff is 16.6 GEL/1,000 m^3 or approx. 7 EUR, so adding the storage cost element will increase the transmission tariff by between 114% (Case I) and 185%. It is important to remember, however, that today's residential consumer tariff is approx. 230 GEL/1,000 m^3. This means that the consumer tariff will increase by 3.5% (Case I) and 5.7% (Case III).

Establishing a storage facility at Ninotsminda may open for reductions of the import price of gas from Russia. It is, however, uncertain and difficult to estimate how much lower the price for gas purchased during summer for injection into the storage may be. But if the summer gas price is approx. 5% lower than the winter price, the total effect on the consumer tariffs after establishment of gas storage may be neutral.

8. CONCLUSION AND RECOMMENDATIONS

The proposed UGS project at Ninotsminda is technically and economically feasible but in the prevailing economic conditions in Georgia it has so far proven difficult to attract the necessary investors with relevant expertise in establishing and operating such large-scale UGS facilities.

RAMBOLL has conducted a SWOT-analysis[11] of the project and here follows the most important observations:

- Strengths
  - Meets the need to increase the security of gas supply

[11] SWOT = Strengths; W=Weaknesses; O=Opportunities; T=Threats

    - Estimated unit costs may be comparable or lower than fee rates charged by EU UGS operators for similar services
- Weaknesses
    - Tariff adjustments are dependent on regulator approval
    - Low level of interest from potential investors with relevant UGS experience
- Opportunities
    - Enables optimisation of benefits from transit fee gas and of price differentials between winter and summer
    - Gradual development of storage capacity from national needs only to out-of-region needs as well
- Threats
    - Financing may prove difficult in prevailing economic conditions
    - Vulnerable towards recent efforts, on populist grounds, to undermine regulator rulings

At the moment, the most fundamental drawback for developing the UGS project in Georgia is that it seems impossible to create a strong and competent consortium of companies with relevant experience for establishing and operating a UGS facility.

It remains a hard fact that the BP and Statoil, the two key companies behind the Shah Deniz gas field development involved in and the associated project to ship gas to Turkey and possibly beyond to Europe by means of the South Caucasus Pipeline project so far have expressed that they are not interested or willing to become involved in establishing underground gas storage in Georgia.

It is, however, possible that the plan for establishing underground gas storage in Georgia may benefit form the implementation of the Shah Deniz and SCP. The effects on Georgia from a successful implementation of those 2 mega-projects are two-fold:

1. The SCP project will increase focus on Georgia as a 'beneficiary' and 'host country' because a large part of the investment in the SCP pipeline will be put on Georgia's territory.
2. The gas from Shah Deniz represents a new opportunity for Georgia to diversify its gas supply in the way that Russia will no longer remain the sole supplier of gas to Georgia.

The latter represents an important step towards increasing the technical security of gas supply and it also holds the potential of Georgia gaining commercial benefits from the fact that Georgia is entitled to purchase supplies of 'supplemental gas' in addition to the transit gas payments in the forms of 5% of the transported volumes in the SCP. Establishing a UGS facility has the potential of optimising the benefits of those new supplies of gas in addition to the gas that is imported from Russia and the transit fees that are received from transiting gas through Georgia to Armenia.

The former represents an opportunity for Georgia to demonstrate that it is willing and able to improve the conditions for foreign investors so that they feel more welcome and confident than is presently the case.

Decision makers in Georgia have to realize that the combination of de facto insolvency of most operating energy companies and uncertainty of the macro-economic trends in Georgia makes it very difficult to attract investors such as international financing institutions and international energy companies. Investors are simply not sufficiently confident that the situation will improve and investment risks will decrease.

RAMBOLL is convinced that more than anything else it is important that decision makers in Georgia realise that it is necessary to start a process towards increasing the commercial discipline in all parts of Georgia's energy sector.

Restoring commercial viability and creditworthiness of the energy utilities, one-by-one and the sector as a whole, cannot no longer be avoided if Georgia sincerely wants to attract the interest of competent investors in the form of international energy companies and international financing institutions.

The need for policy changes also includes the need to secure that the regulatory function of GNERC[12] can be exercised without interference of any kind.

During the EC TACIS project RAMBOLL has sought to clarify whether the current tariff regulation practise actually discourages the undertaking of investments aiming at improving the reliability of the gas transmission system.

Unfortunately, the reaction received from GNERC is not very helpful is this respect. This is disappointing and according to RAMBOLL it seems that the current regulatory practises also discourage investments in the energy systems.

It is obvious that it poses an important barrier to having someone undertake investments in the energy sector – if investments are discouraged due to difficulties in having higher tariffs approved.

This kind of 'self-censorship' by the regulator in order to avoid political or populist pressure against its rulings to allow tariff increases in order to secure cost recovery is clearly not acceptable.

Furthermore, it represents a misunderstanding of what really serves the medium-long term interests – not only of the gas consumers in Georgia because it will prevent the consumers from enjoying the benefits of a more

[12] Georgian National Energy Regulatory Commission

reliable gas system with a higher security of supply. This attitude is also against the interests of GGIC and the gas distribution companies and not least the interest of those energy companies, which are considering to enter into Georgia to make investments.

UNDERGROUND GAS STORAGE: PLANNING AND MODELING WITH SIMSIM, A NEW RESERVOIR ENGINEERING SOFTWARE TOOL

Andras GILICZ and F. PACH
ER-PETRO LTD.HUNGARY

Abstract: A novel, simplified reservoir simulation software tool has been developed for reservoir engineers called SimSim. It combines the advantages of the simple material balance technique and complex numerical reservoir simulation: it keeps the simplicity and speed of the former, and has results like the latter, i.e. it can calculate pressure, saturation, fluid flux and hydrocarbons in place distributions. An example for underground gas storage modeling is presented.

Key words: numerical reservoir modeling, simulation, underground gas storage, material balance, water influx

1. INTRODUCTION

In certain areas in Hungary several hydrocarbon reservoirs are located close to each other, like in the Algyő region[1]. Because of this and due to long production periods of several decades and significant production volumes, reservoirs interfere with each other. This means, that reservoirs influence their pressures and productions mutually. Such systems can rapidly get very complicated hydrodynamically, and it was always a problem for engineers how to model them adequately. One solution is numerical reservoir simulation, but to set up and manage such large numerical models is very tedious and expensive. Another, simpler solution had to be found. Developments started with reviewing material balance techniques, and think about, how they could be further developed in order they can model

J. Hetland and T. Gochitashvili (eds.),
Security of Natural Gas Supply through Transit Countries, 227–235.

interfering reservoirs. It turned out, that the water influx calculation method or term had to be improved. A novel transient flow equation was found, which proved to be much more flexible to calculate water influx, as previous, earlier methods (e.g. Van Everdingen-Hurst[2], etc.). With ongoing application it turned on, that more and more benefits result from the this new approach. All this resulted finally in a software tool called SimSim. SimSim stands for simplified simulation. It is not a replacement for reservoir simulation, but it is a good starting point for a fully detailed reservoir simulation study. SimSim is a certain kind of pre-processor that allows rapidly (in a matter of minutes) decide which parameters are best to evaluate in a full simulation, perhaps saving many hours or days of simulating data, that produce ambiguous results.

2. SIMSIM'S ABILITIES

SimSim fills the gap between material balance techniques and complex reservoir simulation yet keeping the simplicity and speed of the material balance but providing reservoir simulation like results, i.e. pressure, saturation, hydrocarbons in place and fluid flux distribution within the reservoir.

SimSim can be used in a twofold way:

1. for interfering reservoirs
2. for single reservoirs where the reservoir is subdivided to compartments (the compartments forming the simulation grid)

The basic novelty in SimSim is the new way of water influx calculation. SimSim applies a new transient flow equation for waterinflux calculation. This has several benefits over conventional methods[2]:

- the aquifer is characterized with two parameters only: water quantity in the aquifer and its so called eigentime (see below)
- no superposition in time is needed, which speeds up calculations significantly
- time step size is not hooked to pressure measurement time points
- nearly arbitrary time steps can be used, which allows high timely resolution
- there is no need to assume aquifer shape (e.g. radial or linear)
- there is no need to assume aquifer type (steady state, pseudo steady state, transient, infinite acting, etc.)
- the material balance equation can be written in a Taylor series form which allows very fast solution for reservoir pressure

- there is much less cross-correlation between the parameters of SimSim than for the conventional methods.
- experience shows, that found optima is mostly unique and more sensitive to parameters
- interference to neighboring reservoirs or aquifers can be accounted for easily
- compartmentalized reservoirs can be readily investigated

The eigentime is defined with the following integral:

$$\tau = \int_V \frac{\Phi \mu c}{kh} dV$$

where:
Φis the porosity in the aquifer
μ is the viscosity of the water in the aquifer
c is the effective compressibility of the aquifer
k is the permeability of the aquifer
h is the thickness of the aquifer
Vis the volume of the aquifer

The eigentime of the aquifer (τ) is very similar to the inverse of the hydraulic diffusivity. The higher the eigentime, the slower the aquifer reacts to production/injection impacts and vice versa. The eigentime is:

- ***Directly proportional*** to the aquifers volume, porosity, water viscosity and average compressibility,
- ***Inversely proportional*** to the average permeability and thickness of the aquifer.

As can be seen, the eigentime concentrates several important petrophysical and geological information in one parameter, which makes its practical application very convenient. There are no more esoteric dimensionless time, water influx constant, etc., only the eigentime.

In fact the above formula for the eigentime is rarely used (although it coud be.) Instead an estimation is made initially for the eigentime, and it is then updated during the simulation. A good rule of thumb is to take 1 day for 1 million m^3 aquifer pore volume. So a "usual" aquifer of several hundred million m^3 pore volume has an eigentime of approximately several hundred days.

SimSim needs conventional material balance input, i.e.:

- rock and PVT properties

- production/injection data
- initial reservoir pressure
- measured average reservoir/compartment pressures, which will be matched
- initial estimates for reserves and aquifer properties, which will be optimized

As can be seen, the input data are completely conventional which can be found in every petroleum company's files, and no additional information needs to be researched in order to run the software.

SimSim performs a pressure match of measured and calculated reservoir or compartment pressures with an automatic, non-linear optimization technique, called the Nelder-Mead simplex algorithm[3]. During pressure matching SimSim's parameters (e.g. hydrocarbons in place, aquifer size and eigentime, etc.) are varied in a systematic manner according to the simplex algorithm to achieve pressure match. In mathematical terms the residuals sum of squares (least squares) between measured and calculated pressures is minimized. The parameters to be optimized can be freely selected by the user.

3. SIMSIM'S LIMITATIONS

- SimSim has less resolution than a conventional numerical reservoir simulator.
- SimSim does not count for capillary and gravity forces.
- Due to the above limitations SimSim can neither be used for gas or water coning modeling, nor for gravity drainage modeling.

4. APPLICATION AREAS OF SIMSIM

SimSim can be used to

- estimate initial hydrocarbons in place and their distribution
- estimate initial reservoir pressure (if not known exactly and an estimation is needed)
- estimate aquifer size and its activity
- calculate water influxes
- calculate theoretical reservoir/compartment pressures
- calculate drive indices and average reservoir/compartment saturations
- calculate interfering fluid volumes between reservoirs or compartments

- estimate strength of interconnectivity between reservoirs or compartments
- get an estimate of the internal reservoir structure
- get information from distribution of aquifer water volumes and aquifer strength
- get information from the main fluid flow directions
- etc.

Due to these abilities SimSim can be placed somewhere between material balance methods and numerical reservoir simulation. It is more complex than material balance, yet simpler as a numerical reservoir simulator. But it has a great power to investigate complex interference or flow phenomena within a reservoir with the simplicity of a material balance, and it is by far much faster than a reservoir simulator. Running time is in the range of a few seconds even if several reservoirs/compartments interfere. Depending on the difficulty of the problem, running time is usually in the range of a few (ten) minutes in case of parameter optimization. In this way SimSim is an ideal tool for reservoir engineers to provide fast, quick look answers within shortest time with relatively simple data input. SimSim can be used as a pre-stage for reservoir simulation, giving a first impression of main fluid flows, and reservoir drive mechanics. SimSim should not be viewed as a competitor or replacement of conventional reservoir simulators, but it is rather a complement of them. In addition SimSim's philosophy bears some new ideas, which will be the basis of a new family of reservoirs simulators having some striking new features, especially concerning speed, stability even with very high number of grid cells.

5. GAS STORAGE IN HUNGARY

The significance of natural gas beside oil and coal in the Hungarian power supply industry started in the early 60-ies, and gained since then more and more importance. Initially gas had only 1÷3% share in the energy consumption, which however grove over 30% by now.

Up to the 70-ies gas consumption was dominated mainly by industry (heavy industry, chemical industry, electricity, etc.), but the demand for household gas supply increased permanently. This process significantly has changed the structure of gas consumption, because the earlier consumption was without peak periods and was mainly independent from seasons, whereas current demand is of strongly fluctuating nature having peak consumption periods depending on yearly seasons.

Part of the Hungarian hydrocarbon reservoirs had natural gas, but they are meanwhile depleted. Hungarian gas accumulations are in general small, and are at the current stage mostly depleted, the chance for new discoveries is low, the gas caps of existing oil reservoirs cannot be strongly produced because of technological reasons, and gas import is increasing. This means, that the domestic gas production is unable to satisfy domestic demand. Due to the above reasons it was necessary to establish underground gas storage reservoirs to provide safe supply.

Table 1- Some gas storage reservoirs in Hungary.

Parameter	**Hajduszoboszló**	**Pusztaszöllős**	**Pusztaederics**	**Zsana**
Mobile gas, Mm^3	1400	210	330	1300-2200
Peak Capacity Mm^3/d	20	2.4	2.9	18-24
Number of wells	77	14	23	42-56
Cushion gas, Mm^3	2240	350	290	1800
Number of reservoirs	1	4	2	1
Average depth, m ss	1000	1000-12000	1300	1700

First underground gas storage plans in Hungary born in 1973, and after design and realization underground gas storage started in 1978-1981. First gas storage reservoirs were that of Hajduszoboszló, Pusztaszöllős and Pusztaederics. Later Zsana has followed. These underground gas storage reservoirs were actually depleted gas fields. Some characteristic parameters of these are summarized below.

Hungarian domestic gas demand in winter has achieved a daily rate of up to 70÷80 Mm^3/d, so the development and new establishment of underground gas storage has an increasing importance. Underground gas storage is the main source of providing peak capacity, because as mentioned, most of the Hungarian gas reservoirs are already depleted, production of gas from gas caps of oil reservoirs is limited. This fact is also enhanced, that the structure of gas consumption changes permanently, i.e. industrial consumption declines, household consumption is increasing.

At the end of the 80-ies all domestic gas reservoirs were investigated to find out which ones are appropriate for underground gas storage, and what has to be done to convert them.

The adequate knowledge of the parameters (such as volume, petrophysical properties, aquifer size and activity, their determination, and forecast, etc.) of the underground gas storage reservoirs is very important, because this guaranties the safety of gas supply in the future, especially in critical periods like January and February, where the consumption is high, and the gas storage reservoirs are mainly discharged.

The distribution of the domestic gas reservoirs is not even in the country, they are concentrated in a relatively small area. This raises another problem: the reservoirs interact with each other, they mutually influence their productions and pressures. This is a serious problem, and has to be considered in the planning of underground gas storage. SimSim was able to tackle the problem.

6. CASE STUDY: THE ZSANA RESERVOIR

The exploration of this reservoir started in 1977. After a serious blowout in 1979 production started in c. 1980. Original gas in Place was estimated to be about 5.5 billion m^3. After a production of about 4 billion m^3 gas up to the end of 1992, reservoir pressure has significantly declined, and it became clear, that there is limited aquifer support. So the decision was made, to convert the reservoir to underground storage. A detailed integrated study project was initiated to clarify reservoir geology, structure, petrophysics and dynamics[4,5]:

- The reservoir structure was updated by shooting a 3D seismic survey.
- Detailed reservoir geology clarified sedimentology.
- Core analyses established petrophysical interpretation.
- By modeling with SimSim original gas in place and reservoir dynamics was updated, clarified.
- A detailed reservoir simulation study served finally for the realization of the project.

7. SIMULATION RESULTS WITH SIMSIM

The Zsana main gas reservoir has two satellite reservoirs. It was suspected, that the three reservoirs interfere with each other, but the extent of interference was not known.

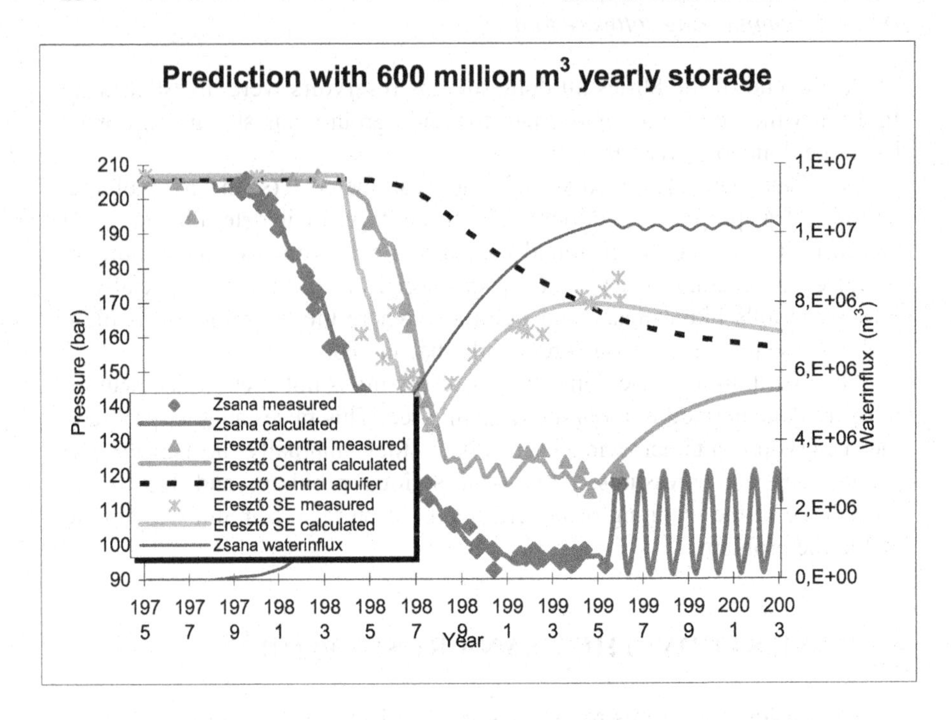

Figure 1: - Prediction of the behavior through time of the Zsana (Hungary) underground storage reservoir.

As a first stage, SimSim was used to clarify reservoir dynamics. Results were then used later to build the detailed numerical model. The investigation performed with SimSim revealed the following important facts:

- The Original Gas in Place in the reservoir is around 5.8 billion m^3
- The two satellite reservoirs around the main gas reservoir have little influence on the main reservoir, but influence each other strongly.
- One of the satellite reservoirs is pumping up the other satellite reservoir.
- One of the satellite reservoirs can be used for water disposal, which will have a favorable effect on the gas storage.
- Thermal effect are neglible compared to hydrodynamical effects in the reservoir.
- Future reservoir pressures could be predicted for different storage scenarios.
- Predicted reservoir pressures were confirmed by later, independent pressure measurements.

Figure1 depicts the pressure matches and a prediction case with 600 million m^3 yearly storage. The Zsana underground gas storage reservoir currently operates well, and contributes to the safe energy supply of the country

8. CONCLUSIONS

A novel software tool has been elaborated for reservoir modeling, called SimSim. The software tries to fill the gap between material balance techniques and conventional reservoir simulation, trying to combine the advantages of both. It needs relatively simple input, runs fast as a material balance, but provides simulation like results, i.e. it calculates pressure, saturation and fluid flux distributions. A case study of a Hungarian underground gas storage reservoir is presented, where the software played a major role to clarify the complex hydrodinamics of three interfering gas reservoirs.Design an prediction of the expected pressure conditions during operation could be obtained with the software with high confidence.

References

1. Pach, F., Gilicz, A., Palasthy, Gy.: „The Algyő Field in Hungary: Modelling Fiveteen Reservoirs with Interfering Aquifers" paper SPE 38929 presented at the 1997 SPE Annual Technical Conference and Exhibition in San Antonio, Texas, 5-8 October.

2. Craft, B.C and Hawkins, M.F.: Applied Petroleum Reservoir Engineering" Prentice-Hall Inc.1959.
3. J.A.Nelder – R.Mead: A Simplex Method for Function Minimization. Computer Journal, 7. 308-313, (1965).
4. Bessenyei, I., Lumdsen Horváth, G., Paulik, D., et. al: „An Integrated Evaluation of a Future Underground Gas Storage Site" 22nd Petroleum Itinerary Congress and Exhibition, Tihany, Hungary, 6-9. October, 1993.
5. Gombos, Z., Farkas, É., Papp, I. et al:"Subsurface Gas Storage in the Zsana Field" 22nd Petroleum Itinerary Congress and Exhibition, Tihany, Hungary, 6-9. October, 1993.

ASSESSMENT STUDIES FOR THE CREATION OF UNDERGROUND GAS STORAGE FACILITIES IN GEORGIA

David ROGAVA
Georgian Technical University, Candidate of geological sciences, assistant professor

Abstract: Georgia experiences a significant lack of energy resources and the country is basically oriented to the import. Therefore, the problem of energy storage and creation of reserves is an urgent problem. Creation of such reservoirs undoubtedly will be connected with implementation of those big projects which provide for the construction of major gas pipelines on the territory of Georgia – with the capacity of approximately 50-60 bcm.

The article reviews in detail the tectonic division of Georgian territory, geographic locations of the areas allocated for underground gas storage (UGS) facilities and their lithological and structural features and peculiarities

Key words: Natural gas, Underground storage, Energy reserve, Geophysical exploration, Survey

1. INTRODUCTION

One of the main principles of any country's security is energy supply that would be sufficient to provide energy for facilities that are of strategic and vital importance.

According to the international regulations the energy independence of a country is calculated as the total primary power resources (E own) and total power supplied to the national economy ratio (E sup), i.e.

J. Hetland and T. Gochitashvili (eds.),
Security of Natural Gas Supply through Transit Countries, 237–245.

$$K = \frac{E\ own}{E\ sup}$$

According to 1999 data for Georgia this ratio equaled to K=0.27 [1].

Generally, the construction and exploitation of underground gas storage facilities got started at the beginning of the previous century. The first reservoir of this kind was built in 1915 in Canada (Welend – Kaunts). Since then underground storage facilities have been built in several countries such as the USA, the Former Soviet Union, Germany, Poland, Czechoslovakia, France, Great Britain, Italy, Austria. In 1977 already 204 bcm gas was stored underground in 385 reservoirs in the USA [2].

According to the rules for strategic reserves control and in anticipation of possible emergency situations the EU member states are obliged to hold mandatory reserve of energy resources in order to provide average daily consumption up to 90 days. The countries that provide energy sources are entitled to reduce the volume of these reserves by only 15%. The European legislators consider the possibility of creating these reserves within the territory of one country for the usage by second and third countries, which may also include the owners of the reserves. This, in brief, is the rationale behind a pronounced need for creating gas storage facilities in Georgia as they could also be of interest to the European countries as well. Furthermore, the geological structures of some of the European countries will not be suitable for underground gas storage.

If we manage to find appropriate structures to keep more than 2 bcm gas, a underground gas storage project would probably be considered commercially attractive for investors. [3,4] That is why this issue is deemed urgent in Georgia today (2003).

Ever since 1966 such activities, i.e. identification and assessment of geological structures for constructing UGS facilities have been in progress in Georgia. By virtue of geological survey and prospecting works two suitable fields were discovered [5]. The first is located in the Khashuri district, 100 km from Tbilisi. By means of seismic exploration the upland of Mokhisi with 300 m amplitude, was mapped. When drilling 330 m deep an appropriate strong water-bearing rock area was identified. Besides, according to the geophysical exploration data, water-proof packets within 650-1200 meter interval were picked out. They were worth considering as UGS facilities. A further exploration of this area is put on halt.

The other area suitable for UGS facilities was mapped in the Rustavi district near the territory of western Azerbaijan. Due to seismic exploration two anticlines were identified here, which are located near the Karadag-

Tbilisi gas pipeline. According to the author [5] it was suggested and considered to be reasonable to drill two parametric boreholes:

1. The first one is to the South of the #1 Karaiaz borehole, at the area of Upper Eocene sediments spreading. The aim would be a precise identification of the Upper Eocene sediment conditions in one section with seismic survey data.
2. The second one is within the scope of East Marneuli anticline after carrying out detailed seismic survey works.

In 1969 V. Likholetnikov together with A.Vasserman [5,6] identified the so-called trap types in Kolkheti central and eastern parts in Bajocian, Bathanian, Neocomian-Aptian, Cenomanian, Turonian-Paleocene sediments and also natural reservoirs – in Upper Cretaceous and Paleocene sediments.

Owing to lacking energy resources in Georgia the country is rather import-oriented. Therefore, the problem of energy storage and the creation of reserves are of strategic importance. Undoubtedly, the creation of such reservoirs would be connected to the large projects that provide for the construction of the major gas pipelines over Georgian territory [7].

In the 1990-ies Georgian geological institutions were carrying out survey studies in order to determine and investigate some favourable structures for UGS facilities. These appropriate structures of brachyanticlines and anticlinal domes within folds consisting also of such rocks can be used as gas reservoirs (mainly cracked sandstones, tuff sandstones and tuff breccias). They are displayed at depths around 1000-1500 meters - sometimes even more than 2000 meters - and are covered with waterproof and gas-proof limestone rocks, with capacities estimated at 1-1.5 bcm [8].

The understanding and the interpretation of tectonic problems today are quite different from three decades ago. The Caucasus with its long and complicated evolutionary development, though it sounds paradoxical, is the classic polygon for its exploration both from fixism as well as a mobility point of view [9,10]. The mobility approach gives far more reasonable explanation of the latest facts. However, researchers think its tectogenetic model is still under the process of sophistication. Therefore, the determination of the tectonic essence of this territory - and the specification of its structures – constitutes the priorities of their study.

Using the already published map of the tectonic division of Georgian territory [10], the areas allocated for UGS facilities can be viewed in the following way:

2. UDABNO

This area is situated in Gardabani and Sagarejo regions, between the rivers of Mtkvari and Iori, and near the villages Udabno and Krasnogorsk at the Azery border, 50 km away from Tbilisi. Within this area the recommendation is to construct a UGS facility comprising two domes – northwest and southeast, which resulted from undulation in the folds dome.

According to the above-mentioned chart the survey area is within Marneuli block (4), which is to the East of Artvin-Bolnisi block of Anticaucasus folded system (III2/2). It belongs to Bolnisi subzone and (II) is near the border of (III) Anticaucasus folded system of the Transcaucasus intermountain trough (east immersion), which is itself known as interzone dividor continuously developing deep break.

Based on the structural and tectonic data the Sakharetbi anticline is presented in the dome by Oligocene-Lower Miocene Maikop suite rocks and flanks are covered with Middle and Upper Miocene sediments.

The lithological and structural features of building rocks are as follows: for collectors – there are sandstone packs in the shale rock layers of Maikop stratum Sakaraulo horizon with about 10mm capacity. Its gas-proof lid is more than 500 meter strong Maikop stratum upper Kotsakhuri horizon shale rock, with rare streaks of finest rock sandstone.

3. UPLISTSIKHE

This area is in the Gori region in the river Mtkvari gorge 80 km away from Tbilisi. According to the tectonic segmentation chart the Gori-Uplistsikhe area is within the Mukhrani-Tiriphoni block and the river Mtkvari intermountain trough (I) and belongs to the (II) Trans-Caucasus intermountain zone. It is also near the same deep break (to the north), which is regarded as the border of the Trans-Caucasus intermountain massif and the Minor Caucasus folded system.

Two areas are discussed herein. The first area is the so-called Khidistavi dome. It is displayed in Tsedisi-sincline (right bank) mould built by the Oligocene-Lower Miocene Maikop layers. The second area is the Gori-Uplistsikhe anticline eastern part (left bank), which is separated from the western part by a break, and is presented by sediments of Oligocene and Miocene. In both structures sandstone packs in the Maikop stratum Sakaraulo horizon limestones are considered as collectors. The Maikop stratum Kotsakhuri horizon limestones with sandstone interlayers - and the Miocene sediments situated above them - can serve as a lid, as these structures provide a gas-proof screen.

4. ARDJEVANI-TSVERI

This area is situated in the Manglisi-Kobura sector of Tsalka and Tetri Tskaro regions, on the south slope of the Trialeti range, 60 km away from Tbilisi.

According to the tectonic map, it is within the Aspindza-Tbilisi sector (3) and located in the South sub-zone of the Ajara-Trialeti zone (III3), at its border to the North. Based on the main structural and tectonic characteristics it belongs, first, to Arjevan-Sevan anticline, which is sublatitudal with the dome inclined to the undulation (doubledomed in the West sector, braccianticlinal in the East). Second, to Toneti and Tsveri brachyanticlines, which are made of Paleocene-Lower Eocene (flish), Middle Eocene (tufogenes), Upper Eocene and Oligocene – Lower Miocene (mainly containing clay) layers of rocks. Periodical breakings are hardly recorded.

The lithological and the structural features of rocks according to the purpose of their usage are as follows:

- As collectors - tuffs, tuff breccias, tuff sandstones of Middle Eocene age; thickness of this stratum is above 1000 meters;
- As for a gas-proof lid, there is the stratum of 1000-1500 meter thickness, with Upper Eocene limestone, argillite and sandstone alternatively.

5. SAMGORI SOUTH GUMBATI OIL DEPOSIT

This is the area where according to the government order in 1992 foreign specialists (UGS Gmbh Mittenvalde) prepared a preliminary technical project for the construction of UGS facilities. It is located within the Aspindza-Tbilisi Sector (3) and belongs to the South sub-zone of the Ajara-Trialeti zone (III4/1). It is situated at the northeast edge, probably exactly at the deep fault, which by itself is the border between (II) Transcaucasus intermountain sector and the (III) Minor Caucasus folded system (Anticaucasus). According to P. Gamkrelidze this is approximately to the West of this fault in the Sartichala sub-zone of sinking at the East edge of the Georgian block.

The deposit is situated in the Gardabani region, on the right bank of the river Iori, 30 km to the East of Tbilisi, and 7 km from the Vaziani railway station. It is situated on the East end of the Teleti structure lowland and is made up by sediments of the Upper Cretaceous, the Lower, Middle and Upper Eocene and Oligocene. In the arch of the fold there are dome-shaped brachy-anticline, where oil was found and quarrying was carried out.

In 1991 it was planned to prepare this deposit for exploration as a UGS facility. The middle Eocene aged tuff-breccias and tuffs with hollow gaps caused by 550-600 m thick cracks. A gas-proof lid was created by the Upper Eocene "Tbilisi" and "Navtlugi" strata of shaly clay, intermittent argillites, clay sandstones and marls, at 1000-1100 meter thickness.

6. SAMGORI-PATARDZEULI OIL DEPOSIT

This survey area is situated in the Gardabani region, on the left bank of the river Iori, 30-40 km away from Tbilisi. It is in (II) Transcaucasus intermountain sector, (II3) eastern molassa zone of sinking – within the river Kura intermountain trough, sligtly-folded Sartichala sub-zone, near the deep fault, to its East, or, as the author of the map calls it: in the molassa zone of Gare Kakheti (II2/3) the so-called of torn molassa blankets.

Structural and tectonic features of Samgori-Patardzeuli deposit are determined by big latitudinal brachyanticline fold, which is connected by three domes: Samgori, Patardzeuli and Ninotsminda though the specialists [11] now assume that Sartichala will emerge domelike. Thick sediments of Upper Cretaceous, Eocene and Oligocene-Miocene, make up the survey area. At present Ninotsminda Brachy-anticline of the Samgori-Patardzeuli oil deposit is being considered (in the Sagaredjo district), which is connected with the Middle Eocene volcanogenic rocks, and is of rather asymmetric structure [12].

According to the lithological and the structural data among the above-mentioned rocks the Middle Eocene tuff-breccias, the tuff sandstones, and tuffs with argillite interlayers, which also have hollow spaces of cracking origin can be used as collectors. Their thickness reaches 600 meter. Based on the same data, the "Tbilisi" and the "Navtlugi" strata with layers of alternating carbon clays, argillites, clay sandstones and marls of the Upper Eocene age, can be used as a gas-proof screen. The thickness of this stratum is between 900 and 1400 meter. The Georgian National Oil Company(GNOC) is reluctant to use this deposit as UGS facility as there are over 80 boreholes for the deposit utilization and it is technically very difficult to plug them, although it is vital from the safety point of view.

A special expedition of "Sakgeologia" has conducted survey-drilling works in order to identify useful structures for UGS construction in western Georgia. On the basis of data analysis from the boreholes situated within the Kolheti artezian basin, Dranda, West Chaladidi and Kulevi anticlines were considered perspective for carrying out hydrogeological exploration works[4].

7. DRANDA

This area is situated 30 km to the southeast of Sukhumi. It is located in (II) the Trans-Caucasus intermountain sector tectonic unit, within the scope of Ochamchire-Kulevi block (3), (II1) at the edge of the West molassa sinking zone, near the border of (2) the Samurzakano block belonging to the same zone. Under the Quaternary cover, based on the seismic data, and also the drilling data, a number of local folds have been revealed, which are almost latitudinally spreading. These are anticline brachyforms [2], which are isolated by the flat bed of synclines. Drilling revealed that Pliocene folds are less shrunk than those situated underneath the folds of the Cretaceous, the Paleocene and the Miocene, and in some places the Lower Pliocene sediment rocks. According to the main structural and tectonic data, the Dranda anticline might be spotted in Upper Cretaceous sediments, the Upper, Middle and Lower Miocene with over 1400 meter thickness.

The lower Miocene clay rocks with sandstone blocks in thickness are considered as collectors, and as a gas-proof lid – the Upper and Middle Miocene hard clay rocks. As for the collectoral rocks, they are located at 1000-1400 meter depth.

8. WEST CHALADIDI

This area is 20 km northeast of Poti, to the West Chaladidi anticline of the Pontain-Meotian (Miocene) sediments. It is in the lower part of the above-mentioned Ochamchire-Kulevi block (3), near the south border of (II1) the intermountain cavity zone, at the southwest edge of the Poti-Abedati and the Senaki-Tsaishi. On the Poti-Abedati fail line the Abedati, Nokalakevi and the Eki mountain over thrust sheet folds are displayed intermittently, which corresponds to the lower structural floor disjunctive break zones [9]. On the mentioned fail line, from the northeast to southwest, the relocation of epicentres of the Samegrelo earthquake (1941) is specified.

In these structures the Pontian basal conglomerates are regarded to constitute gas collectors, which were opened within 1182-1273 meters. Their lid is made of from Pliocene and limestone sand Quaternary system gas-proof rocks.

9. KULEVI

This area is to the northwest of the Chaladidi structure, in the area of the river Khobistskali mouth 20 km from Poti. Among the Pontain-Meotic sediments the Kulevi anticline is considered as the main structural and tectonic unit. It is also in (II1) west molassa zone of sinking – Rioni intermountain cavity, in Ochamchire-Kulevi block (3), which P. Gamkrelidze separates as sub-zone of Kolkheti – in the most sunk part of the Georgian block to the West. The tectonic structure of this zone is not yet sufficiently explored. The Neogenic and Cretaceous sediments are not clearly dislocated. The blanket anticlines, which are more northwest oriented, (Kulevi, Chaladidi, Kvaloni etc) are identified. In the upper part of the Meotic sandstone packs are situated on a 1050-1135 meter depth from the surface, and they are probably 1200-1500 meter deep in the Meotic basal part. The rocks with corresponding litological and structural features are regarded as collectors. A gas-proof screen is made up from the same rocks as in the Chaladidi UGS area, i.e. Pliocene and limestone sand quarterly system rocks. There are no recommended UGS facility structural measurements, though the depth of collector location is 1100-1300 m. The capacity of the UGS facility is probably up to 1 bcm.

Based on the data bank Georgian geologists do not exclude other areas than those described in this survey. Similar sites can be identified in various parts of the country. For instance has the Dgvebi anticline (in the Khobi region) not been researched [4]. It would be desirable in future projects that all geological institutions, authorities and relevant bodies participate actively in surveying the natural structures deemed suitable for UGS. They definitely have the needed potential.

REFERENCES

1. Rogava D., On Construction of UGSs in Georgia. Mining Journal. #1 (6), Tbilisi, 2001.
2. Mining Eniclopedia. V.1, Moscow, 1984.
3. Preliminary Technical Project of UGS in Southern Dome of Samgori, Republic of Georgia. Paper. Undergrungspeicher- und Geotechnologic – Systeme Gmbh. April, 1992.
4. Survey of Using Republic of Georgia Hydrogen Energy and Technology in Fuel and Energy Complex within 1991-2010. Underground Storage of Hydrogen and Oxygen. Author Mobitski A.K. Tbilgidroproekt, 1991.
5. Identification and Exploration Works for Underground Storage of Gas in USSR. Issue 10, Moscow, Nedra Publishing. 1971.

6. Likholatnikov V.M. Structural and Tectonic Peculiarities of Western Part of Kolketi Lowland with the Aim of Holistic Evaluation of Perspectives of Oil and Gas Capacities and the Problem of Construction of UGS. Postgraduate Dissertation Abstract. Moscow, 1976.
7. Gochitashvili T. Perspectives of Main Gas Pipelines and Gas Transit. Tbilisi, 1994.
8. Rogava D. On the Construction of UGS in Georgia. International Scientific and Engineering Conference. "New Technologies and Georgia", Transactions of Section of Mining and Geology. Tbilisi, 2002.
9. Gamkrelidze I. P. Technical Structure and Alpine Geodynamics of the Caucasus. Tectonics and Metallogenia of the Caucasus. Transactions of Geological Institute of the Academy of Sciences of Geology. New Series, Issue 86, Tbilisi, 1984.
10. Gamkrelidze I. P. Once More on the Tectonic Zonning of the Territory of Georgia. Materials of the Scientific Session on the 110th Birth Anniversary of Djanelidze A.I. Tbilisi, 2000.
11. Papava D., Jashi O., Takaishvili A. On Some Issues of Samgori-Patardzeuli oil depository Depth Structure. Georgian Oil and Gas. #1 (4), Tbilisi, 2001.
12. Beraia G., Oviani B., Sidamonidze M. Increase of Oil Exploration on Middle Eocene Depositories and Enhancement of Construction Processes of UGSs by Implementing New Technologies. Georgian Oil and Gas. # 2 (6), Tbilisi, 2002.

GEOTECHNICAL SURVEY OPTIMIZATION FOR PIPELINE CONSTRUCTION IN HIGH MOUNTAINOUS REGIONS

Omar KUTSNASHVILI and Iveri KUTSNASHVILI
Omar Kutsnasvhiliv: the Georgian Technical University and GeoEngineering LLC. Iveri Kutsnashvili: GeoEngineering LLC

Abstract: This article summarizes years of scientific and practical experience in geotechnical and other survey work for the optimisation of main pipelines in high mountainous regions. It also demonstrates the strong interdependence of environmental and technical components of the pipelines.

Specialists will find useful information regarding the characteristics of the survey; design, construction and operation of main pipelines that are crossing high mountains as far as climate and intensive spread of dangerous geodynamic processes are concerned. In particular, in contrast to the discretional method, the article underlines the importance of the continuous engineering method in order to obtain appropriate information while conducting geotechnical surveys.

Key words: Geotechnical mapping, Environment-facility Binary system, Continuous engineering

1. INTRODUCTION

Data covering the design, construction and operation practice of the Trans Caucasian main gas pipeline have been collected and accumulated since the beginning of the 1980-ies. The pipeline design was drafted by the Kiev Survey-Design Institute without properly considering the complex geotechnical conditions, and the possible adverse consequences. Rather

J. Hetland and T. Gochitashvili (eds.),
Security of Natural Gas Supply through Transit Countries, 247–270.

traditional approaches were applied. Site study results were addressed for the selection of crossings of rivers, roads and the like. Design facilities – such as compressor stations and the deployment of valves along the gas pipeline were addressed at the construction site. Furthermore, no longitudinal geotechnical profile of the pipeline route was drafted, neither was any in-depth geotechnical mapping provided at detailed scale (1:5000-1:2000).

Owing to this significant methodological weakness instability and erosion became apparent during the construction phase. A perilous geological process got started, and a critical deterioration of the geo-ecological condition developed as soon as the construction was accomplished. In brief: 30 new landslide formations were generated, erosion was observed at 60 sites, land plots were destroyed on 850 hectares that required urgent reinforcement and rectification [1].

The inaccurate pipeline design and construction was followed by dangerous geological events that resulted in public complains. For this reason, the government was obliged to carry out additional survey and relevant improvement measures.

From 1987-88 the geological events escalated, as landslides, avalanches and mudflow are observed every year – sometimes entailing catastrophic outcome. For example in 1987 the pipeline was cut off in the Baidari ridge under the pressure of a 1.3 million m^3 mudflow. In 1993 a 5 million m^3 landslide cut the 1200 and the 700 mm diameter gas pipelines at the village Geneso and instigated fire.

In 1996 an adverse mudflow affected the pipeline at the river Tergi, in the Gveleti area, and resulted in pipe exposure of the main gas pipeline in several areas. In 1999 the mudflow strongly damaged the 1200 mm pipeline in the Kazbegi region. And, in the year 2000 the mudflow provoked a pipe exposure of the 1200 mm pipeline in the Kuro Ridge that was buried at a depth of 8 meter, and revealed a scouring and protection design failure.

Generally, a distance of approximately 200 km of the Trans Caucasian main gas pipeline is considered a dangerous zone in terms of geodynamic events. Currently the complex geodynamic conditions are observed at the river Amala, Baidar and the Kuro area (in the Kazbegi region), where intense erosion and mudflow occurrences are present.

At a distance of 100-120m the pipeline is deemed dangerous in terms of linear erosion that is apparent at the steep slope north-east of the Jvari peak (2246 m above sea level) in the vicinity of the Jvari Pass. In addition, the area from the Kaishauri Plateau (the Dusheti region) towards the river Tetri Aragvi and the river Khadonistzkali is located in the erosion-tended area. Above the village Kvesheti a 700m gas pipeline crosses a mudflow formation. At the village Chiriki, the lateral erosion along the river Tetri Aragvi has generated an instable slope that provoked landslide at a distance of 90-100m. In the vicinity of the village Patara Pasanauri the river Aragvi washed out the construction line, and formed of colluvial material accompanied by a pipeline exposure. In addition a small landslide was generated at the construction line. At the village Kvavili the pipeline passes through a potential landslide formation that may cause movements that may cause adverse incidents.

Thus, the Trans Caucasian main gas pipeline is located in a highly hazardous zone that is in need of urgent improvements.

Difficulties created during the pipeline construction and the operation of the Georgian section of the North-South Caucasus Trans-Caucasian main gas pipeline justify the urgency of applying an improved geotechnical survey methodology.

2. ENVIRONMENTAL CONDITIONS AND MAJOR TECHNICAL PARAMETERS OF MAIN GAS PIPELINES IN HIGH MOUNTAINOUS REGIONS

The construction of a main pipeline in terms of complexity and reliable operation is stipulated by technical solutions that are adjusted to environmental factors characterized by:

- Diverse natural conditions that are typical of high mountainous regions combined with seasonal fluctuations, and
- The interference of the construction, installation and operation with the environment, and the consequent need for environment-facility synchronization and optimisation.

Factually, pipeline construction requires accurate information with regard to the natural-techtogenic binary system, comprising the natural environment and the pipeline, for the design and construction. This also includes geotechnical data, as well as process optimisation for the selection of technical and technological parameters that become vital for the normal and safe operation of the pipeline.

The following physical environmental components of the main pipeline construction contribute to the complexity of nature:

- Physical-geographical conditions;
- Geological structure, and seismology and tectonics;
- Geomorphologic and geotechnical peculiarities.

A brief description is provided below of the environmental factors that pertain to the main pipelines in Georgia – either existent or pipelines under construction (such as the Baku-Tbilisi-Ceyhan Main Export Oil Pipeline and the South Caucasian Gas Pipeline system) (Table 1 and 2).

The analyses of the tables reflect the complexity, the diversity of components and the fluctuating natural conditions, as well as the characteristics of the high mountainous folded Alpine system. The Caucasus and its central part, Georgia, is a typical representative of such mountainous system.

At existing main pipeline routes, and pipelines under construction, environment complexity is conditioned on the high fluctuation of physical-geographical, geological, seismic, tectonic, geomorphologic, and geotechnical factors. The selection of technical and technological parameters plays a decisive role in the pipeline construction, and also in the design of environmental protection and the engineering-infrastructure that are vital for the normal operability of the pipeline.

Table 1 Trans-Caucasian main gas pipe physical environment characteristics.

	Physical environment characteristics		**ROUTE SECTIONS**					
			Gardabani	***Saguramo***	***Mleta***	***Jvari passing***	***Kazbegi***	***Vladikavkaz***
1	***2***		***3***	***4***	***5***	***6***	***7***	***8***
Physical-geographic conditions.	Hypsometric elevation m		265	510	1580	2395	1744	668
	Precipitation average annual mm		422	636	1315	1733	786	837
	Snow cover winter maximal cm		In winter time 98% snow cover is not generated	6	202	334	96	16
	Air temperatureC^0	Average	12.9	10.8	6.7	-0.2	4.9	10.2
		Absolute minimum	-25	-29	-30	-38	-34	-28
		Absolute maximum	41	39	32	27	32	38
	Climate		The area is characterized by normally warm air with hot summer and two minimums of sedimentation.	The area is characterized by normally humid air, with two minimums of normally cold winter and summer sedimentation	The area is characterized by normally humid air, with cold winter and short-term summer, with minimal sedimentation in winter.	The area is characterized by high mountainous, normal humid air, with absence of real summer	The area is characterized by normally humid air, with cold winter and short-term summer, with minimal sedimentation in winter.	Normally humid air is characteristic of the area

	Physical environment characteristics	ROUTE SECTIONS					
		Gardabani	*Saguramo*	*Mleta*	*Jvari passing*	*Kazbegi*	*Vladikavkaz*
1	2	3	*4*	5	*6*	*7*	*8*
	Vegetation cover	Vegetation is characterized by eastern Georgian plain field, chestnut soils, with salty soil complex.	Vegetation is characterized by bushy field with forest elements. Here brown soil of eastern Georgian forest lower belt is observed.	Cover is presented by Eastern Georgian sub-alpic forests, with soils of forest upper belt	Cover is presented by Eastern Georgian sub-alpic field with cord mountainous-field soils. .	Cover is presented by mountain pine and cord mountainous-field soils.	Vegetation is presented by alluvial and field soil.
Geological structure and seismotechtonic.	Geo-structural – morphological units	According to geo-structural – morphological division, the region belongs to Artvin-Bolnisi zones (belts) Bolnisi lower zone Marneuli block (weakly folded sediment layer)	Area is Caucasian mid mountainous with molasses eastern sinking zone.	Caucasian folded system is characterized for this area, Mestia-Tianeti zone, Shovi-Fasanayri lower zone with strongly compressed, south-directed isoclinals dissection and overthrusted folds.	Area is characterized by Caucasian folded system; Kazbegi-Lagodekhi zone is (folded) south directed with lightly isoclinals folds.	Area is characterized by Caucasian folded system. Kazbegi-Lagodekhi zone (folded) layers in Tergi gorge are upside down to south.	Caucasian anticlinorium's Tergi-Kaspi side depression is noted here.
	Stratigraphy, lithology	Late Pleistocene alluvial pebble bed, loams and clays (Q_3) are presented here	In this area Miopleocene and quaternary clays, sandstone conglomerates interchange can be noticed.	Lower volcanic age Mleta-Black marl set, bituminised, partly layered clay slates and slated marls, carbonate sand stones thin mid-layers are observed in this area.	Lower Lavajin (Mleta set) black marl set, slated marls, marl slates and Kimerij- Titonic (Tzipori set) layered lamestoneare are present in this area.	U.Liakhauri (Kazbegi set) black clay lined slides can be observed in this area with quarts like sand stone frequent mid-layers as well as mid liase (Tziklavri set)-clay slate, rarely thin layered, fine grained Quartz-Arkoze sand stone mid-layers.	Quaternary alluvial sedimentation is apparent here

	Physical environment characteristics	ROUTE SECTIONS					
		Gardabani	*Saguramo*	*Mleta*	*Jvari passing*	*Kazbegi*	*Vladikavkaz*
1	*2*	*3*	*4*	*5*	*6*	*7*	*8*
	Neothechnic developments	Tectonic depression is apparent here filled with sea - continental sediments sunk in the Pliocene – quaternary layer to more than 1000 m.	Intermountain depression filled with Pliocene quaternary continental sediments were sunk in the Pliocene –quaternary layer to more than 1000m.	Mostly secondary and Paleogennic structures were lifted to more than 3000 m in the neotechtonic period can be noted here.	Mostly secondary and pre-secondary structures were lifted to more than 3000 m in the neotechtonic period.	Mostly secondary and pre-secondary structures were lifted to more than 4000 m in the neotechtonic period.	Continental fractions were sunk to more than2000m in the neotechtonic period.
Geological structure and seismotechtonics.	Seismotechtonics	Due to relatively stable substratum of Artvin-Bolnisi (belt) zone macro seismic intensity of the region is characterized by potential intensity of 7, with expectancy time 50 sec, exceeding probability 2%. .	Due to proximity to Achara-Trialeti fold system seismogenerating north edge depth erosion, the region macro seismic intensity is characterized by potential magnitude of 8	Region macroseicmic intensity is characterized by potential magnitude of 8.	Due to seismogenerating deep erosion region is characterized by potential magnitude of 8.	Due to surface passing Caucasian major seismogenerating depth folds within the region from standpoint of anticipated seismic intensity it belongs to the potential magnitude of 9.	Region macro seismic intensity is characterized by potential magnitude of 8
Geomorphologic and geotechnical characteristics	Geomorphologic unit	According to the geomorphologic division area belongs to Caucasian mid-mountain Kvemo Kartli plain region, -Gardabani- Marneuli plain sub region	According to the geomorphologic division area belongs to Caucasian intermountain inner Kartli plain (Mukhrani plain)	According to the geomorphologic division area belongs to East Caucasian flysch line	According to the geomorphologic division area belongs to high mountainous slate line	According to the geomorphologic division area belongs to high mountainous slate line	According to the geomorphologic division area belongs to north Osetia inclined plain (400-700m absolute height)

	Physical environment characteristics	**ROUTE SECTIONS**					
		Gardabani	*Saguramo*	*Mleta*	*Jvari passing*	*Kazbegi*	*Vladikavkaz*
1	*2*	*3*	*4*	*5*	*6*	*7*	*8*
	Relief type	Relief is characterized by alluvial plain with relative sinking area.	Relief is characterized by alluvial plain with relative sinking area.	Relief is characterized by Mid and high mountainous erosive-denudation	Relief is presented by less apparent edge frosty, nival old frosty area.	Relief is presented by less apparent edge frosty, nival old frosty area.	Relief is presented by accumulative plain
	Geodynamic condition	Melioration caused swamping and clay soil settlement procedures are apparent in this region. .	Zone belongs to contemporary sinking and excessive accumulation with mudflow events apparent in this region. .	Zone belongs to contemporary lifting and excessive erosion/denudation with mudflow occurrences.	Zone belongs to contemporary lifting and excessive erosion/ denudation with carstal occurrences.	Zone belongs to contemporary lifting and excessive erosion/ denudation: with avalanche, landslide, stone fall, and mudflow.	River divagation and consequent lateral erosion is currently active.

Table 2 BTC and SCP Pipeline Georgia Section Physical Environment Characteristics

	Physical Environment Characteristics		**ROUTE SECTIONS**			
			Gardabani	***Bedeni-Tsalka***	***Tavkvetili-Tskhratskaro***	***Vale***
1	***2***		***3***	***4***	***5***	***6***
Physical-Geographical Conditions	Hypsometric Elevation, m		270	1915-1600	2352-2475	1230
	Average Annual Precipitation, mm		422	736	643	554
	Winter Maximum Snow Cover		-	46	89	76
	Atmosphere, C^0	Average	12.9	5.9	2.6	9.0
		Absolute Minimum	-25	-34	-39	-32
		Absolute Maximum	41	33	30	39
	Climate		The area is characterized by normally warm air with hot summer and two minimums of sedimentation	The area is characterized by cold winter and long summer transforming from normally humid to mountainous steppe	The area is characterized by mountain steppe atmosphere with short summer and cold normal snow sedimentation in winter	The area is characterized by mountain steppe drought atmosphere. Cold, with normal snow sedimentation in winter and long warm summer
	Vegetation cover, topsoil		Vegetation/cover is presented by Eastern Georgian Plainfield, chestnut soils, with salty soil complex.	Vegetation/cover is presented by high mountain valleys; With black topsoil and knobbly mountain ridge topsoil	Vegetation/cover is presented by sub-alpine plains; with black topsoil and knobbly ridge topsoil	Vegetation/cover is presented by secondary-steppe and forest steppe landscape; with brown topsoil

	Physical Environment Characteristics	ROUTE SECTIONS			
		Gardabani	*Bedeni-Tsalka*	*Tavkvetili-Tskhratskaro*	*Vale*
1	*2*	*3*	*4*	*5*	*6*
Geological structure and seismotechtonic.	Geostructural – Morphological Units	According to the geostructural morphological division area belongs to Artvin-Bolnisi zones (belts) Bolnisi lower zone Marneuli block (weakly folded sediment layer)	According to the geostructural morphological division area belongs to Aspindza-Trialeti zone south sub zone, Aspindza-Tbilisi sector	According to the geostructural morphological division area belongs to Adjara-Trialeti zone south sub zone; Aspindza-Tbilisi sector	According to the geostructural morphological division area belongs to Adjara-Trialeti zone south sub zone; Akhaltsikhe sector
	Stratigraphy, Lithology	Late Pleistocene alluvial cobble loams and clays (Q_3 are presented here)	Here can be noticed quaternary continental formations and upper Pliocene-early Pleistocene dolerite lava flows located under them and volcanogenic-continental formation variations (tufo, tufo-like) [b, pк, al N_2^3-Q_1]	Upper Miocene-quaternary Epusives; Dolerites, Andesittes, Andesit-Basalt, Andesit-Datsit lava flows and piroclastilites are present here [aN_1^3-N_2^1; bN_2^3-Q_1; a_1hQ_4]	Oligocene sandstone and clay interchange can be apparent in this area.
	Neothechnic Developments	Tectonic depression is apparent here filled with sea-continental sediments sunk in the Pliocene –quaternary layer to more than 1000 m.	Cover folding in the upper Pliocene-early Pleistocene lavas; 2000m amplitude neotechtonic uplift	Cover folding in the Neogenic lava can be observed; 2000m amplitude neotechtonic uplift	Neogenic depression (500-1200m) under regional uplift; Differentiated sink area

	Physical Environment Characteristics	**ROUTE SECTIONS**			
		Gardabani	***Bedeni-Tsalka***	***Tavkvetili-Tskhratskaro***	***Vale***
1	***2***	***3***	***4***	***5***	***6***
Geological structure and seismotechtonics.	Seismotechtonics	Due to relatively stable substratum of Artvin-Bolnisi (belt) zone macro seismicity of the region is characterized by potential intensity of 7, with expectancy time 50 sec, exceeding probability 2%. .	Region macro seismic intensity according to the sub latitudinal direction seismogeneration Adjara-Trialeti south edge active depth erosion is characterized by potential intensity of 8, with expectancy time 50 sec, exceeding probability 2%.	On junction of the Abul-Samsakhra and Javakheti volcanic morphostructure seismogeneration erosions region macroseismicity is characterized by potential intensity of 8-9, with expectancy time 50 sec, exceeding probability 2%.	On Tskhinvali-Kazbegi across active siesmogeneration depth erosion the region macro seismic intensity is characterized by potential intensity of 8, with expectancy time 50 sec, exceeding probability 2%.
Geomorphologic and geotechnical characteristics	Geomorphology	According to the geomorphologic division area belongs to Caucasian mid-mountain Kvemo Kartli plain region, Gardabani- Marneuli plain sub region	According to the geomorphologic division area belongs to the Beshtasheni (Tsalka) trench	According to the geomorphologic division area belongs to South Georgia volcanic plain Tabatskuri sub region	According to geomorphologic division area Akhaltsikhe trench
	Relief type	Relief is characterized by Alluvial plain with relative sinking region.	Relief is characterized by ridges and trenches	Relief is characterized by volcanic-techtonomorphic; fold, nivelic, with ridges and gorges	Relief is characterized by low mountainous-knob erosion

	Physical Environment Characteristics	**ROUTE SECTIONS**			
		Gardabani	***Bedeni-Tsalka***	***Tavkvetili-Tskhratskaro***	***Vale***
1	***2***	***3***	***4***	***5***	***6***
	Geodynamic Condition	Melioration caused swamping and clay soil settlement procedures are present here.	Intensive physical weathering and depth erosion is present here	Physical weathering is intensive, with landslides, talus and other gravitational developments	Intensive weathering. With landslides, erosion and other developments can be observed here.

3. BAKU-TBILISI-CEYHAN

The Baku-Tbilisi-Ceyhan pipeline, which is under construction (2003), starts from the Sangachal Terminal, adjacent to Baku. It is designed to transport 50 million tons of crude oil per year through the new Ceyhan terminal at the Mediterranean Sea. The pipeline goes through the territory of Azerbaijan, Georgia and Turkey with a total length of 1750 km. The length of the pipeline in Azerbaijan accounts for 442 km, in Georgia 248 km, and in Turkey 1060 km. The pipeline is scheduled to go on stream at the end of 2004. The pipe diameter amounts to 42" (1066.8mm) in Azerbaijan and Turkey, whereas the diameter in Georgia is increased to 46" (1168.4mm) owing to the high hypsometric elevations (ranging from 270 m to 2475m) due to high pressures. The Georgian section of the Baku-Tbilisi-Ceyhan pipeline consists of the following facilities:

- Two pump stations;
- Discharge and receiving stations;
- One measurement station,
- 27 valve stations;
- Cathodic protection system;
- SCADA;
- Temporary facilities such as camp bases for workers, storage areas, and waste placement areas (etc.)

4. THE SOUTH CAUCASUS PIPELINE SYSTEM

The South Caucasus Pipeline system is designed for transportation of 7.3 billion m^3 per annum of gas from the Sangachali Terminal located near Baku (Azerbaijan) via Georgia to Erzrum through the Turkish gas distribution network for domestic consumption in Turkey. The total length of the South Caucasus Pipeline System is to be 690 km. In Georgia a new pipeline will be constructed in parallel with the Baku-Tbilisi-Ceyhan pipeline on the common construction line. The length of the 42" (1066.8 mm) pipe amounts to 248 km in Georgia and 443 km in Azerbaijan. Besides the pipe the South-Caucasus-Pipeline project employs the following facilities and infrastructure [7]:

- One compression and measurement station;
- Three valve stations;
- Natural gas distribution unit;
- Cathodic protection system;

- SCADA;
- Supplementary facilities required for the construction.

The Baku-Tbilisi-Ceyhan and the South-Caucasus-Pipeline projects will be arranged in a 22-meter wide construction corridor for two parallel pipelines buried in a 2.2-meter deep trench. This alone illustrates the complexity of facilitating pipeline systems in regards of environmental issues and the complicated morphological mountainous relief.

The complexity of the binary system (i.e. the environment-pipeline) plays a significant role in identifying the scope and method of an appropriate geotechnical survey that is crucial for the design and construction of the pipeline system. Obviously, the more complex the system, the more extensive the required survey-design works. Thus the scope of the main pipeline survey-design methodology and the optimisation of the scale are of significant economic importance to mountainous regions.

A more accurate way of addressing environment-facility interdependence is to accurately forecast the binary system operational regime. Systems dynamics is determined by the interrelation of forces existing in natural conditions and the forces generated during construction and operation of the pipeline and its related facilities. The interrelation of these forces enables the binary system to operate both in stable and instable conditions. A stable binary system excludes the generation of dangerous events that represent a real threat to the safety of the pipeline when no geotechnical fluctuating condition is observed or the fluctuation is of such a minor extent that it cannot impact on the construction activities or the design stability. According to available data, it has become evident that there are no sufficiently stable binary systems relating to the main pipelines in Caucasus.

The NSC North-South Caucasus (further referred as NSC) Gas Pipeline can serve as a good example for an instable binary system, as its construction and operation, as already mentioned, was accompanied by rather dangerous geodynamic events and pipe damage. Circumstances like this require implementation of specific engineering activities during operation of the NSC that can ensure the balance of the binary system represented by the 'environment-and-pipeline', and its operation in an optimal regime. Also the engineering survey activities should be performed with adequate consideration of the physical environment structure, stability, geotechnical parameters, construction and the operation condition of the facilities.

Three possible strategies could be applied to manage an instable binary system:

1. **Improvement of the physical environment**, its essential realignment or creation of a completely new environment by applying modern melioration methods;
2. **Re-adjustment of the design** and re-alignment of the pipeline route and facilities to the existing physical environment structural, organization and dynamics;
3. **Applying a combined strategy** of the physical environment, design re-alignment and improvement.

Geotechnical surveys aim is to optimise the binary system management. To a great extent a correct decision of the task would depend on the rational selection of the survey method and the optimal identification of the scope of the survey, and also on the design, construction, and operation at different stages.

5. BASIC PRINCIPLES FOR A GEOTECHNICAL SURVEY IMPROVEMENT

The requirement for an essentially new and modern geotechnical survey can be interpreted as developing a consistent scientific conception in consideration of both modern scientific achievements and experience built up worldwide from main pipelines construction projects.

Key questions are:

- What are the main weaknesses of the existing geotechnical survey methodology, and
- Which new ideas and scientific concepts can be considered as bases for new conception?

One response to the first key question is the following:

- The traditional geotechnical survey is characterized by discrete information. It means that old survey methods cannot provide comprehensive data on the geological structure, substance composition and properties of the whole pipeline route. This principle is not observed in all the three main gas pipeline cases. The technical documentation of the Baku-Tbilisi-Ceyhan pipeline, the South-Caucasus Pipeline system, and particularly of the North-South Caucasus Gas Pipeline does not include longitudinal geotechnical profiles of the route and the pipeline corridor in the geotechnical maps. Trial pits and boreholes that are arranged approximately 1,5-

2,0 km apart from each other, cannot provide accurate information on the fluctuating geotechnical conditions.

- Standards and method statements envisage geotechnical survey performance only in the pre-construction period that leads to insufficient forecasting of the geotechnical conditions for the main pipeline construction in mountainous areas. Therefore accurate quality assessment of the survey is excluded which is crucial for the mitigation of adequate environmental risks in the design and development phase.
- The traditional geotechnical survey does not pay adequate attention to the natural environment protection from an adverse impact during pipeline construction and the subsequent operation period. The reason is often found to be some artificial improvement of the economic parameters at the expense of a possible negative ecological impact. The Trans Caucasian Gas Pipeline can serve as a good example for the underestimated ecological condition and the possible changes.
- Geotechnical surveys are often executed without sufficient detailed scientific research. Sometimes also the pipeline design and the condition assessment for construction are rather insufficient.
- Standards and method statements governing pipeline design and construction do not cover clearly the geotechnical work, methods and the scope by the various stages through design and construction.
- No comprehensive geotechnical classification for pipeline construction in mountainous areas exists that could sufficiently address various aspects for the development of the binary system comprising the 'environment-and-pipeline'. This would include quantitative and qualitative characteristics of both the environment and the facilities.

First of all a further development and optimisation of the geotechnical surveys would depend on the appropriate solution of the above mentioned issues aimed at rectifying the deficiencies. The conception set out in this article is based on such concepts that will facilitate the correction of the flaws and to further develop and optimise the geotechnical surveys. Accordingly, the following suggestions are offered:

1. The geotechnical surveys for the design and construction of main pipelines should provide comprehensive data on geological structures, properties and composition of substances of the entire pipeline route. This can be accomplished by applying a "continuous engineering" method which will be explained in the next paragraph.

2. The geotechnical work required for pipeline construction in high mountainous regions is considered to be a unified technological process of survey-research. The investigations are carried out throughout the lifetime of the pipeline, through design, construction and operation. It will ensure the necessary prognostic information on the structure and the characteristics of the natural dynamics of the environment and the pipeline. The sole purpose of these investigations is to prevent the construction and operation interruptions as well as the negative ecological changes in the pipeline area.
3. In contrast to the traditional methods the geotechnical investigations should be aimed at positively solving environmental protection objectives. Only this approach can ensure a successful solution of all the major geotechnical objectives during the design and construction of the pipeline.
4. The geotechnical investigations for the main pipelines in typical mountainous areas should be sufficiently comprehensive. They should be based on the analysis of information that is derived as a result of geotechnical investigations and mapping. The assessment and the prognoses should become a necessary part of the engineering surveys and also be addressed in the recommendations.
5. Depending on the design and project development the geotechnical investigations should develop from an overall regional assessment to a concrete local assessment with increased accuracy and detailed level.
6. The complexity of the geotechnical conditions should be assessed according to a binary system approach. Based on appropriate environmental and design parameters a similar tactic should be applied for the selection of the design and solutions. It is important to note that the results of the geotechnical assessments will be probabilistic. Therefore geotechnical monitoring and/or observation can only achieve the comprehensive information that is necessary to manage the stability of the binary system.

6. DISTINCTIVE FEATURES OF GEOTECHNICAL SURVEYS AT DIFFERENT STAGES OF A MAIN PIPELINE PROJECT DEVELOPMENT

The logical model of geotechnical surveys to be performed at different development stages of a main pipeline project is given in Figure 1. The main distinctive feature of the model is that the geotechnical survey is considered as a complete technological process that is performed during the entire period of or the pipeline project - through planning, design, construction and operation. The project development is divided in five stages each of which contains a survey component. The geotechnical surveys comprise conceptual, basic, detailed, pre-construction and monitoring survey phases. The project development includes phases of conceptual, feasibility, front-end engineering and design (FEED), and also construction and operation.

Direct links exist as well to feedback between the different survey stages and the relevant stages of project development. Forecast information that is received at each survey stage represents the basis for the design, construction and operation tasks.

Corresponding to the principle of targeted requirements the content and volume of the geotechnical information that will be received at each phase of survey should strictly respond to the main engendering objectives to be solved at the given stage, namely:

- At **Stage I** conceptual surveys are carried out in order to prepare the conceptual design. The particular objective is to identify the general pattern of the geotechnical conditions for the pipeline construction internationally, in a country or within a certain region, and to make estimation of their possible change in order to select and to make an optimal option for a 10 km wide corridor of interest.
  The task to be solved is mainly accomplished by analysing the literature and archive documents available. At the same time the reconnaissance of the geotechnical surveys are essential. This allows for accurate selection of the most appropriate area for a pipeline. At this stage the accuracy of the total cost estimate for the pipeline construction is within ±30-50%.

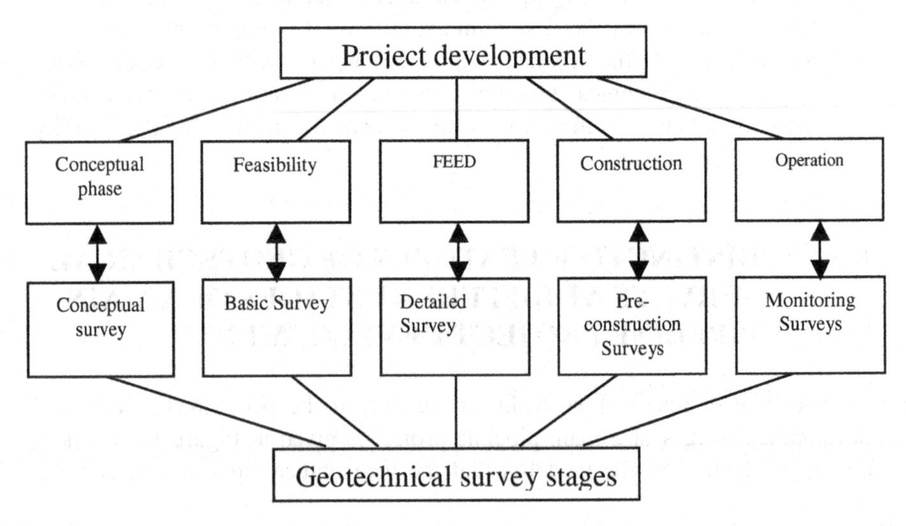

Figure 1 Geotechnical survey stages in the project development

- At **Stage II**, a feasibility study combined with general engineering surveys are conducted with the main objective of selecting the best alternative routes with a corridor of 500 m width within the 10 km wide corridor of interest as identified at Stage I.
  Such preliminary surveys should take into consideration the geotechnical mapping of the construction area as well. At this stage the construction cost would be determined with a total inaccuracy of 15-25%.
- At **Stage III** a detailed design will be developed based on detailed engineering surveys carried out in order to select a 100 m wide corridor within the previous 500 m wide corridor from Stage II.
  This stage is a very important pre-construction phase. Therefore the surveys carried out at this stage are rather comprehensive, and would include large-scale geotechnical mapping, arrangement of boreholes and trial pits of significant volume, filed geophysical, pilot and regime observation; examination of physical environment components. On the basis of these activities the location of the pipe and its related facilities at the area – as identified at Stage II – and determination of technological parameters that are needed for construction. Furthermore, operational conditions for the project and its envisaged binary system are determined, and costs and construction period are identified with an inaccuracy of ±10%. At the same stage all measures are verified to ensure a smooth operation of the facility, including long-term stability and ecologically safe operation.
- At **Stage IV** the pipeline construction starts for which the relevant work documentation is prepared based on additional engineering surveys conducted in a 22 m wide construction corridor.
  At this stage the exact coordinates for the construction facilities are finally determined along with such details of geotechnical and technical solution that can affect the organisation of the construction work and the operational stability of the binary system.
  As usual, the pre-construction survey stage coincides with the construction period. In order to check the issued geotechnical decisions and the recommendations the following should be envisaged for control purposes: authors' supervision, geotechnical documentation of trenches, construction trenches and other boreholes and trial pits as well as the regime and survey works in order to accurately identify the engendering construction activities.

- **Stage V** denotes the operation that includes a long period (decades) during which usually monitoring surveys are carried out aimed at optimised planning and implementation of preventive measures in order to ensure a smooth functioning of the pipeline.

At each stage of the project development the sequence and scope of the survey and the components consisting geotechnical conditions for the pipeline construction will be determined by taking into consideration the above-mentioned theoretical assumptions. While solving the issue the geotechnical conditions for the pipeline construction represent a determined common system of such interrelated system-generating components as:

- geographic,
- geomorphologic,
- hydro geological conditions,
- geological structure,
- seismicity,
- current tectonic movements,
- dissection degree of soils and physical-mechanical properties,
- geological processes and developments,
- inert materials deposits and water supply sources.

These components should be studied in a certain technological sequence with the observation of the general methodological principle, i.e. from the general towards the specific. The scope of the geological and environmental components, and the related accuracy, vary at the different stages depending on the requirements that are determined according to each engineering assignment. In addition, by considering the surveys conducted for pipeline construction in high mountainous regions, the following issues need to be taken into consideration:

- All components of physical environment are to undergo an in-depth study at each geotechnical survey stage
- For a studied area the volume, accuracy and quality of the survey at similar natural conditions are strictly in line with the geotechnical survey stages and the consequent tasks. The studied area is narrowing from a conceptual and feasibility survey to a monitoring survey. As for the physical environment, here the detail level of the in-depth study will increase.

The physical environment structure and the properties determining the pipeline construction require adequate survey and method application. The table 3 provides multiple surveys for various stages. This scheme reveals the library material and literature study, relevant experience generalization, and constant office study, and acquires a key role at each geotechnical survey stage. In addition the whole material is subject to expert approval.

6.1 Scheme illustrating geotechnical activities at various construction stages required for the pipeline construction

From the geotechnical activities as proposed in table 3, the geotechnical mapping suggests a new method for providing complex combined data derived from the applied information. (A detailed description of this method could be obtained from scientific research papers published by O. Kutsnashvili and B. Iurovski (see reference list) Due to the complex relief dissection and rock exposure the geotechnical mapping in high mountains and hardly accessible areas actually represents one of the most effective forms of geotechnical surveys. Geophysical surveys and aerial-space photo materials complement the geological mapping in term of acquiring relevant, technically and economically justified information for the design and construction of pipelines.

Geotechnical mapping, as a constant geological environment-related information source the effectiveness is much dependent on the scale of mapping and the area selection. The scale is selected in line with the project development stage, the complexity of the geotechnical condition and the study area. The most rational high-scale feasibility stage is considered (1:25000-1:10000), while for detailed design a much more detailed geotechnical mapping is required, hence a scale of 1:5000 or 1:2000 is suggested. For a draft construction the survey documentation requires a rather complex and local area mapping at scale 1:500.

The territories for the geotechnical mapping are to be selected in consideration of the geological events formation, their development and the localization zones that affect the construction in particular areas.

Table 3 Geotechnical survey stages

Types of Geotechnical activities	***Geotechnical survey stages***				
	Conceptual	**Feasibility**	**Detailed**	**Construction**	**Monitoring**
Existing library, design/constructive and scientific material obtaining					
Main pipeline design, construction and operation study and generalization					
Geotechnical reconnaissance					
Geotechnical mapping					
Borehole drilling, trial pit arrangement and geotechnical investigation					
Geophysical survey					
Field trial activities (dynamic and static probing, plate load test, compaction quality control, etc)					
Rocks, soil and water composition and properties determination under laboratory investigation					
Regime stationary observations					
Inert construction material and power supply sources determination.					

Types of Geotechnical activities	Geotechnical survey stages				
	Conceptual	Feasibility	Detailed	Construction	Monitoring
determination.					
Field and laboratory survey data office study, geotechnical reports, conclusions and recommendations preparation					
Scientific research					
Expertise					

Note: Grey-marked area size characterise certain work relative volume at each survey stage.

7. CONCLUSIONS

The importance of geotechnical surveys is justified by the construction and operation experience of main pipelines – especially in high mountainous regions.

This article proposes the optimal scope and the scale of geotechnical investigations for each and every stage of project development pertaining to main pipelines. Focus is made on the vitality of geotechnical mapping as a method for receiving integral and constant information that is to be applied during the construction of the main pipeline.

Some methodological adjustments and improvements are proposed in the survey standards that apply to pipeline design, and pipeline construction in high mountainous areas in order to ensure safe, reliable and environmentally friendly construction and operation.

REFERENCES

1. E.Gamkrelidze, Sh.Javakhishvili. Map for Georgian Seismic Risk, Tbilisi 2000
2. L.Varazashvili, O.Kutsnashvili, E.Tzereteli, Z.Tatashidze. Problems of natural-technologic hazards sensitivity decrease in Eurasian Transport System functioning area", Moscow general regional conference "Assessment of Natural Risk Management" 2000, P 5.
3. O.Kutsnashvili. "Bases for Geotechnical Survey" Tbilisi. 1997. 76 pages.
4. O.Kutsnashvili. New general rules of engineering geology, Material of Georgian Technical University Scientific Technical Conference, Tbilisi 1993
5. L.Maruashvili. Scheme for Georgian Geomorphologic dissection. Tbilisi 1969
6. Georgian SSR atlas Tbilisi 1964.
7. Environmental and Social Impact Assessment. Baku-Tbilisi-Ceyhan: Georgia. Non-technical guiding summary. Project documentation for public advertising April 2002.
8. Environmental and Social Impact Assessment. South Caucasian pipeline Georgia. Non-technical guiding summary. Project documentation for public advertising April 2002.

PART IV: ECONOMICS AND LEGISLATIVE ASPECTS OF GAS TRANSPORT AND GEOPOLITICS

PROSPECTS FOR GAS SUPPLY AND DEMAND AND THEIR IMPLICATION WITH REFERENCE TO TRANSIT COUNTRIES AND THEIR POLICY

- drawing upon recent experiences from European countries

Boyko NITZOV
Senior Expert (Investment), Energy Charter Secretariat[1]

Abstract: *This paper examines some risks that affect decisions for the construction and operation of major transborder pipelines in the context of Eurasia, including the Caspian Sea and Transcaucasia. An attempt is made to provide highlights on the linkages between microeconomic and macroeconomic determinants of such undertakings, as well as ways and means for the development of policies conductive to the implementation of projects of such nature.*

Key words: Oil and gas pipeline economics, pipeline investment risks, risk mitigation strategies, pipeline transit countries' policies

1. GAS SUPPLY: FACTORS AFFECTING THE WELLHEAD COST OF GAS

1.1 Finding costs

Finding costs are exploration and development expenditures in current dollars (excluding expenditures for proved acreage) divided by reserve additions (excluding net purchases). When reported, finding costs for gas

[1] The views expressed in this paper are those of the author and not necessarily of the Energy Charter Secretariat.

J. Hetland and T. Gochitashvili (eds.),
Security of Natural Gas Supply through Transit Countries, 273–306.

are converted to barrels of oil equivalent on the basis of 0.178 barrels of oil per thousand cubic feet of gas [1].

In essence, finding costs are an indicator of how expensive it is to replace reserves in proved acreage, or "proved reserves". These are commonly defined as oil and gas that have been demonstrated to exist beyond reasonable doubt and can be economically extracted with existing technology and at the prevailing market conditions. Besides, "proved reserves" may refer, in addition to "conventional" crude oil and natural gas, to natural gas liquids, heavy oil and certain other types of hydrocarbons, but generally exclude data on "unconventional" hydrocarbon resources, such as tar sands, shale, oil and gas from coal, deepwater and polar oil and gas, very heavy oil, etc. "Proved reserves" are, as a consequence, limited to conventional hydrocarbons: oil that is easily movable and can be produced and refined with relative simplicity, and natural gas that is mostly methane; both must be relatively free of unwanted or harmful substances, such as dissolved salt, heavy metals, carbon dioxide, hydrogen sulfide, etc. As far as finding costs are concerned, they are related to the replacement cost of proved reserves *with reserves of similar properties* (for example, conventional hydrocarbons with conventional hydrocarbons).

1.2 Lifting costs

Lifting costs are "production costs", i.e. costs incurred to operate and maintain wells and related equipment and facilities, including depreciation and applicable operating costs of support equipment and facilities and other costs of operating and maintaining those wells and related equipment and facilities. They become part of the cost of oil and gas produced. Production costs include the following *sub-categories* of costs:

- Well operations and maintenance;
- Well work-overs;
- Operating fluid injection and improved recovery programs;
- Operating gas processing plants;
- Ad valorem taxes;
- Production or severance taxes;
- Other, including overhead.

The following are *examples* of production costs (lifting costs):

- Costs of labor to operate the wells and related equipment and facilities;
- Repair and maintenance costs;
- The costs of materials, supplies, and fuel consumed and services utilized in operating the wells and related equipment and facilities;

- The costs of property taxes and insurance applicable to proved properties and wells and related equipment and facilities;
- The costs of severance taxes.

1.3 Other costs

Depreciation, depletion, and amortization (DD&A) of capitalized acquisition, exploration, and development costs are not production costs, but also become part of the cost of oil and gas produced along with production (lifting) costs identified above[2]. Finding, lifting and development costs constitute the major elements of the cost of supply of hydrocarbons. Table 1 provides a sample of these costs as reported by a major company over 1998-2002.

1.4 Upstream supply factors

1.4.1 Cost of supply

It is obvious that an operating company has little control over many factors negatively affecting the cost of supply[3]. For example, host country government policies may influence effective tax rates, the cost of material needed to develop the field (especially import material where duties are levied), the terms of access to foreign exchange, the inflation rate in the host country, etc. Market prices of inputs such as labor, power, insurance, etc., may change over time in a way that negatively impacts cash flow. Upstream operations are also subject to specific risks related to the performance of the reservoir, well flow rates and some other geological and technical risks inherent to the petroleum industry. The combined effect of all these factors is graphically illustrated by the wide margin of short-run variation in the cost of supply. In the instance illustrated in Table 1, cost of supply varies between a low of $6.40 and a high of $7.80 per boe[4] for BP (median $6.60), and a low of $6.30 and a high of $9.40 per boe for other industry majors (median $7.65).

The long-run cost of supply is a topic beyond the limits of this paper. Suffice to mention that a debate, fueled by concerns about an impending

[2] *Cf.* Noresco glossary of terms.

[3] A "project risk" is the probability of an event that negatively impact the project cash flow. Greater cost of supply is one of these risks.

[4] Barrel of oil equivalent.

peak in conventional oil[5] production, the global environment and other issues of similar magnitude, is as intense as ever. From the point of view of cost of supply, the ground taken by one of the two main factions in the dispute[6] [2,3] boils down to an expectation for a rapid and irreversible increase in *both* the *cost* and the *price* of hydrocarbons, possibly already within ten years or so. The other faction believes that such a forecast is based on some wrong premises and does not properly account for a number of factors (e.g. advances in technology, demand-side factors, etc.). We only note here that, while there is little past evidence in support of the pessimistic view, emerging concerns have prompted the energy industry to endeavor in directions that provide greater flexibility and hedge against some risks in a would-be world with a declining supply of conventional hydrocarbons. By definition, these steps mean that a higher cost of supply is regarded as a quite realistic scenario.

Table 1: Cost of supply (an example)

			1998	1999	2000	2001	2002
Result and oil price							
Replacement cost operating result[7]		$ billion	3.56	7.28	15.71	14.5	12
BP average oil realizations [8]		$/bbl	12.1	16.74	26.63	22.5	22.69
Finding and development costs							
BP		$/boe	4.7	3.21	3.29	3.68	4.14
Range of other oil majors -	Maximum	$/boe	12.84	6.57	5.11	8.34	n/a
	Minimum	$/boe	3.17	2.86	3.21	4.34	n/a
Finding costs							
BP		$/boe	1.33	1.02	1.22	0.54	0.79
Range of other oil majors	Maximum	$/boe	6.55	2.99	2.8	3.76	n/a
	Minimum	$/boe	1.53	0.58	1.18	1.9	n/a
Lifting costs							
BP		$/boe	3.2	2.7	2.6	2.7	2.6
Range of other oil majors -	Maximum	$/boe	3.7	3.7	4	4.3	n/a
	Minimum	$/boe	2.6	2.4	2.6	2.6	n/a
Cost of supply[9]							

[5] Conventional oil excludes natural gas liquids, condensate, oil from coal and shale, bitumen, extra heavy (<10 API) and heavy oil (10-17.5 API), polar oil, deepwater oil (>500 m water depth), enhanced recovery, gas-to-liquids (GTL).

[6] For a sample of both views: Campbell, C.J.: Industry Urged to Watch for Regular Oil Production Peaks, Depletion Signals [2]; Lynch, M.C.: Petroleum Resources Pessimism Debunked in Hubbert Model and Hubbert Modelers' Assessment. Oil and Gas Journal, July 14, 2003 [3].

[7] Replacement cost operating result adjusted for special items and acquisition amortization.

[8] Crude oil and natural gas liquids.

[9] Cost of supply comprises exploration expense, lifting costs and depreciation adjusted for special items and acquisition amortization.

BP		\$/boe	7.8	6.4	6.4	6.6	7.3
Range of other oil majors -	Maximum	\$/boe	9.4	7.9	8.2	8.8	**n/a**
	Minimum	\$/boe	7.4	7.3	6.3	6.4	**n/a**

Source: BP (2003).

1.4.2 Supply: drivers and reserve requirements for long-range gas pipelines

Oil and gas reserves have similar distribution by size: few giant fields hold the bulk of hydrocarbons, while thousands of smaller ones contain just a fraction of reserves. Figure 3 shows gas reserves and reserve-to-production (R/P) ratios as of the end of 2001 by country. Potential piped gas exporting countries are generally those in tiers 1-3, with exportable reserves exceeding 0.8 Tcm and R/P ratio exceeding 20 years. Size of reserves alone does not make a country a potential exporter: despite significant reserves the US and EU have low R/P ratios and are relatively "gas poor". In both instances, domestic supply is increasingly insufficient as compared to demand; this is a major driver for gas pipeline projects.

Gas supply diversification is another driver for transborder gas projects, especially using LNG. Examples of this kind of driver are the US and the EU, where the cheapest per-unit gas is supplied by pipeline. In addition to diversification and security of supply, LNG has the advantage of being smaller-scale and more flexible in comparison with transcontinental pipelines.

A third type of driver (lower cost of supplies from another country) can be illustrated by the case of Turkmenistan and Iran, both of which are among the most richly gas-endowed countries in the world. However, in northeast Iran it is cheaper to bring gas from just across the border in Turkmenistan, rather than all the way from the Gulf. In the past, gas has been exported from Iran to the former Soviet Union's Transcaucasia under similar circumstances.

Transit countries have the geographical advantage of being located between the "gas rich" and "gas poor" countries, and may themselves be "gas poor" (Ukraine) or "gas rich" (Kazakhstan).

Figure 1 indicates some "gas rich" countries, i.e. those that enjoy both large gas reserves and high R/P ratio. Such countries are usually net gas exporters[10]. In fact, the Asian countries and the Middle East are in the top

[10] Of the countries listed in Tiers 1-3, only the Ukraine is a net gas importer. However, reserves are reported for the Ukraine in a system that differs from the one generally in use in the international petroleum industry. Instead of listing proved reserves, the Ukraine reports A+B+C_1 reserves. The Ukrainian government itself points out that if reserves were

three tiers of gas-rich countries worldwide (with reserves exceeding 0.85 Tcm and R/P greater than 35 years) hold 70% of the global proved reserves. At current rates of extraction, these reserves combined would last some 125 years, i.e. 3-4 times as long as Norwegian reserves and 12-13 times longer than reserves in the U.S. (*cf.* Table 2).

The *minimum* required proved reserves base to consider constructing a large transborder pipeline (a 36+ inch, 3,000+ km pipeline) is in the range of 250 billion cubic meters gas for 20 years of operation. The minimum required reserve base goes sharply up when larger diameter pipeline or longer lifetime is expected, and can exceed 1 trillion cubic meters for 56 inch pipelines. The only countries that can conceivably support such projects for very large diameter pipelines are those included in Tiers 1-3 in Figure 1.

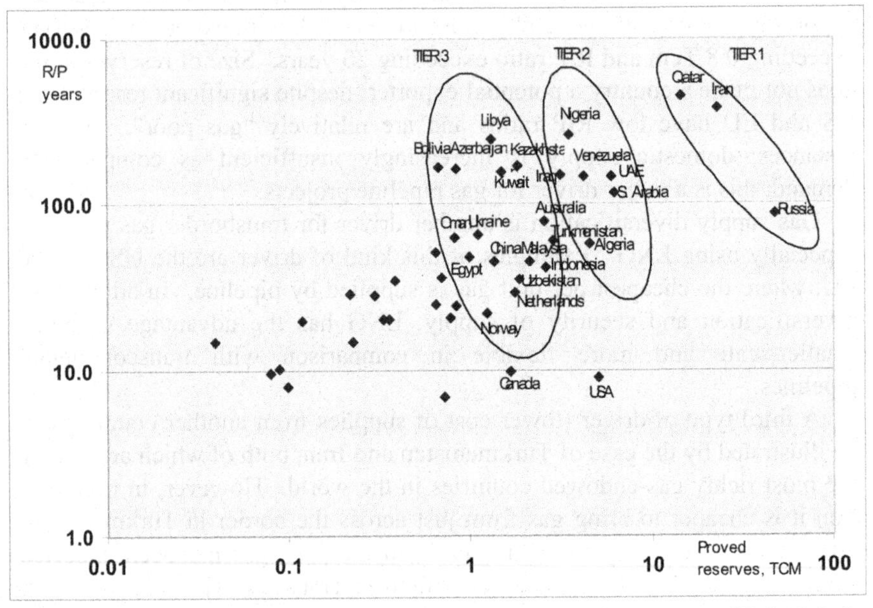

Figure 1 Gas reserves and R/P ratios as of end 2001 (log scale). Data source: BP Statistical Review of World Energy (except R/P for Iraq, which is an estimate).

Table 3 illustrates basic interdependencies between pipeline diameter, required market size, reserve base and feasible distance to market at an average cost-of-transportation index (ACTI)[11] of \$35 per 1,000 m^3.

to be reported as "proved", they may have to be reduced by 50% (*Cf.* Report on the Investment Climate and Market Structure in the Energy Sector of the Ukraine, ECS, Brussels, 2002, p. 62). In this case, the Ukraine will fall out of Tier 3.

[11] ACTI is an indication of total cost of transportation (capital and operational cost) at a given load factor of the pipeline (usually 60-80% of maximum capacity).

Table 2 Reserves and R/P ratios of selected countries. ***Source: BP Statistical Review of World Energy 2002 (except R/P for Iraq, which is an estimate).***

	Reserves Tcm at end 2001	R/P Years at end 2001
Azerbaijan	0.8	163
Iran	23.0	380
Iraq	3.1	150
Kazakhstan	1.8	170
Qatar	14.4	443
Russian Federation	47.6	88
Saudi Arabia	6.2	116
Turkmenistan	2.9	60
U.A.E.	6.0	146
Uzbekistan	1.9	35
Subtotal	**107.7**	**124**
World	155.1	62

Table 3 Pipeline sizing, gas reserves, incremental demand and distance constraints at constant average cost-of-transportation index (ACTI)[12].

Pipeline diameter, inch	Throughput, billion cubic meters p.a. at 80% utilization	Required reserves for 20 year operation, billion cubic meters dry gas	Required incremental market size (offtake), MW electric equivalent	Distance to market at ACTI $35 per 1000 cubic meters[13]
28	5.5	110	2,700	2,400
32	7.5	150	3,650	2,900
36	10	200	4,900	3,100
42	13.5	270	6,600	3,500
48	21	420	10,300	4,000
56	31	620	15,200	4,500

In locations where the basic constraints cannot be resolved (an efficient route to market is not available), "stranded" gas reserves occur. Associated gas has in many instances such a fate, since an oil project may be justifiable even if there is no market for the co-produced gas. In such instances, gas is either flared or injected in a suitable structure outside the oil-bearing reservoir, in the hope that later on it would be possible to bring it to use. Flaring is progressively becoming less of an option due to environmental constraints, and there may not be a geological structure suitable to inject the

[12] *Cf.* part 1.5 for models of pipeline production and cost functions. The model is used to calculate values in this and other tables and in some charts.

[13] At $50 per 1,000 cubic meters at the wellhead.

gas into. The marginal cost of gas under such circumstances tends to be negligible or even negative.

2. GAS TRANSPORTATION: COSTS, CONSTRAINTS, RISKS

2.1 Production, total and marginal cost functions of oil and gas pipelines

For an oil pipeline[14]:

$$\Delta P = 587.76 \frac{Z}{Q}$$ *Equation 1*

Where Q is the quantity of oil in barrels per day (bpd), Z is the horsepower and ΔP is the pressure differential, and

$$\frac{dP}{dL} = \frac{0.241 fsQ^2}{D^5}$$ *Equation 2 (Shell-MIT equation for moving oil through a pipe)*

Where dP/dL is the pressure drop in psi, f is a friction factor (roughness of the pipe), s is the specific gravity of oil, Q is the quantity of oil in bpd, D is the internal diameter of the pipe in inches, and L is the length of the pipeline section in miles. Pressure drop in psi:

$$\frac{dP}{dL} = \frac{\Delta P}{L}$$ *Equation 3*

Therefore:

$$\Delta P = \frac{0.242 fLsQ^2}{D^5}$$ *Equation 4*

From Equation 1 and Equation 4:

$$Q = 13.46 \left(\frac{ZD^5}{fs}\right)^{\frac{1}{3}}$$ *Equation 5, production function of moving oil through a pipeline*

[14] The derivation of the production function and marginal cost function of moving oil through a pipeline is by Brito and Sheshinki (Alternatives to the Strait of Hormuz, by Dagobert Brito and Eytan Sheshinski, The Energy Journal, **19(2)**, 1998, 135-147).

Solve *Equation 5* for horsepower:

$$Z = 0.00041\frac{fsQ^3}{D^5}$$

Equation 6, pump station horsepower

Construct the cost function (*TC* – total cost):

$$TC = 0.00041c_1\frac{fsQ^3}{d^5} + \rho(r,T)(c_2D + c_3Z)$$

Equation 7, total cost function of moving oil through pipe

Where c_1 is the operation cost per horsepower/day, c_2 is the construction cost per inch/mile, c_3 is the construction cost per installed horsepower, $\rho(r,T)$ is the imputed daily cost to the pipeline as a function of interest rate r and project life T.

Marginal cost *MC* is given by:

$$MC = 0.00123.C_1\frac{fsQ^2}{D^5}$$

Equation 8, marginal cost function of moving oil through pipe

Similarly, for moving gas through a pipeline compressor horsepower *HP* is:

$$HP = \frac{Z_gRT}{\left(\frac{k-1}{kE_p}\right)}\left[\left(\frac{P_2}{P_1}\right)^{\left(\frac{k-1}{k}\right)} - 1\right]$$

Equation 9, compressor station horsepower

Where Z_g is average gas compressibility factor, R = 1,544, T is temperature (R), P_1 is inlet pressure (psia), P_2 is outlet pressure (psia), k is adiabatic exponent and E_p is adiabatic efficiency, and

$$Q = 737\left(\frac{T_0}{P_0}\right)^{1.02} D^{2.53}\left(\frac{P_1^2 - P_2^2}{G^{0.961}T_fLZ_g}\right)E$$

Equation 10, modified Panhandle volumetric flowrate of gas in pipe (production function of moving gas through pipeline)

Where T_0 is standard temperature and P_0 is standard pressure, G is the relative density of gas to air, T_f is flowing gas temperature (R), L is length of pipe (miles) and the other parameters are as described above. Other formulas[15] may also be used to assess the production function, for example:

$$Q = \left(0.03393\frac{T_0}{P_0}\right)\left[\frac{\left(P_1^2 - P_2^2\right)D^5 R}{M_g T_s Z_g L f}\right]^{0.5}$$

Equation 11, Weymouth volumetric flow of gas in pipe (production function of moving gas through pipeline)

Where M_g is the molecular weight of gas, T_s is the ambient (flowing) temperature of gas, R is the gas constant and f is a friction factor[16].

Rewrite *Equation 9* for horsepower by using Weymouth (*Equation 11*)[17]:

$$HP = \left\{\left[\frac{P^2}{\left(\frac{Q}{a}\right)^2\left(\frac{b}{c}\right) + P^2}\right]^{\frac{(2k-2)}{k}} + 1\right\}\left(\frac{d}{k-1}\right)$$

Equation 12

Where $a = 0.03393\frac{T_0}{P_0}$, $b=M_g T_s Z_g L f$, $c=D^5 R$ and $d=Z_g RTkE_p$

Construct the cost function:

$$TC = C_1\left\{\left[\frac{P^2}{\left(\frac{Q}{a}\right)^2\left(\frac{b}{c}\right) + P^2}\right]^{\frac{(2k-2)}{k}} + 1\right\}\left(\frac{d}{k-1}\right) + \rho(r,t)\left(C_2 D + C_3 HP\right)$$

Equation 13

Marginal cost is given by:

$$MC = C_1\frac{\Delta HP}{\Delta P} = C_1\left[\left(\frac{\Delta HP}{\Delta Q}\right)\left(\frac{\Delta Q}{\Delta P}\right) + \left(\frac{\Delta HP}{\Delta P}\right) + \left(\frac{\Delta HP}{\Delta Z}\right)\left(\frac{\Delta Z}{\Delta P}\right)\right]$$

[15] If the pressure drop in a pipeline is less than 40% of P_1, then the Darcy-Weisbach incompressible flow calculation may be more accurate than the Weymouth or Panhandles A and B for a short pipe or low flow. In main pipelines, compressible flow calculations are generally used.

[16] Equation 11 parameters in SI dimensions where applicable. *Cf. Dahl et al.* in this volume.

[17] Conversion from SI to British units needed.

Equation 14

2.2 Scale effects on cost of gas transportation by pipeline

Figure 2 illustrates the effects of gas pipeline scale on the average cost of transportation index (ACTI)[18]. The case shown is that of 28, 40, 48 and 56 inch pipelines operating at high pressure (100/140 bar) delivering gas at a distance of 3,000 km[19]. Larger diameter pipelines are capable of delivering gas at a lower cost-of-transportation than small diameter pipelines.

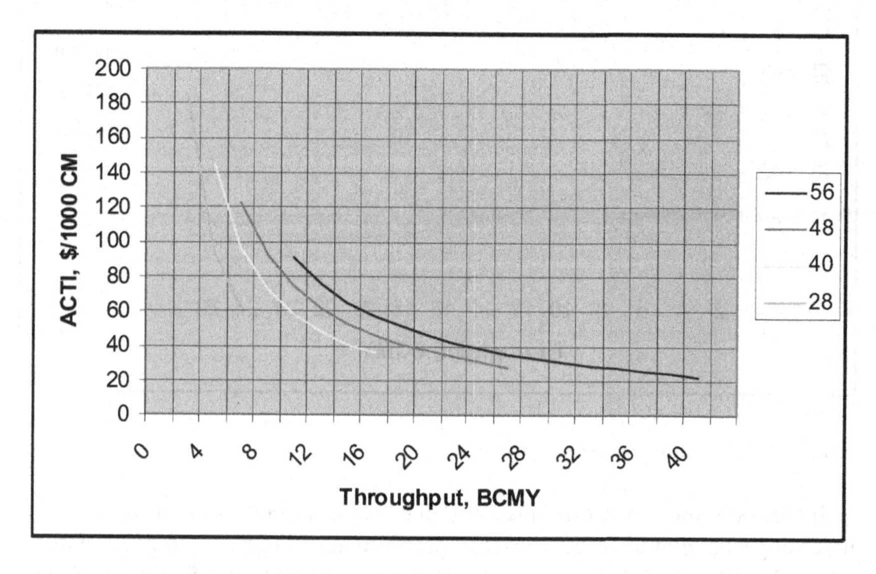

Figure 2: Gas pipeline scale effects on average cost of transportation (ACTI, representative values)

For any given diameter ACTI is strongly dependent on pipeline load factors. ACTI tends to increase sharply if operated below a load factor of around 60% of the maximum capacity of the pipeline. For example, a 48 inch pipeline operating at around 90% of capacity would deliver 24 bcm p.a. at ACTI ~$31/1000 CM, while a 56 inch pipeline operating at 60% of capacity would deliver the same quantity at ACTI ~$38/1000 CM. At

[18] The average cost-of-transportation index is total cost divided by volume of transported gas at various utilization levels (load factors) of the pipeline, i.e. unit cost of transportation at varying throughput.

[19] Other assumptions include 30 years to scrap, 12% ROR, etc. Model input parameters values are generally representative for the year 2000.

similar load factors, however, the cost-of-transportation advantage is unequivocally enjoyed by larger pipelines (for example, other things being equal, at 80% load factor the ACTI for a 48 inch pipeline would be ~$35.00 and the ACTI of a 56 inch pipeline would be ~$29.50 at a distance of 3,000 km).

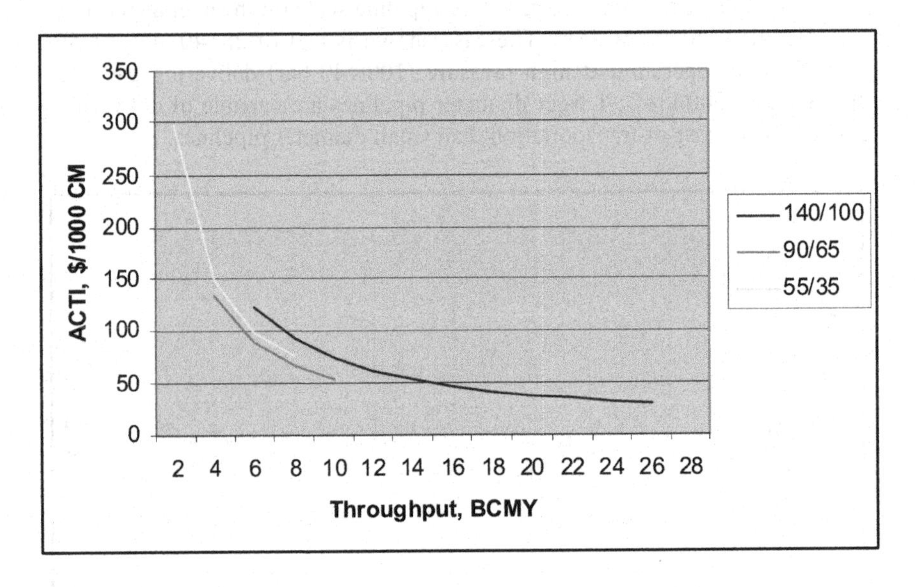

Figure 3: ACTI at various operating pressures.

Increasing the operating pressure has scale effects similar to that of increasing the diameter at constant pressure (*cf.* Figure 3, representative values for 48 inch, 3,000 km long pipeline at 140/100, 90/65 and 55/35 bar discharge/suction pressure). Note that a low-pressure pipeline would not be able to achieve low ACTI at long distance even at full load (*cf.* Table 4 for effects of higher pressure on feasible length of pipeline).

Table 4 Effects of advanced pipeline technology (higher pressure) on feasible supply distance at constant throughput, wellhead cost, ACTI and market price.

Operating pressure, suction/discharge at compressor stations, atm	Throughput, billion cubic meters p.a. at 80% utilization	Pipeline diameter, inch	Wellhead cost of gas, US$/1,000 cubic meters	ACTI, US$/1,000 cubic meters	Delivered cost of gas, US$/1,000 cubic meters	Maximum feasible pipeline length, km
35/50	7	48	50	50	100	2,700
45/65	7	42	50	50	100	3,100
70/100	7	36	50	50	100	3,500
100/140	7	32	50	50	100	3,850

Other scale effects concern required reserves and markets. In their totality, the cost and scale parameters of a gas pipeline act as constraints that have to be met in order to even consider endeavoring in the venture.

3. GAS DEMAND

3.1 Market constraints (netbacks and incremental absorption capacity)

The cost of transportation constitutes 30-50% of the border price of gas on transcontinental routes (e.g. in Eurasia) and a pipeline of a given diameter can generally only assure reasonable delivered cost of gas over a certain distance, if upstream netback constraints are to be also met. For example, if wellhead cost of gas is \$50 per 1,000 m^3, and gas is competitive at a delivered cost (or border price) of \$100 per 1,000 m^3, then the maximum feasible lengths of high-pressure pipelines would be those listed in Table 5 (representative values).

Table 5 Distance constraints at constant wellhead and market gas prices and ACTI.

Diameter, inch	Throughput, billion cubic meters p.a. at 80% utilization	Wellhead cost of gas, \$/1,000 cubic meters	Marginally efficient gas price at pipeline destination, \$/1,000 cubic meters	ACTI \$/1,000 cubic meters	Maximum feasible pipeline length at listed ACTI, km
28	5.5	50	100	50	3,500
32	7.5	50	100	50	4,000
36	10	50	100	50	4,400
42	13.5	50	100	50	4,900
48	21	50	100	50	5,800
56	31	50	100	50	6,500

The ACTI and delivered cost-of-gas equivalence of pipelines of various diameters is therefore only one of the major factors that affect the feasibility of a solution. Other factors of paramount importance are wellhead cost, wellhead netbacks, and the ability of a market to absorb the incremental increase of supply, which for large diameter pipelines is in the range of 20 to 30 billion cubic meters per year. Lower wellhead cost extends the "feasibility radius" of a gas field; large diameter pipelines may be able to reach further at reasonable delivered cost. However, markets may set limits on the absorption of the massive increase in supply that is inherent to large diameter pipelines. The absorption capacity of a market may spell doom even for projects that are otherwise within the constraints: a pipeline would need an "anchor" market of a certain minimum size to be built (*cf.* Figure 4).

Recent advances in technology (other than higher pressure pipelines as discussed) have opened up some previously stranded gas reserves for both energy and non-energy (feedstock) uses. Of particular importance are:

- Gas liquefaction plants of new design that allows single-train operation (about 2.5-3 million tons per year or more) as opposed to earlier designs that required two trains (about 2.1-2.5 million tons each, 4.5-5 million total per plant). Other advances in LNG technologies (e.g. transportation and re-gasification) have also contributed to lowering the delivered cost of LNG;
- Emerging gas-to-liquid (GTL) technologies;
- Advances in the development of mobile (barge-mounted) methanol and fertilizer plants using natural gas feedstock.

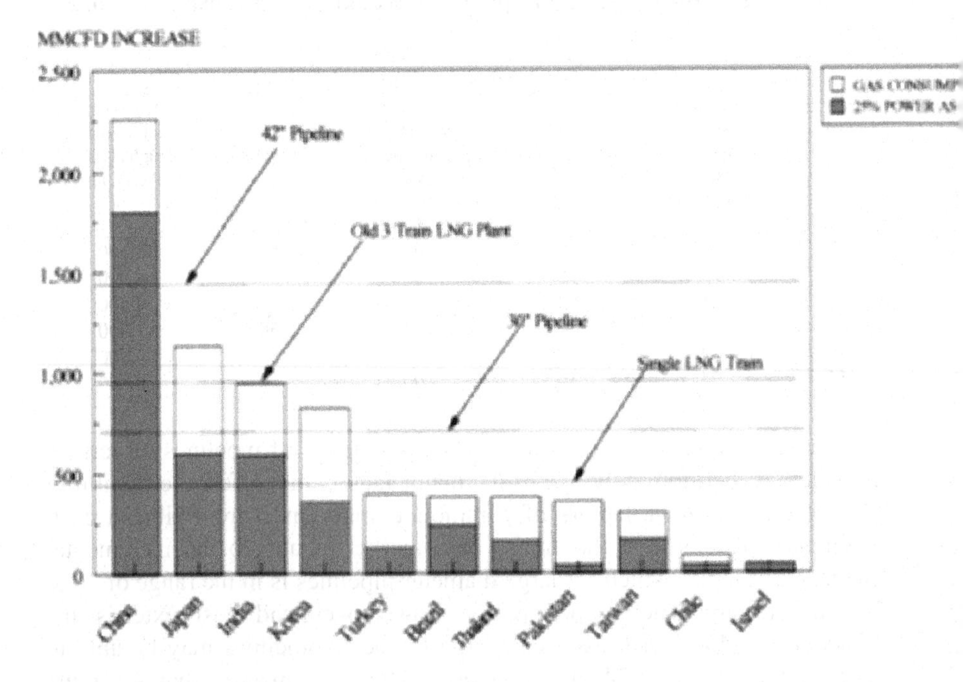

Figure 4 Required market incremental size vs. gas transportation solutions (Source: Jensen Associates, Inc.).

The technical constraints and technology/market size interdependencies are project risks that have historically had a combined impact resulting in the segmentation of natural gas imports and generally of gas shipments across international borders. For many years, the pattern has tended to be one of inherently rigid gas transportation systems that take gas from discrete sources to the particular markets they are specifically designed to cater to. Accordingly, project legal frameworks and contract structures have been dominated by long-term sales and purchase agreements that, unlike oil contracts, usually deal with a host of specific project factors. The need of such contractual arrangements was rooted in the absence of international non-captive gas market that could provide cost/price markers; therefore risk mitigation instruments had to be bilaterally designed and reflected in commercial contracts.

In the absence of competitive markets, to minimize risks gas prices have tended to be indexed to other fuels, particularly oil products, and the gas is sold under long-term contracts. Since gas could only be supplied to a predetermined market, and the pipeline has to operate with a minimum of fluctuations, from the point of view of both seller and buyer such a contractual structure served the purpose best. Another factor shaping the choice of the long-term contract approach is the high transportation cost. Cost of gas transportation is generally 6-7 times higher than that for oil per unit of energy equivalent transported at the same distance. Therefore disruptions on the transportation route generally have greater negative impact (carry a higher risk) than for other fuels.

3.2 Competition: gas *vs.* other fuels and gas *vs.* gas (same-fuel competition)

Overall energy demand is usually presented as a function of income Y and energy prices P_e. Models often take in consideration other factors, such as population growth, dwellings parameters, end-use energy technologies, weather, etc. In a general form, energy demand Q_e as a function of Y is typically represented as:

$$Q_e = K_1 Y^{\alpha} \qquad \textit{Equation 15}$$

where K is a constant and income elasticity of demand α (valid for small changes in Y) is

$$\alpha = \frac{\frac{\delta Q_e}{Q_e}}{\frac{\delta Y}{Y}} \qquad \textit{Equation 16}$$

Similarly, energy demand as a function of price and price elasticity would be:

Equation 17

$$Q_e = K_2 P_e^{\beta}$$

Demand is said to be "elastic" if elasticity is greater than unit, "inelastic" when elasticity is less than unit, and "unit elastic" when elasticity equals one. When substitutes are available, demand would be also dependent on prices of alternative fuels, so

$$Q_e = K(P_{e_1}^{\beta_1}, P_{e_2}^{\beta_2} ... P_{e_n}^{\beta_n})$$ *Equation 18*

where P_{en} and β_n are the price and cross-elasticity of fuel[20] type n.

Long-run elasticities have been found to be higher than short-run elasticities, which should be expected, given the fact that switching from one fuel source to another often requires switching to a different midstream and end-use technology that involves considerable investment. For example, to switch from fuel oil to natural gas in heating, it would be necessary to build gas pipelines and install gas or dual-firing equipment. The process requires time and capital outlays that may not be undertaken at all.

Fuels that are hard to substitute with each other may have elastic demand with respect to changes in own price or income, but will have low (close to zero) cross-elasticities of demand. Such fuels will "not be in the same market" and would not compete with each other; within their respective markets, however, there may still be strong "same-fuel" competition. To illustrate, let quantity demanded of gas Q_g be determined by its own price P_g, prices of oil P_o, power P_p, coal P_c and income Y:

$$Q_g = \phi(P_g, P_o, P_p, P_c, Y)$$ *Equation 19*

Hence:

$$\frac{\delta Q_g}{\delta P_g} P_g + \frac{\delta Q_g}{\delta P_o} P_o + \frac{\delta Q_g}{\delta P_p} P_p + \frac{\delta Q_g}{\delta P_c} P_c + \frac{\delta Q_g}{\delta Y} Y = 0$$

Equation 20

Re-write Equation 20:

[20] "Fuel" in the case should be understood as any energy source that is an alternative (e.g. heating up with electricity produced at hydropower plants as opposed to heating up with natural gas).

$$-\varepsilon_{gg} = \varepsilon_{go} + \varepsilon_{gp} + \varepsilon_{gc} + \qquad \textit{Equation } 21^{21}.$$

When no substitutes to gas exist, ε_{gg}=-ε_{y} , when perfect substitutes exist (e.g. $\varepsilon_{gp}=\infty$), $\varepsilon_{g}=\infty$. Most short-run cross-elasticities have been found to be <0.15, and majority long-run elasticities for various fuels have been found to be well below unit (*cf.* Tables 6-8). On the other hand, "good substitutes" are perceived to have cross-elasticities >>1.

Table 6: Cross elasticities of fuels in the industrial sector of the US, 1970-1985. Source: Considine (1989), quoted by Waverman and Watkins.

Cross-elasticity	All fuels	Stationary fuels
Oil-natural gas	0.1	0.05
Oil-coal	0.03	0.17
Oil-power	-0.04	0.11
Natural gas-oil	0.22	0.04
Natural gas-coal	-0.32	-0.2
Natural gas-power	0.67	0.63
Coal-oil	0.11	0.63
Coal-natural gas	-0.62	-1.01
Coal-power	0.44	1.38
Power-oil	-0.05	0.04
Power-natural gas	0.34	0.32
Power-coal	0.4	0.14

What this means is that, in a sense, there is no single market for energy where different fuels compete to gas or to each other, but a series of overlapping markets that exhibit very significant barriers to entry in the short run and considerable barriers in the long run. For example, natural gas tends to be "substitutable" almost exclusively in heat applications (*cf.* shaded cells in Table 8) across several sectors, but not all: in transportation, it is exceedingly difficult to find substitute to liquid fuels. Besides, the pattern of overlap is widely divergent in various countries: natural gas has proven an excellent substitute to coal in Germany, but not in Japan, where coal consumption is much lower in the first place due to a dedicated earlier policy of substituting coal by gas (*cf.* elasticities of gas to coal in these countries in Table 7).

Table 7: Cross-elasticities of fuels. Source: See Table 6

	US	Germany	Italy	Japan
Natural gas-oil	0.14	-1.85	0.53	-
Oil-natural gas	0.28	-0.36	0.13	-

[21] Derivation of Equation 21 is by Waverman and Watkins.

Coal-natural gas	-0.43	1.74	-0.42	0.06
Natural gas-coal	-0.09	5.99	-0.20	-
Natural gas-power	0.48	-1.13	-0.57	2.10
Oil-power	-0.69	-	0.21	0.19
Coal-power	1.04	0.97	-	-0.49
Power-oil	-0.15	-	0.15	0.14
Power-coal	0.09	0.43	-	-0.11
Oil-coal	-	1.46	0.12	-

Table 8: Fuel "substitutability" matrix. Source: See Table 6.

	Residential			Commercial		Industrial		Transport
	Lighting	**Heating**	**Appliances**	**Heating**	**Lighting**	**Boiler heat**	**Process heat**	
Oil - natural gas	n/a	yes	negl.	yes	n/a	yes	some	low
Oil - power	negl.	yes	negl.	yes	negl.	yes	some	low
Oil-coal	n/a	yes	negl.	n/a	n/a	yes	negl.	n/a
Natural gas - power	negl.	yes		yes	negl.	yes	some	n/a
Natural gas - coal	n/a	yes	negl.	n/a	n/a	yes	negl.	n/a
Power - coal	negl.	yes	negl.	n/a	negl.	yes	negl.	n/a

Elasticities are a basic tool of models used to forecast gas demand. However, apart from the issues mentioned so far, they have other drawbacks that become part of the models. For example, demand models often presume reversible elasticities: high prices reverse the effect of low prices and *vice versa*. However, it may be difficult to convince a user to switch back to oil, once a shift has occurred to gas, even if gas prices are higher than oil prices in energy equivalent terms for a long period of time – such a switch-back would require scrapping of equipment, re-purchase of equipment for use in oil heating, and some loss of value/comfort. When used without qualifications, elasticities in models presume no shift in demand curve, equal response to equal change in the long run and other bold assumptions that are just not validated.

Apart from these factors, many additional considerations apply to modeling gas demand: changes in technology that open new niches and entire new markets to gas penetration, physical bottlenecks on transport routes, weather patterns, perceptions about reliability of supply (and accordingly policies designed to address security of supply issues), externalities (pollution, climate change), to name but a few. The outcome is that gas demand has proven itself tough to accurately predict, as evidenced, for example, by the widely inaccurate forecasts of gas demand in Turkey – and in many other countries. The demand-side uncertainties are an additional factor compounding risks involved in gas pipeline projects. A common technique to deal with these uncertainties and risks is to conduct detailed project feasibility studies to assess a particular geographic market and then aim for fast market penetration ("ramp-up" of deliveries) under long-term contracts with inbuilt price formulas tying gas prices to prices of other fuels, usually oil. The result is the absence of true gas-to-gas competition.

4. TRANSIT COUNTRIES: RISKS AND RISK MITIGATION POLICIES IN TRANSBORDER PIPELINE PROJECTS

4.1 Country risks *vs.* project risks

Moving gas over long distance involves significant risk and expense, regardless of country risk levels. The fact has prompted some analysts to point out that the major issue in gas pipelines is the "tyranny of distance" [4] which makes the undertaking less attractive as the length of the route grows. This very circumstance makes risk-reducing policies a major factor in increasing the "feasibility radius" of gas reserves. The longer the pipeline and the larger its diameter, the stronger the need to assure a comprehensive risk mitigation framework that may involve bilateral, multilateral and commercial instruments.

The types of risks relevant to gas pipeline projects are related to supply, market (commercial), financial, engineering, and regulatory and political factors. Wherever risks are seen as lower and whenever risk mitigation instruments are successfully deployed, the chances of a project to succeed improve. This paper deals with contract risk mitigation strategies in some more detail below. However, regardless of the degree to which the technical, economic and operational risks have been addressed in the contractual structure, *not all types of risks can be mitigated by contractual instruments*. Of particular importance are risks that are not project-specific or pipeline route-specific, but are related to the general framework conditions for the gas sector and the energy sector as a whole. This includes also the financial, regulatory and political framework of the economy. *Those risks are part of the general investment climate,* and often entail additional costs. They are discussed in this part of the paper.

The link between country and project risks[22] is quite often straightforward: in "permissible" countries, project assessments use discount

[22] A number of sources provide information about various country risks. In this paper, sources consulted are [5,6,7,8,9]. Some indices were not available for all countries on the list. Wherever available, most recent value of index is used. Risk ranks in each category

rates that may be three to four times lower than the ones used for "limited exposure" and "prohibited countries"[23]. A composite risk ranking for a number of both "gas rich" and "gas poor" countries in Asia is provided in Table 9. Figures 5 and 6 illustrate the rank order correlation between country risk ratings and target lender returns; data suggests it is in the range of -0.98. Based on these data, **for every unit increase in country risk rating, the target rate of return increases by about 0.26 percentage points in "limited exposure' countries and by about 0.33 points in "prohibited countries".**

Table 9 Illustrative composite risk levels in some Asian countries (100=least risky). Sources: [5,6,7,8,9,10]. See Footnote on this page.

	Risk rank	Risk category
Turkmenistan	12.8	Very high
Iran	17.0	Very high
Iraq	20.5	Very high
Bangladesh	21.2	Very high
Uzbekistan	22.4	Very high
Indonesia	25.9	Very high
Azerbaijan	26.8	Very high
Ukraine	30.6	Very high
Pakistan	31.0	Very high
Russian Federation	35.0	Very high
Kazakhstan	37.9	Very high
India	40.6	Very high
Thailand	47.4	Very high
Malaysia	51.1	High
Oman	54.5	High
Saudi Arabia	57.3	High
China	59.5	High
Qatar	63.5	Moderate
United Arab Emirates	71.0	Low
Australia	79.3	Low

Table 10 illustrates how reducing country risk levels affect basic gas pipeline project parameters at a constant length (representative values).

were normalized (lowest risk rating = 100). Moody's ratings were transformed to numerical values by using logistic transformation pursuant to [10]. Composite risk rank is average of available normalized rankings, may not be fully comparable across listed countries because of different weights due to lacunas in data, and is intended for illustrative purposes only.

[23] This is just one of the possible ways to take into account country risks when assessing projects, and not necessarily the best or most widely used. *Cf.* here, Table 9-10, and Fig. 5-6: Nitzov, B: Building Better Energy Markets: What Next? Proceedings of the 26th Annual International Conference of the IAEE, Prague, 2003.

Other things being equal, greater country risk significantly reduces the chances of long pipelines to be implemented, increases the cost of gas and limits the potentially accessible markets. In the case illustrated in Table 10, each point increase in the rate of discount (equivalent to about 3 points increase in country risk ratings) corresponds to an increase of the gas transportation tariff by about $3.40 per 1,000 cubic meters, or to an increase of the minimum required gas sale price by the same amount. Many of the gas-rich countries that have both the required reserves and R/P ratios to support at least one large-diameter pipeline belong to high-risk categories.

Table 10 Impact of increased (political) risk on minimum acceptable tariffs and prices (representative values).

Pipeline length, km	Throughput, billion cubic meters p.a. (32 inch high pressure line)	Wellhead cost of gas, $/1,000 m^3	Target rate of return (project discount rate for NPV calculations)	Minimum acceptable transportation tariff to secure required NPV and loan coverage ratios, $/1,000 m^3	Minimum price of delivered gas to secure required NPV and loan coverage ratios, $/1,000 m^3
3,850	7	50	10	~48-50	~100
3,850	7	50	15	~65	~115
3,850	7	50	20	~82	~132
3,850	7	50	25	~100	~150
3,850	7	50	30	~117	~167

The cost of large-scale cross-border energy projects typically runs into billions. The energy sector of transition economies has investment needs that exceed approximately 10 times those of OECD countries as a percentage of GDP (7%-9% as compared to 0.8%-1.1%). A great number of investment barriers were (and continue to be) related to market structure and the ability to attract private investment. Restructuring and privatization are often seen as means to enhance competition (for example, by increasing the number of both suppliers and buyers of oil and gas) and improve the investment climate (*inter alia*, by providing clear market signals to investors).

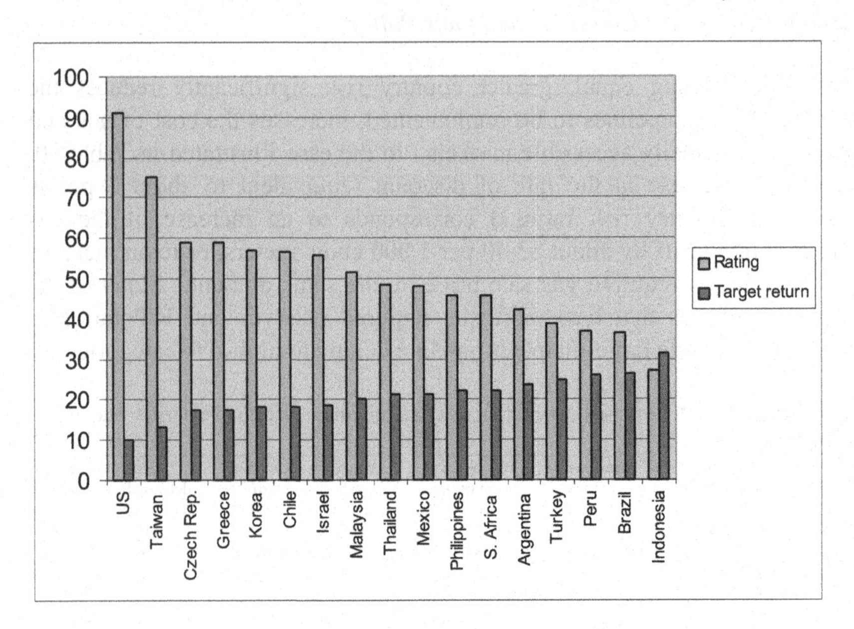

Figure 5 Target investor returns in "limited exposure' country risk ratings. Note: US listed for comparison. Risk rating 100 = best. Target return in discount percent points. Source: [11].

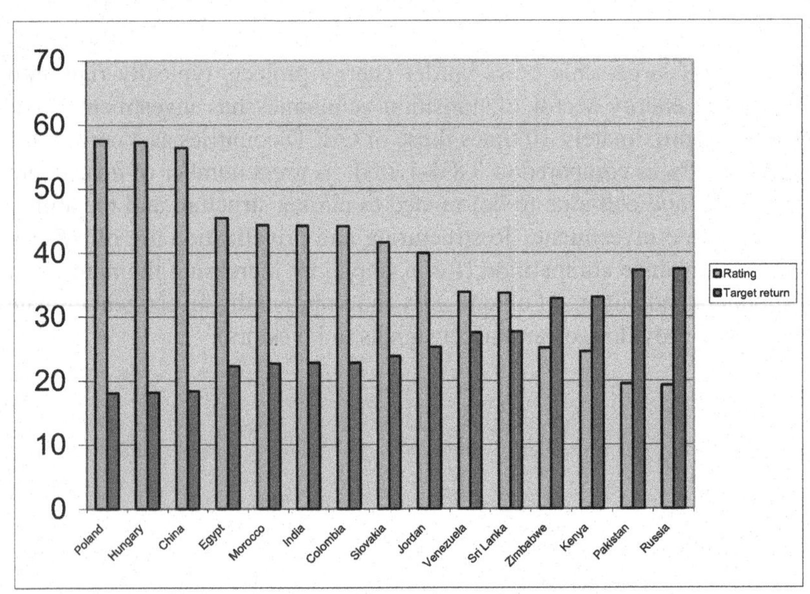

Figure 6 Target investor returns in "prohibited exposure" country risk ratings. Note: Risk rating 100 = best. Target return in discount percent points. Source: [11].

Data suggest that *gas and power are the sub-sectors where the issue of risk reduction is particularly urgent, and especially gas*[24]. This is primarily due to the fact that, while investment needs for gas are much the same as for oil, the presence of state industries is far more pronounced in the gas sector. In gas transportation, essential reforms prior to privatization have yet to be carried out in many countries. While no country has a legal monopoly on gas production, a number of countries have legal or *de-facto* gas transportation monopolies and no unbundling has been implemented in many instances. All of these factors discourage private investment.

Table 11 Estimated direct cost of risk in the gas industry and the economy of some countries[25].

	Composite risk score	GDP, 1999, billion 1990 $	Minimum required annual investment in the gas sector infrastructure, million $	Annual cost of capital at 12% discount	Risk-adjusted discount rate, percent	Annual cost of capital at risk-adjusted discount rate, $ million	Annual cost of unmitigated risk, $ million	Annual loss as share of GDP, percent
Turkmenistan	12.8	2.8	70	8.4	42%	29.4	21.0	0.75%
Uzbekistan	22.4	21.4	535	64.2	37%	197.95	133.8	0.63%
Azerbaijan	26.8	5.2	130	15.6	32%	41.6	26.0	0.50%
Ukraine	30.6	63.5	1588	190.5	28%	444.5	254.0	0.40%
Russian Federation	35	360.6	9015	1081.8	24%	2163.6	1081.8	0.30%
Kazakhstan	37.9	25	625	75.0	22%	137.5	62.5	0.25%
Total		**478.5**	**11,963**	**1,435.5**		**3,014.55**	**1,579.1**	**0.33%**

The high-risk investment environment associated with most gas-rich countries in Eurasia adversely affects the chances of particular projects being funded. An inadequate investment climate exacerbates concerns about security of gas supply, on the one hand, and increases the cost of capital for host countries, on the other hand. Gas-rich countries that are rated "high risk" pay an additional price due to decreased investment; gas-poor countries pay additional price in the form of higher energy prices due to the increased cost of supply. In Table 11, an attempt is made to illustrate what the extra

[24] Oil, gas and power infrastructure typically absorbs about 90% of energy investment.

[25] This illustration does not account for the varying weight of energy sub-sectors in different countries (e.g. in Turkmenistan the sector is mostly dependent on natural gas, coal has a greater importance in the Ukraine, etc.). The same is valid about the relative size of the sectors of different countries compared to each other, e.g. the gas sector in Russia and in Azerbaijan. The aggregated estimate, however (bottom line) is representative to a sufficient degree and may be regarded as an indication of the magnitude of the problem.

cost incurred directly may be in key gas-rich countries that fall in the category of "highly risky" places to do business in[26].

Table 11 illustrates what the extra capital costs might be. The true measure, however, would include the opportunity cost. Unrealized gains, such as those from additional oil and gas exports that did not occur because of lack of investment, are a better measure of the costs to the society's welfare. This paper does not try to provide a comprehensive estimate of what the negative impact may be, but rather just an illustration (Table 12).

Table 12 Illustrative values of losses/gains from greater risks in some countries (gas sale proceeds).

	Average production in 1988-1992, billion cubic meters per year	Average production in 1993-2001, billion cubic meters per year	Annual loss/gain in production, billion cubic meters	Value loss/gain at sale price of $50 per 1,000 cubic meters, million $ per year	Annual value loss/gain as share of 1999 GDP (1990 Dollars), percent
Azerbaijan	9.2	5.7	-3.5	-176	-3.4%
Kazakhstan	6.9	7.5	0.7	33	0.1%
Russian Federation	583.9	553.5	-30.4	-1521	-0.4%
Turkmenistan	76.6	33.2	-43.4	-2170	-77.5%
Ukraine	25.5	17.1	-8.4	-419	-0.7%
Uzbekistan	38.5	48.2	9.7	485	2.3%
Total	**740.6**	**665.3**	**-75.4**	**-3,768.7**	**-0.8%**

The next section provides an analysis of contractual structures and risk mitigation instruments in transborder pipelines.

4.2 Practices and Options in Risk Mitigation

The technical and economic constraints and the legal environment affect the choice of the contractual framework for gas deliveries via pipeline. For example, depending on the number of suppliers and clients, and the regulations governing the construction and operation of a gas pipeline, the role of the pipeline can be that of a gas merchant, a hybrid pipeline or a gas transporter. A "merchant" pipeline typically buys all the gas at the inlet point, transports it and re-sells it at outlets to different clients; it does not provide third party access, unless specifically required to do so by law. In the latter case, it becomes a "hybrid" pipeline. A "hybrid" pipeline would engage in both gas sale and purchase and gas transportation activities, as well as in any of the associated services. Hybrid pipelines are a particularly difficult case to regulate and to ensure competitiveness and non-

[26] *Cf.* here and Tables 11-12: Nitzov, B: Building Better Energy Markets: What Next? Proceedings of the 26th Annual International Conference of the IAEE, Prague, 2003.

discrimination. A "gas transporter" pipeline does not engage in gas purchase and sale, it only provides the service of transportation and related services (e.g. balancing, storage, swap platform, etc.) to both upstream and downstream clients. A "gas transporter" pipeline may provide third party access under both long-term and short-term contracts, albeit some preference may be given to longer term contracts. Regulations are needed to assure that the "gas transporter" does not discriminate and provides a fair playing field to all parties that technically qualify.

Further contractual complications arise when the pipeline crosses an international border. In this instance, there is a split of delivery point and pipeline integration; in this respect, the following options exist:

- Delivery at the border where separately owned and operated exporting and importing pipelines interconnect;
- Cross-border delivery where an integrated pipeline operates in both the exporting and importing countries without transiting;
- Trans-national transportation where an integrated pipeline either transits or delivers gas to several countries along its route [11].

The first scenario is the simplest one. Potential complications arise when <u>both</u> the type of pipeline ("merchant", "transporter" or "hybrid") <u>*and*</u> regulations differ in the countries along the route (e.g. TPA). As a minimum, this solution *requires operational dispatch coordination.* The second scenario usually *requires a bilateral agreement* to resolve issues related to regulation, taxation, force majeure and insurance. The third scenario is the most complicated and usually cannot assure efficient operation without a multiparty international convention or treaty [11]. Where interconnected networks exist (e.g. in Eurasia), the options partially overlap, *requiring the establishment of a coherent hierarchy of bilateral and multilateral treaties.* In an integrated network environment, such instruments have already become a *sine qua non* for transborder gas operations - little or no investment occurs where integrated network pipeline regimes are not bilaterally and multilaterally established and *harmonized.*

Table 13 provides a comparative listing of some transborder oil and gas pipeline projects in Asia. Project frameworks almost invariably feature an intergovernmental agreement, long-term off-take contracts and absence of TPA. Upstream access is more often than not either on PSA terms or is part of the activities of the relevant national oil/gas company. Finally, transit is typically regulated by an *ad hoc* host government agreement. Overall, the framework is rather removed from the idea of a liberalized gas market, where the role of a government would be by-and-large to oversee the fairness, absence of discrimination and competitiveness of the business environment.

Figures 8-9 illustrate what it takes to complete the project structure of an international pipeline in a relatively simple case, where the upstream part of the project is on PSA terms and there are only two countries involved. The case is that of Unocal's proposed pipeline from Bangladesh to India. Under the adopted approach, a gas export pipeline agreement and a gas transportation agreement would have to be negotiated, and title on gas would pass to customers at the border. This structure allows national regimes to apply for the sections of the pipeline on the territory of each of the two countries, but leads to a necessity to involve a host of parties and negotiate multiple agreements of several types (*cf.* Figure 8).

In cases where transit country(ies) is/are involved, the contractual structure needed to resolve potential conflicts of law on the territory of the involved countries. Addressing major investor risks may be even more complicated. A main drawback is the dearth of international law norms dealing with the right to construct a transit *onshore* pipeline across the territory of another country. Some analysts believe that in this instance an intergovernmental agreement becomes a *sine qua non* where transit pipelines are expected to be built [13]. We note that in simpler cases where only two countries are involved, this is not the case. For example, Unocal's project in Bangladesh and India is not underpinned by an intergovernmental agreement. For *offshore* pipelines, the right to construct transit pipelines is granted to investors on the continental shelf by the Law of the Sea[27], subject to the requirement to obtain the littoral country's consent to the pipe route. We note, however, that issues related to the *operation* of the pipeline, e.g. third party access, are less clear cut even in the case of offshore pipelines located on the shelf.

Table 13: Project structure of some transborder oil and gas pipelines in Asia.

Project	Upstream	Intergovernmental agreements	TPA	Long-term offtake contract	Transit	Operator
BTC	PSA	Yes	No	Yes	Host government agreement	Controlled by upstream partners
Azerbaijan-Turkey gas pipeline	PSA	Yes	No	Yes	Host government agreement	Controlled by upstream partners
Bangladesh-India gas pipeline	PSA	No	No	Yes	n/a[28]	Controlled by upstream partners
China-South Korea gas pipeline	Undecided	Yes	No	Yes	Under discussion	National companies
Iran-Pakistan gas pipeline	National oil / gas company	Yes	No	Yes	n/a	Independent

[27] By the 1958 Geneva conventions and the 1982 UN convention on the law of the sea.
[28] Not applicable.

	(government controlled)					
Iran – Armenia gas pipeline	National oil / gas company (government controlled)	Yes	No	Yes	n/a	National companies
Kazakhstan-Iran oil pipeline	PSA and national company	No	No	Yes	Under discussion	Controlled by upstream partners/national companies
Malaysia – Philippines gas pipeline	National oil/gas companies (government controlled)	?	No	Yes	n/a	National companies
Qatar – Pakistan gas pipeline	PSA and national company	?	No	Yes	n/a	Independent
Turkmenistan – Pakistan gas pipeline	National company, PSA	Yes	No	Yes	Host government agreement (proposed)	Undecided
Vietnam - Thailand – Malaysia gas pipeline	PSA, national company	?	No	Yes	?	Controlled by upstream interests
Russia (Kovykta) – China gas pipeline	PSA or other national regime	Yes	No	Yes	Host government agreement or n/a, depending on route	National companies in China, undecided in Russia
Kazakhstan – China oil pipeline	National company	Yes	No	Yes	n/a	National companies
Russia-China oil pipeline	Private national companies	Yes	No	Yes	n/a	Controlled by upstream interests in Russia, national company in China

The absence of harmonized practices in establishing a contractual and/or regulatory regime for transborder routes for pipelines translates into a general perception of higher entry barriers and risks for business. Like in the case of the general business climate, there is a directly attributable cost associated with these barriers and risks. As a matter of illustration only, assuming that the cost of the projects listed in Table 10 is around $40 billion and interest rates are 5% p.a., a one-year delay in building the required interlaced legal and contractual system fore these transborder projects would mean a loss of around $2 billion, or a bit less than $150 million per project on the average. We would not even bring in the picture the missed opportunities and gains from oil and gas exports and tradable services and the negative impact on the return of upstream investments that go with this loss[29].

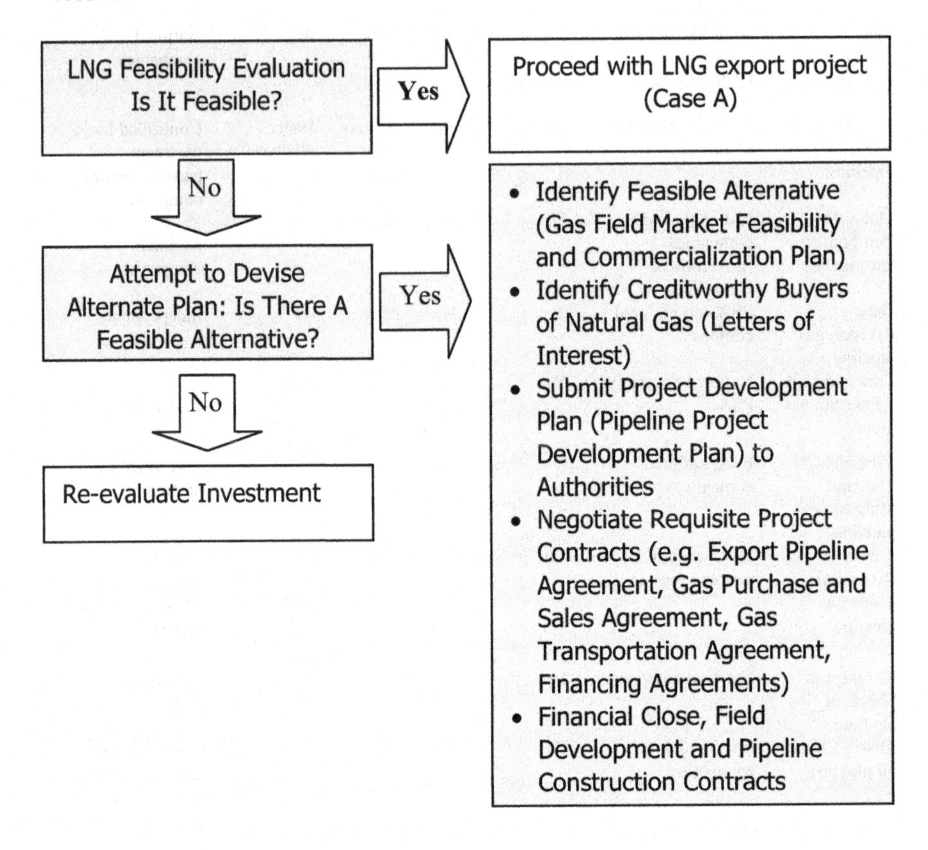

Figure 7: Transborder gas pipeline decision tree outline. ***Source: Unocal (with minor amendments by author).***

[29] In economic terms, the non-competitive environment for the design, implementation and operation of transborder pipeline projects in Asia results into a deadweight loss consisting of two elements (loss due to deficient investment climate in particular countries and loss due to barriers and risks specific to the pipeline business across international borders). We try to demonstrate in this paper that the first element could be in the range of 0.3% of GDP on an annual basis, and that the second element may be of similar magnitude.

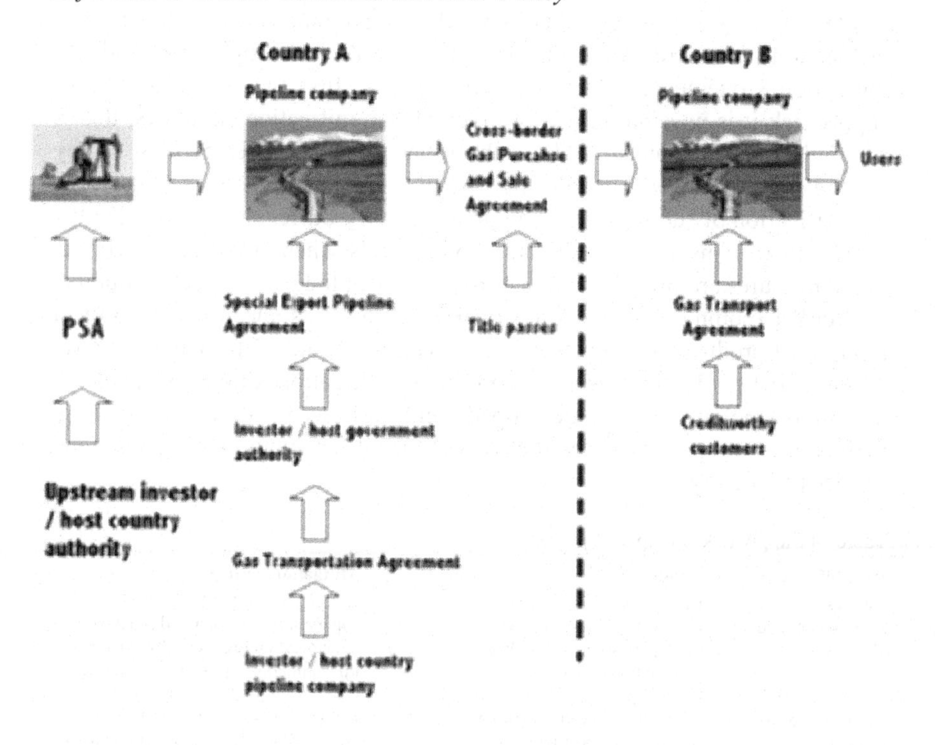

Transaction	Contract Type	Party A	Party B
Transmission on upstream country territory	Gas transport agreement	Upstream investor and host government authority	Pipeline company in country A
Title pass (at border or other location)	Gas purchase and sale agreement	Upstream investor and host government authority	Multiple parties
Transmission in downstream country	Gas transport agreement	Multiple parties	Pipeline company in country B
Users	Gas purchase and sale agreement	Aggregator	Users

Figure 8: Contractual structure for the commercialization of the Bangladesh-India gas pipeline project. *Source: Unocal (with minor amendments by author).*

Typical project-related risks are supply, market and economic environment, financial, construction and completion, regulatory and political risks[30] [11,12]. Supply risks mean that there is a probability[31] of non-

[30] The following discussion is based on [11] and [12] with amendments and additions.

[31] Risks are generally defined in terms of the probability of an event that adversely affects the cash flow and other economic parameters of an undertaking. An economic agent

availability of gas or oil at the point where the pipeline enters the country of destination (downstream). Table 14 lists some typical supply risks and risk mitigation strategies[32]. Note that while in the first two subcategories of supply risk it is possible to use contractual risk mitigation strategies, this is generally not adequate to address the other two risk subcategories. They require taking steps outside the project framework *per se*, such as diversification of supplies, building up emergency buffer stocks, assuring the capability of consumers to switch to other fuels when supply is short, etc. Such measures are often part of energy policies in the gas importing country.

Market (commercial) risks are related to the costs and prices of project inputs and products and the size of the market. Table 15 provides a matrix of these risks. In this instance, too, some mitigation strategies require a consistent policy to be developed and followed country-wide (e.g. maintaining proper balances between current accounts and goods and services accounts).

Table 14 Supply risks in pipeline projects.

Risk subtypes	Comment	Mitigation strategy
Upstream reservoir-related	Reserve base may have to be adjusted downwards due to reservoir underperformance; wellhead cost of gas may be higher than expected, etc.	Independently certify reserves, assure supervisory access, avoid schemes dependent on few fields (provide back-up suppliers), cap costs via hedging, etc.
Competitive calls on reserves	Supplier unable or unwilling to match life of gas infrastructure to life of supply contract	Contractually restrict third party sales Assure first call on future discoveries Define quantity reduction terms and conditions in supply contract
Single/few sources, import dependency		Diversify Dual firing capability Storage/reserves
Midstream bottlenecks	Delays in construction or expansion of capacity Competing calls on gas along the route Tariffs set at levels that make delivered cost of gas or wellhead netback unacceptable	Dual firing capability Storage/reserves Diversify

Financial risks significantly overlap in their origins with regulatory and political risks (all of them can be caused by government policies or

(company) is rarely in a position to affect the degree of a risk (e.g. to reduce the probability of a certain undesired event), and, in a sense, has to operate in the risk environment "as is". From this point of view, risks exist objectively and cannot be "reduced". However, it is possible to provide protection against some (but not all) types of risk and to pursue a strategy which avoids the taking of certain risks.

[32] No attempt is made to be exhaustive of all risks and mitigation strategies, or to suggest particular solutions.

intervention), but are somewhat more manageable. Table 16 provides an overview of financial risks.

Table 15 Market (commercial) and economic risks in transborder pipeline projects.

Risk subtypes	Comment	Mitigation strategy
Inability to achieve required ramp-up	Required load factor of pipeline to assure acceptable upstream and midstream NPV must be achieved shortly after commissioning; any delays have profoundly negative effects on the project.	Secure alternative off-take Take-or-pay clauses
Upstream cost overruns	Affects wellhead cost of gas; cost increase flows down through entire gas chain and may reduce demand to unacceptable levels.	Design production split formulas in PSA contracts allowing greater cost gas allocations or other relief
Delays in project kick-off due to absence of buyers	Absence of creditworthy buyers negatively impacts ramp-up rates.	Build in PSAs and pipeline agreements sufficient flexibility to allow additional time for gas sale contracting
Inflation	Inflation in target market country has an adverse impact that is stronger than that in the gas export country.	Tie prices to international markets
Debt service ratio	Reduces the ability to secure funding for import project components (typically pipe, compressors, SCADA, some know-how)	Use non-recoursive finance
International liquidity ratio	May negatively affect upstream and midstream project components	In transit countries, pay fees in kind Forex hedge
Foreign trade collection experience	May make supply of equipment and material more difficult or expensive or jeopardize sale receipts flow.	Train local personnel in collection techniques Secure right to disconnect for non-payment
Current account balance vs. goods and services	In countries where hydrocarbons are the main export item, may have transient negative implication during the early stages of the project.	Assure measures are taken to avoid the Dutch disease
Parallel forex markets	Usually lead to cost increase upstream.	Minimize the need to convert to and from local currency up- and midstream

Table 16 Financial risks matrix in pipeline projects.

Risk subtypes	Comment	Mitigation strategy
Increased exposure due to delays	Most lenders will require a minimum coverage of 2:1; delays are likely to adversely affect financing capability and in turn lead to snowballing delays.	Assure minimum cash flow via alternative outlet sales in upstream country Loan restructuring options
Gas-to-other fuels and/or gas-to-gas competition depresses price	Linking gas prices to that of other fuels (formula-based pricing) exposes cash flow to same kind of volatility as that of, for example, crude oil. Once downstream infrastructure is in place, it may become efficient to bring gas from other sources.	Hedge Long-term contracts Right of first refusal
Loss from forex control	Forex control may lock in upstream and midstream investment, reduce returns for investors and hinder debt servicing.	Option to change production split in PSAs in case forex controls are introduced Pay for transit country services in kind
Loan default,	History of such events in the gas export country are	Minimize local components

unfavorable loan restructuring	a severe drawback, especially when local partner participation is required.	
Delayed payments on suppliers' credit	This is a severe drawback when local contractors with such history are deployed on the project, e.g. due to "local component" requirements.	Minimize local components
Repudiation of contracts by government	Mostly affects upstream and midstream elements of the project.	Bilateral and multilateral investment protection agreements
Expropriation of private investments	Mostly affects upstream and midstream elements of the project	Bilateral and multilateral investment protection agreements

Completion risks are primarily related to commissioning delays, contractor default and cost overruns. Typical outcomes of such events would be the loss of revenue, difficulties in servicing debts, the calling-in of bonds, the need to refinance or seek new loans. Prudent strategies would involve the use of reputable contractors, early triggers of performance bonds, etc.

Regulatory risks generally fall in the subcategories of government action or inaction with regard to economic, environmental, labor, and tax regulations. Typically, such risks lead to increased cost which has to be passed on to the consumer, increasing prices and potentially lowering demand, thus causing lower pipeline load factors, loss of market share and shortfalls in sales revenue.

- Economic regulatory risks may involve the inept introduction of TPA or inefficient pricing mechanisms;
- Environmental risks may mean an increase in compliance cost, which would have to be passed on to clients, increasing prices;
- Labor risks may involve higher payroll costs, requirements to maintain long-term pension funds liabilities, higher compliance costs, etc.;
- Tax risks may involve the need to pass on new or higher taxes, causing demand shifts.

Table 17 provides an overview of political risks. A mitigation strategy for such risks would by default involve the development and the implementation of long-term consistent policies for risk reduction, which is impossible without the participation of the governments concerned. Bilateral and multilateral instruments and other forms of cooperation are required, often with considerable input and advice from industry.

Table 17 Political risks matrix in cross-border pipeline projects.

Risk subtypes	**Comment**
Economic expectations vs. reality	General economic instability increases, increased cost of business
Economic planning failures	Delay construction, interruption of supply
Political leadership	Refusal to lend

External conflict	Refusal to lend, need to reroute pipeline to avoid certain countries, sabotage
Corruption	Increased cost of doing business, additional fees in transit countries
Military in government	Refusal to lend, need to reroute pipeline to avoid certain countries
Organized religion in politics	Refusal to lend, need to reroute pipeline to avoid certain countries, reduced competition increases cost
Law and order tradition	Spurious judicial proceedings/awards, inability to protect rights, supply disruptions
Racial and national tensions	Refusal to lend, need to reroute pipeline to avoid certain countries, sabotage
Political terrorism	Sabotage, supply interruption
Civil war	Sabotage, supply interruption, loss of supply
Political party development	Reduced competition, delays due to successful lobbying by various groups
Bureaucracy quality	Construction delays, extra cost of doing business

At present, there seem to be three distinct approaches to dealing with the concerns of investors related to transborder pipelines. The first one is to try to limit the number of stakeholders in a project and use in the financial engineering phase "executable" agreements that spell out in great detail the liabilities of the parties. This applies equally to corporate and project finance and combinations thereof. While helpful, this approach does not overcome the need to repeat the entire process over and over again, resulting in custom-crafted solutions. As is the case with crafted products, their cost tends to be higher than the cost of standardized items. The second approach is to try to provide "umbrella" or "framework" bilateral and, in certain instances, trilateral agreements between the governments of the countries concerned. As demonstrated in the case studies of Turkmenistan, Kazakhstan, China and Russia, such agreements are often just non-binding political declarations that do not have much of an impact on national pipeline laws and regulations. In this instance, too, the need to hand-craft the legal and contractual structure for *each* project separately is outstanding. The third approach, which was pioneered by the BTC pipeline, is to develop more or less harmonized host government agreements. This approach also includes the development of "soft law" instruments available on a multilateral base, such as model agreements, best practices guidelines, etc. Within the Energy Charter Treaty process, a number of such documents have been developed or are under consideration.

Based on the foregoing, it is recommended to consider the benefits that the development of bilateral and multilateral understandings regarding transborder pipelines may have in terms of reducing the perceived risks, simplifying the project structure, and helping harmonize the legal base and contractual arrangements. It seems worthwhile to remember that due to the scale of the projects even a modest improvement in implementation could translate into considerable gains for the parties involved and the public good.

REFERENCES:

1. Mark Rodekohr. Financial Developments in '96-'97: How the U.S. Majors Survived the 1998 Crude Oil Market Storm. EIA presentation, London, May 27, 1999 (web version).
2. Campbell, C.J.: Industry Urged to Watch for Regular Oil Production Peaks, Depletion Signals; Oil and Gas Journal, July 14, 2003.
3. Lynch, M.C.: Petroleum Resources Pessimism Debunked in Hubbert Model and Hubbert Modelers' Assessment. Oil and Gas Journal, July 14, 2003.
4. Thomas Stauffer "Caspian Fantasy: The Economics of Political Pipelines", Brown Journal of World Affairs, Summer/Fall 2000 - Volume VII, Issue 2.
5. Heritage Foundation's Index of Economic Freedom.
6. Transparency International's Corruption Perception Index.
7. A.T. Kearney's Foreign Direct Investment Confidence Index.
8. Moody's Sovereign Ratings List.
9. PRS Group's International Country Risk Guide.
10. Guillermo Larraín, Helmut Reisen and Julia von Maltzan: Emerging Market Risk and Sovereign Credit Ratings, 1997, OECD's Technical Paper No. 124
11. Erb, Harvey and Viscanta: Expected Returns and Volatility in 135 Countries, The Journal of Portfolio Management, Spring 1996. *Cf.* also by the same authors: Political Risk, Economic Risk, and Financial Risk. Financial Analysts Journal, November/December 1995, and Do World Markets Still Serve as a Hedge? The Journal of Investing, Fall 1995.
12. Stickley, Dennis: Risk Management in Gas Imports and Gas Project Risk Management, s.d., s.p., p. 5.
13. M. Bannikov (Elrus Consulting) quoted in NGV, 1/2003.

IMPACTS OF THE EUROPEAN GAS DIRECTIVE ON GAS EXPORTERS

Recent experiences with the new Norwegian regulatory regime for accessing the natural gas transportation system

Hans Jørgen DAHL, Sondre DYRLAND and Thor BJØRKVOLL

Dahl: Adjunct Professor at the Norwegian University of Science and Technology (NTNU), Trondheim – Norway. Dyrland: Research Fellow at Centre for European Law, University of Oslo – Norway. Bjørkvoll: Post-Doc Research Fellow at the Norwegian University of Science and Technology (NTNU), Trondheim – Norway, and Senior Researcher at The Foundation for Scientific and Industrial Research at the Norwegian Institute of Technology (SINTEF) - Norway [1]

Abstract: The paper analyzes the EU Gas Directive and the new Norwegian statutory instruments for regulating natural gas transportation services. Such services are of great importance for gas sellers in their capacity as shippers in the system. The paper indicates that the regulation complies with economic theory on how to regulate a natural monopoly. Core elements are the establishment of an independent operator, Gassco AS, a new transparent tariff regime, and non-discriminatory access rules for booking of capacity rights.

Key words: EU gas directive, Norwegian natural gas transportation systems, economic and legislative regulation, natural monopoly, tariff regulation, security of supply

[1] Dahl is the main author of the paper. Dyrland is the main author of Chapter 2 and partly Chapter 3, and Dahl and Bjørkvoll are the main authors of Chapter 4. Bjørkvoll is also the main author of Chapter 4.5.3

The analyses, opinions and conclusions expressed in this work are entirely those of the authors and should not be interpreted as reflecting the views or positions of the supporting organizations and institutions.

J. Hetland and T. Gochitashvili (eds.),
Security of Natural Gas Supply through Transit Countries, 307–342.

1. INTRODUCTION

Although the natural gas industry has been a familiar business worldwide for decades, the natural gas industry is a fairly new trade for Norway. The first exploration and exploitation of hydrocarbons from the Norwegian Continental Shelf (NCS) was carried out as late as in the early nineteen seventies. [2] Since then however, remarkable growth has taken place and Norway is now a major European natural gas supplier. In 2002 Norway supplied approximately 12 % of the consumption in the OECD Europe, and the Norwegian market share is expected to grow in the years to come.

The natural gas reserves on the NCS are significant and current production level can be maintained for many years. In year 2002 the total volume of gas landed was 62.7 billion standard cubic meter (bcm). Within a few years this figure is expected to rise to approximately 75 bcm/year.

A vital and important means for facilitating the gas export is the natural gas transportation system. This system has been continuously developed over the last two decades and it is now the world's largest sub-sea natural gas pipeline system. Among it's major components we find huge processing plants such as those sited at Kårstø and Kollsnes, several offshore riser platforms, onshore terminals and of course, a large number of pipelines interconnected to form a comprehensive network, amounting to more than 6,500 km in total length. In Figure 1 below a drawing is presented showing the outline of the transportation system.

Much invention, innovation and many technological changes have taken place over the years in order to facilitate this significant development. But innovative solutions have not been sought for technical issues alone; innovative solutions have been a pre-requisite in the Norwegian, and lately also the European Union's regulation and organization of the industry.

This paper examines some of the major events and regulatory issues that have been introduced and implemented by this industry during recent years. It starts out with a review of the main provisions of the EU Gas Directive; followed by a section describing the implementation of these requirements in Norwegian statutory instruments. Finally, the paper outlines some basic state-of-art economic theories applicable for regulating natural monopolies and

[2] Royal Norwegian Ministry of Energy and Petroleum: "Fact Sheet 2003 Norwegian Petroleum Activity", see web page: http://www.odin.dep.no/oed/engelsk/p10002017/p10002019/026031-120012/index-dok000-b-n-a.html

also analyses, to some extent, how the regulatory authorities have applied such theory in the design of the new regime

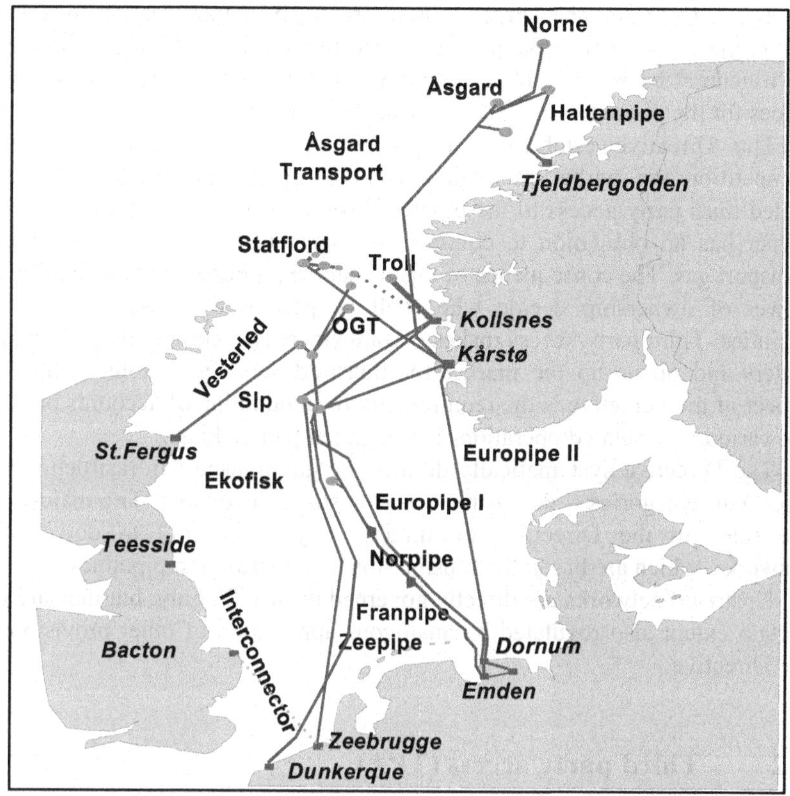

Figure 1. The Norwegian natural gas transportation system

2. PRINCIPLES OF THE EU GAS DIRECTIVE*

2.1 Introduction

The Gas Directive[3] imposes on EU Member States to gradually open their national gas markets to competition. The Directive regulates the *transport*

* This chapter 2 is to a large extent based on parts of Dyrland, Sondre and Ketil Bøe Moen: *Market opening and third party access: An overview of the EU Gas Directive,* SIMPLY 2002, Scandinavian Institute of Maritime Law Yearbook 2002. For a more updated and thorough interpretation of the Gas Directive in Norwegian, see Moen, Ketil Bøe and Sondre Dyrland: *EUs gassmarkedsdirektiv*, Bergen 2001. For a thorough analyses in English of some essential aspects of TPA, see Moen, Ketil Bøe: *The Gas Directive and*

market only, but the aim is to develop competition in the market for the *sale* of natural gas.

The objective of the Directive is to create an internal market for natural gas in the EU as part of the process of creating an internal market for energy on a general basis.[4] The hope is that an internal market will provide lower prices for the consumer and greater security of supply.

The Directive establishes a number of rules intended to promote competition and market integration. The most important provision is the so-called third party access to the pipeline system. This means that the pipeline owner has an obligation to contract in respect to a third party wishing to transport gas. The consequence of this is that the pipeline owner cannot by the power of ownership decide who shall be permitted to use the transport facilities. Third party access makes possible increased competition among gas sellers and opens up the market for transport services. Another important aspect of the Directive is the requirement for unbundling of accounts between the various gas-related operations in integrated undertakings.

The Directive systematically divides the gas transport infrastructure into two main categories – downstream and upstream networks. The majority of the rules in the Directive are directly applicable only to downstream networks, which are basically transmission and distribution pipelines.

Upstream networks are directly governed by Art. 23 only, but they are to a certain extent also regulated by analogous application of other provisions in the Directive.

2.2 Third party access (TPA)

2.2.1 Models and parties

There are several ways of implementing a system of third party access to the pipeline network. One option is that tariffs and the other terms and conditions concerning the pipeline owner's obligation to transport the gas of other suppliers are determined in advance of any request for access (regulated access). Alternatively, it could be left to the parties involved to negotiate the price and the other conditions (negotiated access). According to the Gas

---

Third Party Transport Rights – What Pipeline Volumes are Available, Journal of Energy & Natural Resources Law, Vol 21 No 1 February 2003, pp 1-125.

[3] "Directive 98/30/EC of the European Parliament and of the Council of 22 July 1998 concerning common rules for the internal market in natural gas", based on the EC Treaty, particularly Arts. 57 No. 2, 66 and 100 A (new Arts. 47, 55, and 95).

[4] See Preamble (3) and (5).

Directive Art. 14, any Member State can choose to operate with either negotiated access (Art. 15), regulated access (Art. 16) or with a combination of the two.

Regardless of whether Member States operate with negotiated or regulated third party access, the same parties will be involved. The pipeline owner has traditionally handled both the sale and the transport of gas. In the case of third party access, the pipeline owner will always be involved, but only as providing the transport service. Those wishing their gas to be transported by the pipeline owner constitute third parties in the sense of the Directive. These might be both purchasers and sellers of the gas. The Directive refers to both these parties in Art. 15 No. 1 and Art. 16, as "natural gas undertakings and eligible customers".[5] The underlying gas sales agreement will decide which of the parties is responsible for the transport of the gas and thereby who has to request third party access in order to fulfil the gas sales agreement.

Arts. 15 and 16 comprise customers and natural gas undertakings either inside or outside the "interconnected system". According to the definition in Art. 12 No. 13, cf. No. 12, an interconnected system consists of several interconnected pipeline networks. This means that both those who geographically are situated in an area with an existing pipeline network and those who are not linked to such a network, have the right to negotiate, according to Art. 15, or the right to regulated access, according to Art. 16. Those who are not already linked to such an interconnected system must, however, handle the necessary connection to the system themselves. Consequently, they also become part of the interconnected system.

2.2.2 Negotiated access

Under a system with negotiated third party access, the pipeline owner, according to Art. 15, is obliged to negotiate with those who wish to obtain access to the system. However, the Directive does not require that any particular outcome should result from the negotiations. A number of factors are subject to discussion, for instance, price, volume, duration of the agreement, timeframe in respect to intake and off-take (balancing), etc. Hence, there is a need for rules ensuring that actual third party access will be the outcome of the negotiations.[6]

The Directive requires pipeline undertakings "to publish their main commercial conditions for the use of the system" (see Art. 15 No. 2). It is not entirely clear what the concept "main commercial conditions" means. A possible interpretation is that it is related to the information necessary for potential users of the system to assess whether they should start negotiations

[5] See below, section 2.3, on the definition of those allowed to require third party access.

[6] On similar provisions in the Electricity Directive, see Hammer, p. 322.

in respect to access. For instance, the transport tariff is a key issue. However, it can hardly be required that the pipeline undertaking operates with a totally fixed tariff, since a number of factors, such as volume, time of year, distance, etc. can determine the costs of the transporters.[7]

The Directive also has certain requirements in that the negotiations. Art. 15 No. 1 generally demands that the parties should negotiate in good faith. This means that they should enter into negotiations with the intention of concluding a third party access agreement. When the undertaking's main conditions are published, it is reasonable to assume that the same conditions must apply to all potential users of the system. Further, a more general prohibition against discrimination follows from Arts. 7 No. 2 and 10 No. 2, where it is stated that transmission and distribution undertakings "shall not discriminate between system users or classes of system users."

2.2.3 Regulated access

Under a system with regulated third party access (Art. 16), the pipeline owner is obliged to transport the gas of the third party without further discussion, provided the terms and conditions established in advance, either by the state or by the transporter, are met. Hence, the state has a possibility for greater direct influence on the actual competition by regulated rather than by negotiated third party access.

Access to the system should be granted on the basis of "published tariffs and/or other terms and obligations for the use of the system". Both tariffs and other conditions should, according to Art. 14, be "objective, transparent, and non-discriminatory". The Directive does not offer guidelines beyond these, and it is up to the individual Member State to establish further methods of calculation and concrete terms.[8]

2.3 Eligible customers

As mentioned, not all players in the gas market can request access to the pipeline system. According to the Directive, a gradually increasing number of sellers and buyers may request access. In this way, the competition in the market is gradually increased. This opening up of the market is directly linked

7 The Commission has published an interpretation on what is meant by "main commercial conditions", available on the Internet at: http://europa.eu.int/en/comm/dg17/elec/ldochome.htm

8 The Commission has pronounced on the various methods for calculating tariffs and has provided certain guidelines for the way in which calculations should be made.8 It is emphasized that tariffs should be balanced, in order to give the pipeline owner a fair profit, and at the same time encourage potential customers to make use of third party access. About tariffs, see discussion in chapter 4.

to the concept of "eligible customers", who are those customers who have "the legal capacity to contract for natural gas in accordance with Articles 15 and 16" (Art. 18 No. 1).

The Directive does not clarify which natural gas undertakings in their capacity of *sellers* of gas have the right to third party access. However, access to the system is directly related to the agreement with an eligible customer. Therefore, only natural gas undertakings intending to enter into a gas sales agreement with an eligible customer have the right to access or to negotiate for access.

In this way, the concept "eligible customer" becomes a key issue in respect to the liberalization of the gas market. The relevant Articles are not particularly straightforward and are relatively complicated, but the system seems well suited for a gradual opening up for competition, while at the same time securing balance between the Member States.

In principal, the Directive gives each Member State the responsibility to define who should be considered eligible customers (Art. 18 No. 1). However, the Directive requires both a qualitative and a quantitative minimum for those who should be considered as such. These minimum requirements are met as long as all eligible customers are presented with an *offer* of, or the right to *negotiate*, system access. The requirements concern only the degree of competition, and not the number of customers who would actually make use of third party access.

2.4 Unbundling of accounts

Another important measure is the obligation to unbundle the accounts between the various activities carried out by so-called integrated undertakings. An undertaking can be integrated either vertically or horizontally, or both.

A vertically integrated natural gas undertaking is, as defined in Art. 2 No. 16, an undertaking that performs "two or more of the tasks of production, transmission, distribution, supply or storage of natural gas". A horizontally integrated natural gas undertaking is an undertaking that performs "at least one of the functions of production, transmission, distribution, supply or storage of natural gas, and a non-gas related activity" (Art. 2, No. 17).

Integrated undertakings are to keep "separate accounts for their natural gas transmission, distribution and storage activities" (Art. 13 No. 3). These accounts shall comprise both their consolidated balances and a separate balance sheet for each area of operation.

The underlying reason for the requirement of unbundling of accounts is to avoid discrimination, cross-subsidization and distortion of competition (cf. Art. 13 No. 3). The price of the gas delivered to the user consists of several components: the seller's purchase price, transmission, distribution, and storage costs, as well as profit. One would assume that integrated undertakings de facto keep separate accounts for the various operational areas. However, the *published* accounts will normally be consolidated, so that the results and balances of the various operations are not disclosed. Obvious reasons for this include competition considerations.

In a system with third party access, it is of great importance that the transport service component is "correctly" priced in order for competition to be genuine. For instance, it is reasonable to believe that a transmission undertaking would prefer to establish a high price for the transmission component in order to protect itself against rival competition resulting from third party access. When the undertaking has to publish separate accounts, such practice can more easily be countered. Hence, the obligation to carry out unbundling becomes an important means of promoting competition in the gas market. About the publishing of accounts, see section 7.2 below.

2.5 Upstream pipelines

The Gas Directive defines upstream pipelines as those which are part of a production project or which are "used to convey natural gas from one or more such projects to a processing plant or terminal or final coastal landing terminal" (Art. 2 No. 2). A primary question is if all Norwegian offshore pipelines are covered by this definition.

Firstly, according to the wording, pipelines leading directly from the gas field to a coastal terminal abroad are upstream pipelines. Secondly, pipelines from the field to the processing plants on land are also covered. It does not exclude the application of the Gas Directive that parts of the processing of the gas take place onshore. The definition presupposes that transport of gas, which is not fully treated, is regulated by the Directive.[9]

It could be claimed, strictly according to the wording, that pipelines from processing plants on land to a coastal terminal are not covered by the definition "to a processing plant ... *or* a final coastal terminal" (our italics). It would, however, have been somewhat odd if these pipelines were not treated in the same way as other offshore pipelines. Alternative solutions have obvious weaknesses.

Accordingly, all Norwegian offshore pipelines probably fall within the definition of upstream pipelines in the Directive. This is also the position of

[9] It is possible that the Norwegian authorities have had something else in mind, cf. the statement in the report to the Storting No. 46 (1997–98) "Olje og gassvirksomheten" p. 11: "Offshore pipelines to final terminal should, in accordance with established practice, still be regulated as part of the production sector" (our translation).

the Norwegian government.[10] The Directive has one provision only – Art. 23 on third party access, which directly relates to upstream pipelines.

According to Art. 23 No. 1, Member States shall take the necessary measures to ensure that natural gas undertakings and eligible customers are able to obtain access to upstream pipelines. The objective, according to No. 2, shall be "fair and open access" and achieving "a competitive market in natural gas". Regard should be given to security and regularity of supply, existing and potential capacity, and to environmental protection. Further, regard should also be given to various practical difficulties, for instance, those relating to technical specifications and marginal gas fields. The list of situations to keep in mind must be regarded as exhaustive, but the wording still leaves considerable discretion for national authorities when implementing Art. 23.

Art. 23 does not provide further regulations governing how access should be implemented out. It would therefore be up to each Member State within the existing framework to determine its methods for implementation. Neither is there a list of possible reasons for the derogations similar to those given in Art. 17. However, the rationale for refusal of access to the remainder of the pipeline system, implies the possibility of refusal of third party access to upstream pipelines according to Art. 23 if the reasons for refusal are relevant to upstream access.

2.6 Dispute settlement authority

According to the Directive Art. 21 No. 2, a competent authority shall be appointed to achieve a rapid resolution of disputes arising between undertakings in connection with negotiations for third party access. Disputes on denial of access are especially mentioned. The authority shall be "independent of the parties". Member States can make use of existing bodies, for instance, competition authorities or national courts.[11] However, if the State in question is an important player in the gas market, the demand for impartiality indicates that the dispute authority shall be administratively independent of the state authorities.

Art. 23 No. 3 presents similar requirements with respect to impartiality in the solving of disputes relating to access to upstream pipelines.

[10] Ot. prp. nr. 81 (2001-2002), kap. 2.
[11] Hancher, p. 57.

2.7 Derogations from TPA

According to Art. 17 of the Gas Directive, three reasons can justify a natural gas undertaking refusing a third party's wish for access to the system. These exceptions to third party access are fixed, regardless of which type of access (negotiated or regulated) the Member States choose, and "and duly substantiated reasons" must be given for each refusal (Art. 17 No. 1).

The three reasons for refusal of access are lack of capacity, public service obligations (PSO) and financial difficulties due to take-or-pay commitments. PSO's are governed by Art. 3. No. 2, which practically corresponds to Art. 86 No. 2 EC. Derogations from TPA on the basis of economic and financial difficulties due to long-term take-or-pay commitments are regulated in Art. 25.

An important characteristic of third party access according to the Gas Directive is that access may be refused due to "lack of capacity" (Art. 17 No. 1). If the pipeline is being used at full capacity, new customers can therefore not require access. Consequently, existing users of the pipeline, in most instances the pipeline owner himself, would not be subjected to reducing their use when new players wish to enter the market.

In order not to prevent an actual market opening, Member States may, according to Art. 17, take the steps necessary to ensure that natural gas undertakings make "the necessary enhancements" in order to increase capacity. The extent of this provision is, however, not clear. First, Member States have no duty to require enhancements, but they "may" so decide. Second, enhancements may only be imposed if these are economical or if "a potential customer is willing to pay for them". We presume that an order to extend capacity is most timely in cases of so-called bottlenecks in the system, that is, for short distances or interconnections where capacity constraints are a particular problem.

3. NORWEGIAN GAS TRANSPORT REGIME

3.1 Implementation of EU gas directive

In October 2001 the EEA Committee decided to include the Gas Directive in the EEA-agreement [12]. The implementation of the directive in Norwegian law falls in two parts. The downstream rules – which constitute the major part of the directive – are implemented separately by a new act on common rules

[12] European Economic Area

for the internal gas market. The upstream rules - primarily expressed in article 23 of the directive – are implemented by an amendment to the Petroleum Act[13] (PA) and an amendment to the Petroleum Regulation[14] (PR).

In a Norwegian context, the most important issue is the regulation of the upstream network, which comprises all six export pipelines landed in UK (1), Germany (3), Belgium (1) and France (1). The regulation of the upstream network will be discussed in 3.2 to 3.4 below.

To fully understand the new regulation of the upstream networks in Norway, it should be kept in mind that the sector has undergone major changes during the last few years. Without going into details, the key words are as follows: The Norwegian State's oil company, Statoil AS, was partly privatized and listed in 2001. Beforehand the State established a new state owned company, Gassco AS, to operate the upstream network, which until privatization had been operated by Statoil. From 2003 the many pipeline joint ventures also have formed into one JV called Gassled. Last but not least was the Gas Negotiating Committee (GFU), which since 1987 have made joint sales of gas from the Norwegian continental shelf, abolished from 2002. The licensees are now selling all gas individually. The amendments to the PA and the PR discussed below were initiated not only by the obligation to implement the EU Gas Directive, but also by the need to facilitate for the described changes of the organization of the sector.

3.2 Amendment to the Petroleum Act (PA)

The amendment[15] to the PA of June 2002 implements the upstream network provisions in the Gas Directive. In a few words the amendment determines the main rule about TPA for eligible customers and natural gas undertakings and gives a frame for further regulations from the Ministry of Oil and Energy.

Three definitions were added to section 1-6:

- Upstream network, section 1-6 m);
- Natural gas undertaking, section 1-6 n); and
- Eligible customer, section 1-6 o).

The first two definitions are with one minor difference directly translated from the relevant definitions in the Gas Directive. In the definition of upstream network it is clarified that networks and facilities, which are used locally at the production site for production purposes, are not upstream

[13] Act of 29 November 1996 No. 72.
[14] Regulation of 27 June 1997 No. 653.
[15] Act of 28 June 2002 no 61, in force from 1 August 2002.

networks. This clarification is however, in accordance with the wording of Art. 23 No.1.

The definition of eligible customer does not correspond to a definition in the Gas Directive, but is based on the minimum requirement that follows from Art. 18. To be considered eligible according to the PA, a customer must not only comply with the minimum requirements, but also have access to the system in the EEA-state in which it is resident. In other words it must be considered eligible under national law in its resident state. On the other hand, having status as an eligible customer under national law in its home country does not qualify for status as eligible customer in the Norwegian upstream network if the customer's status as eligible under national law is based on requirements that are less strict than the minimum requirements in the Gas Directive. This could be the case for customers resident in a state which has opened up the market to a larger extent than the minimum requirements, for instance by giving industrial customers with an annual consumption of less than 25 million cubic meters of gas per consumption site status as eligible customers.

The main rule about TPA is laid down in an amendment to section 4-8, first paragraph, second sentence: "(...) natural gas undertakings and eligible customers domiciled in an EEA State shall have a right of access to upstream pipeline networks, including facilities supplying technical services incidental to such access."

Further details are not given in the PA, but the Ministry is empowered to give further regulations: "The Ministry stipulates further rules in the form of regulations and may impose conditions and issue orders relating to such access in the individual case" (section 4-8, first paragraph, third sentence).

In addition to the power to give both general regulations and individual conditions and orders, the Ministry also has an important tool in the approval procedure for agreements concluded according to the described rule.[16] The main rule is that any agreement on the use of upstream networks shall be submitted to the Ministry for approval. The Ministry's power, however, is much wider than just to choose between a simple yes or no to a submitted agreement. In connection with the approval the Ministry may "stipulate tariffs and other conditions which will ensure the implementation of projects with due regard to resource management considerations, and which will provide the owner of the facility with a reasonable profit taking into account, among other things, investment and risks".[17] The Ministry has this power also in cases where no agreement has been reached "within a reasonable period of time", and in the case an order is being issued.

[16] § 4-8, second paragraph.
[17] § 4-8, second paragraph, second sentence.

The practical importance of this approval procedure may at first sight seem minor, as the Ministry as discussed below has passed two regulations about TPA and tariffs. Such a conclusion would, however, be premature, as the standard terms and conditions for gas transport is determined by the operator after consultation with the owners of the upstream network and is still subject to subsequent approval from the Ministry.

3.3 Chapter 9 of the Petroleum Regulation (PR)

3.3.1 Introduction

On the legal basis i.a. of the PA section 4-8 discussed above, the Petroleum Regulation is from January 2003 amended [18] with a new chapter 9, which governs access to upstream pipeline networks. In the regulation a number of rules about TPA is laid down in detail: Principles for access to upstream pipeline networks (section 59), agreements for TPA concluded in the primary market and the tariff[19] (section 61 and 63), allocation of new capacity resulting from expansions (section 62), transfer of capacity rights in the secondary market (section 64), the system operator's responsibility for the system (section 66) and dispute settlement (section 68).

3.3.2 Principles for access

From the PA section 4-8 it follows that only natural gas undertakings and eligible customers have third party access. The PR section 59 introduces the additional requirement of "duly substantiated reasonable need" for TPA. The term is not easily understood, but the concept is interpreted by the industry to reflect that any shipper's capacity reservation shall correspond to an actual, present or reasonably foreseen need for transport and/or processing capacity. In other words, third parties are not allowed to take positions in the capacity market that do not reflect their actual needs. Further it seems reasonable to assume that a need must be based on e.g. own production or purchase of gas upstream of an entry point in the network.

[18] Regulation of 20 December 2002 No. 1618, in force from 1 January 2003.

[19] Section 63 of the PR about tariff for agreements in the primary market is discussed below together with the Tariff Regulation.

3.3.3 Primary capacity market

It follows from section 61 of the PR that the owner of the upstream network "shall make spare capacity (...) in the primary market available to the operator, who shall make it available collectively" and that agreements are "are to be entered into with the operator on behalf of the owner".

From the quote above it is evident that although several oil companies own the upstream network jointly they do no not have any capacity rights in their capacity of owners, neither for their own needs nor for third parties. Unlike a so-called "pipe-in-pipe system", the operator on behalf of all the owners markets the entire system capacity in the network jointly.

In doing so, the operator shall act neutrally and non-discriminating, and also secure commercially sensitive information. Any natural gas undertaking or eligible customer, who can duly substantiate a reasonable need for transport, shall get access to the system.

If the demand for capacity exceeds the spare capacity available, actual available capacity is to be distributed according to a formula determined by the operator.[20] However, the owners enjoy certain preferential rights when capacity is scarce. An owner has relative priority for its needs up to twice its equity interest in the network.[21]

3.3.4 Secondary capacity market

Capacity rights acquired in the primary market may be transferred by agreement in the secondary market.[22]

The requirements in the primary market appertaining to natural gas undertakings or eligible customers, and the duly substantiated reasonable need, apply correspondingly for transactions in the secondary market.

Transactions in the secondary market can take place either bilaterally or through a market place established and operated by the operator of the upstream network. The price in the secondary market is the market-clearing price and it is initially not subject to control from the Ministry.

3.3.5 Dispute settlement

Disputes relating to access may be referred to the Ministry or a Ministry authorized appointee. [23] Due to extensive state ownership in the sector and the organization of these ownership interests with the Ministry, it can be

[20] PR section 61, sixth paragraph.
[21] PR section 61, seventh paragraph.
[22] PR section 64, first paragraph.
[23] PR section 68.

questioned whether the Ministry has the necessary neutrality in relation to the parties in a potential dispute.[24]

3.4 System operator; Gassco AS

The operator of the upstream network, the state owned company Gassco AS, is awarded certain powers under Section 66 of the PR. The operator's responsibilities include operation of the network, maintenance and maintenance planning, and co-ordination of processes for further development of the network.[25] The operator shall also co-ordinate gas quality nominations at inlets and outlets and may under special circumstances even issue binding instructions to operators of production facilities in order to avoid operational disturbances or gas quality problems.[26]

Although the operator of the network under the operating agreement is an agent for the owners, under the PR the owners are partially deprived of their right to instruct their agent. According to section 66, sixth paragraph, the owners "may not instruct the operator in his performance of tasks assigned to him in or pursuant to this chapter, unless otherwise is specifically stipulated in these regulations".

3.5 The Tariff Regulation (TR)

The principles for tariffs in the primary market are laid down in section 63 of the PR. The tariff is based on the capacity reserved by a shipper, irrespective of whether it is actually used or not (ship-or-pay).[27] The tariff consists of a capital element and an operating element.[28] The capital element, stipulated by the Ministry, takes due consideration to resource management, a reasonable return on investments and other special circumstances.[29] The operating element is to be determined so that neither the owners nor the operator has any loss or profit other than the return determined by the capital element.[30]

In the Tariff Regulation[31] (TR) the exact tariffs are laid down by the Ministry. The TR divides the upstream network into 4 zones, A to D. Zone A and B are rich gas pipelines to the Kårstø processing terminal, while zone C is

[24] See Gas Directive Art. 23 No. 3.
[25] PR section 66, first paragraph.
[26] PR section 66, second and third paragraph.
[27] PR section 63, second paragraph.
[28] PR section 63, third paragraph.
[29] PR section 63, fourth paragraph.
[30] PR section 63, fifth paragraph.
[31] Regulation of 20 December 2002 No. 1724, in force from 1 January 2003.

the Kårstø processing terminal itself. Hence, the tariff in zone C is for processing services, while the tariff in zone A and B is for transport of rich gas to the terminal. Zone D covers the six export pipelines to final landing terminals in the UK and the Continent. The tariff in zone D is for transport of dry gas.

New tariff zones are currently under establishment as more systems and facilities will be included in the regime in the time to come. In the figure below an illustration is given showing the different tariff zones.

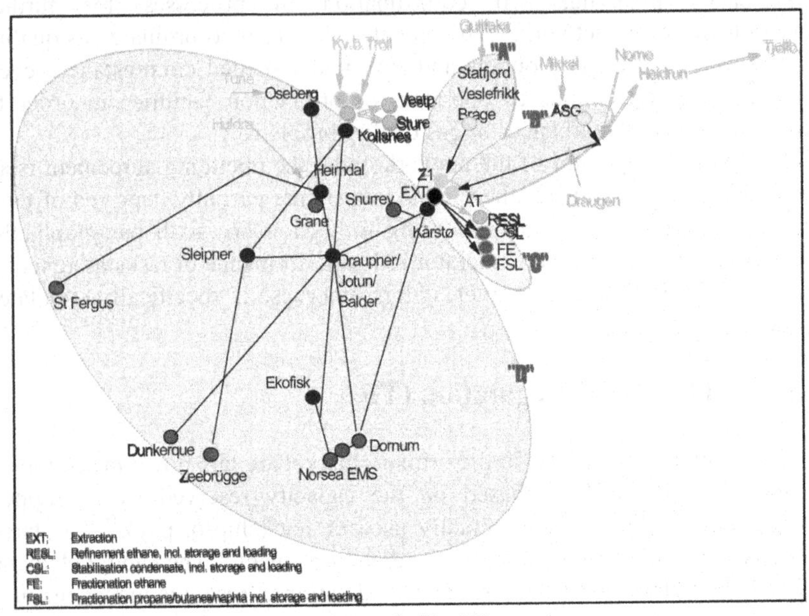

Figure 2, Tariff zones

The TR further determines the tariff formula for each of the four zones.[32] These formulas will not be discussed in this chapter.

3.6 Booking rules and dispatching procedures

3.6.1 Booking rules

The main purpose of the booking rules is to grant shippers access to capacity. The booking rules, including the Gassled terms and conditions for

[32] TR section 4.

transportation, cover principles for how to grant access to initial transportation capacity, capacity in the primary and secondary market. The rules provide a framework for the interface and communication between system operator, pipeline owners and shippers.

The booking rules have introduced new concepts, and some of these are briefly described in the list below.

"Qualification of Shipper" is a term referring to a requirement that a shipper must pass a test of predefined qualification questions related to the companies' credit rating. A qualified shipper will be granted access to Gassco's booking system and the shipper will obtain an electronic and digital certificate that enables him to access the Gassco's web-based software programs for capacity booking. This requirement is derived from the owners' interests to reduce the financial risks involved in providing transportation services in their systems.

"Qualified Need". The Royal Decree is explicit in that shippers must have a duly substantiated reasonable need in order to be allocated available capacity. The "qualified need" may be understood as the volume of gas that the shipper can demonstrate to have available at the Gassled entry points. (In order for the shipper to demonstrate the need, he has to send yearly production forecasts to Gassco or production permits. Further, he shall produce evidence of carry forward balance (i.e. an accumulated carry forward of a specific gas volume equal to the difference between actual production and production permits within a gas year). He shall also demonstrate the volumes of gas sold, purchased and/or lent at entry points. He shall also include potential volumes resulting from development of new production facilities and infrastructure. Gassco will consider the maturity of these fields in collaboration with the Norwegian Petroleum Directorate. Based on such discussions Gassco will decide whether these volumes qualify. The "Qualified Need" test, as described here is developed in order to comply with the Petroleum Regulation.

"Ability to Use" (AtU). In the Norwegian upstream pipeline system the different pipelines are integrated and interconnected. As a consequence of this technical design, many physical dependencies exist between the different pipelines and the many entry and exit points. In order to optimize physical transportation, these relationships must be taken into consideration when booking is done. AtU is thus a term which specifies how much gas a shipper is physically able to transport to a given exit point, taking into consideration all his entry points. The AtU is calculated in order to secure the principle that the shipper's gas shall be delivered with sufficient pressure to enable the system to undertake a transportation service to required exit point independently of how the other shippers chose to use their capacity rights. The AtU test, as outlined here, is developed in order to meet the requirement of the Petroleum Regulations.

"Capacity Allocation Key" (CAK). It is important to note that the AtU

figures, as described above, will in sum far exceed a given exit point's capacity. As a consequence of this fact, the shippers may submit booking requests that in sum exceed the exit point capacity. In such a situation the CAK is the allocation key used by Gassco to share the available capacity amongst the shippers who requested the capacity. The CAK formula is function of "qualified need" and AtU.

"Gross Calorific Value factor" (GCV factor) is calculated to convert between nominations that are done in energy units, and bookings that are done in volume units.

"Time periods" is a term referring to predefined periods of time in which capacity products are defined.

Several new processes have also been developed in order to facilitate capacity allocation and booking. The booking of capacity is applicable at the different entry and exit zones as defined by the Tariff Regulation. The booking is thus booking for entry capacity, process capacity and exit capacity.

If there are no capacity constraints, all shippers will be granted capacity according to their requests but limited by their "qualified need" and their AtU.

The basic principle of distributing capacity when it is scarce, is to let the owners first allocate capacity according to their requests, limited to two (2) times their equity interest in the transportation system [33] and limited by their "qualified need".

Spare capacity is made available in the primary market. In order to obtain capacity in the primary market, shippers have to book capacity in predefined time periods; long-term, medium-term and short-term including booking within the existing Day[34]. For the long-term periods, booking has to be done during an initial booking period. For shorter periods (medium- and short-term) capacity is first offered as an initial product. Any remaining capacity is offered on the "first come, first served" principle.

The secondary capacity market encompasses both the marketplace for transactions of capacity as facilitated by Gassco and the market for bilateral transactions between shippers. Several rules are developed for the use of secondary capacity rights. The core principle is that qualified shippers with existing capacity rights secured in the primary or secondary market can sell this capacity to qualified shippers. Bids and offers are posted on the marketplace and they are valid until acceptance or withdrawal. Invoicing between buyer and seller is done directly on a "net price" basis where the net price is the agreed price less the primary market tariff.

[33] These principles apply for the Gassled system.

[34] Day; in a contractual context the committed gas deliveries shall be delivered within a 24 hours period commencing at 06:00 hours, often referred to as the "Day".

3.6.2 Dispatching procedure

The dispatching procedure regulates information flow between stakeholders such as the production field operators, upstream operators and upstream shippers, Gassco, and downstream operators. The dispatching includes information on available production capacities, forecasts, nominations, instructions, reporting and much more.

4. DISCUSSION; THEORY AND PRACTICE

4.1 Introduction

In this section the authors' intention is to indicate and discuss the extent to which the regulatory transition is complying with relevant and well-recognized principles of economic literature. In order to do so some adequate technical comparisons are made to underpin the analyses.

The first topic looked into is the question; to what extent is it possible to obtain competition among transportation service providers – or in other words: is it possible to have more than one system operator performing the tasks? This question is discussed here by drawing upon some technical and operational aspects of the Norwegian system in particular. This description is also intended to give the non-technically trained reader an insight in some basic physical behavior of natural gas pipeline transportation.

The second issue discussed is the extent to which we can argue that the transportation network is in fact a natural monopoly. Having concluded it is, the third issue assessed is how the actual regulatory transition has adopted the basic principles of regulating a natural monopoly. The latter question is twofold; how do access rules align with recognized principles, and how efficiently does the tariff system work if compared with predefined economic efficiency criteria?

Finally, the last issue touched upon, albeit very briefly, is security of supply in a long-term perspective. The task at stake is how to secure investment initiatives in infrastructure development throughout the coming years. This is a difficult question indeed and no detailed analysis is presented here, merely some few comments.

4.2 Shall transportation services compete?

In this section we discuss the extent to which the integrated Norwegian production and transportation system causes economic advantages. We analyze and discuss in particular how the chosen operational organization minimizes negative economic externalities and creates good incentives for efficient operations. In sum we try to answer the question raised by the title; shall transportation services compete, given the physically integrated Norwegian system?

During the last three decades, the Norwegian transportation system has been developed from a single pipeline system (Norpipe system) into a complex interconnected network, as shown in Figure 1. New transportation- and treatment capacity has gradually been added and the network comprises today rich and dry gas pipelines, compressor stations, riser platforms and two onshore gas treatment plants. The system is by now the world's most comprehensive integrated offshore gas transportation network.

The development is caused by a series of decisions on how to efficiently supply natural gas transportation from a variety of oil and gas production fields offshore to customers in UK and continental Europe. The aim of the oil companies and the Norwegian authorities is to optimize petroleum resource management, optimize the gas value-chain, utilize existing capacity, make transportation reliable, offer flexibility with respect to supply sources and market opportunities, and also take the uncertainties in connection with future supply and demand into consideration. The strength of the Norwegian integrated system is its ability to handle different kinds of gas qualities produced in different regions and simultaneously fulfill changing demands in different markets at a high level of supply security. Hence for both producers and buyers, there are so-called network economics. Simultaneously, and due to the integrated nature of the transportation system, a range of operational aspects and limitations occur. Such issues, well known to pipeline operators, are often related to:

- The need for blending gas from different producers in order to (i.a.):
  - meet GCV and WI [35] delivery requirements
  - meet sulfur requirements
  - meet carbon dioxide requirements

- Varying transportation capacity, as each individual pipeline does not have a well-defined capacity because quantity transported in one pipeline affects the flow in other pipes in the network.

- Optimizing physical gas streams while simultaneously optimizing commercial demands for gas deliveries. Due to the nature of the integrated system, it is generally not desirable or even feasible for the operator to physically transport gas molecules in a routing

[35] Gross Calorific Value and Wobb Index

strictly following the contractual "routing". To take full advantage of the network, the buyer normally thus gets a gas stream originated from different sources.

In an economic context, the facts described above, namely that operations in one part of the network effect the conditions in the rest of the network, are referred to as economic externalities. A clear understanding of externalities is important because both producers and buyers may benefit or suffer from such externalities.

If we assume that the network had been split into several entities and a different operator had operated each entity, in order to introduce competition on transportation services, the identified externalities should preferably have been neutralized by tariff adjustments. Such adjustments are necessary in order to compensate for the disadvantage (or benefits) that a given externality causes on the neighboring entities.

Such neutralization should thus impose penalties on negative externalities and bonuses for positive externalities. However, as the externalities typically will change over time and as overall optimal operation mode depends on a number of discrete decisions such as flow directions and compressor modes, it is generally not possible to develop a tariff scheme that supports the optimal mode.

A "single system, single operator scheme" does not eliminate externalities. These schemes however, internalize some of the externalities as the scheme minimizes the number of boundaries and interfaces calling for explicit and well functioning coordination.

A single operator is also favorable in the case of network failures (for example during a compressor shutdown) as efficient coordination is required in order to secure the integrity of infrastructure as well as minimizing overall losses due to flow disturbances and gas quality deterioration.

Further, and due to economy of scale and scope, we are of the opinion that network management is handled more cost efficiently by a single operator, rather than having the tasks divided between several operators. This means that it is most cost efficient to have one company operating the total network and ensuring fulfillment of delivery and technical requirements rather than having several operators operating pipelines individually with corresponding duplication of systems and organizations.

The recent establishment of Gassco is thus in the view of the authors a promising industry structure that seems to recognize that the creation of one independent system operator is a practical and feasible way of organizing transportation services.

4.3 Is pipeline gas transportation a natural monopoly?

In this section we will discuss some technical issues and relate these to economic considerations in order to illustrate that pipeline transportation of natural gas is recognized as a natural monopoly. The core expression of gas flow in a pipeline can be expressed as follows (Weymouth's equation): [36]

$$Q_{SC} = \left(\frac{T_{SC}}{P_{SC}} \frac{\pi}{8} * \frac{60*60*24}{1*10^6} \right) \left[\frac{\left(P_1^2 - P_2^2\right) d^5 R}{M_{gas} T_S Z_S L f} \right]^{0.5}$$

In this equation the terms have the following meaning:

Q_{SC}	is the throughput (or flow rate),	MSm^3/day
P_1	is the pipeline inlet pressure,	bar
P_2	is the pipeline outlet pressure,	bar
d	is the pipeline diameter,	m
L	is the pipeline length,	m
M_{gas}	is the molecular weight of the gas	kg/kmol
T_S	is the ambient temperature of gas	K
T_{SC}	is the temperature at standard conditions	= 288,15 K
P_{SC}	is the pressure at standard conditions	= 1,01325 bar
R	is the gas constant	= 8314,34 J/(kmol*K)
Z_S	is the average compressibility factor of the gas	
F	is the friction factor	

Based on the flow equation some essential economic aspects can be derived. First, the investment cost of a pipeline is obviously linked to the diameter of the pipeline. The larger the diameter, the more steel is needed, and the more weight the pipeline will have. This implies that investment costs increase with increasing diameter as costs for pipeline material and installation will increase. As we can see from the equation however, the output (flow) will increase in the power of 2.5 for every unit of diameter invested. This economic fact is often referred to as economy of scale.

Another way of illustrating this fact is that an increase in diameter with 50%, keeping pressures and all other parameters fixed, will result in an increase of flow of 175%, indicating in fact that the average cost for natural gas pipeline transportation decreases with an increase in pipeline diameter.

[36] For further details see Dahl and Osmundsen (2002)

If we should consider the total transportation cost formula we must at least include the compressor costs. It is easily demonstrated (see for example Dahl and Osmundsen (2002)) that the compressor costs do not invalidate the demonstrated economics of scale. The work done by the compressors depend on the so-called compression ratio: P_1 (compressor outlet pressure) divided by the pressure at the compressor inlet (the suction pressure). If we keep the compressor ratio fixed, the necessary compressor work per unit gas, and hence the compressor related cost per unit gas, will not increase as the diameter and flow increase.[37]

The above discussion indicates that over a wide range, and due to the pipeline investment cost, the average cost of transporting gas decreases. This is a necessary condition for a natural monopoly.

Although the gas flow equation indicates substantial economics of scale, there are still some "technical constraints" that limit the chosen diameter and hence the lowest possible average transportation cost.

If several pipelines are constructed and operated in order to satisfy the transportation demand in a given geographical region, the "diameter economics of scale" demonstrated above will be of minor importance if we consider the total transportation systems.

However, the "integrated system" and the "single operator" arguments presented in this sub-section take the basic idea of natural monopoly from a single pipeline level to the network level. Although the economics of scale may be exempt on the pipeline level, there are economics of scale and scope at the network level that make the network a natural monopoly. This simply means that the total cost is least when the network is operated as a single, integrated unit.

4.4 The need for regulating a natural monopoly

Prior to recapping the basic principles of regulating a natural monopoly, an illustration can be presented, discussing two important questions. These questions focus on how to approach a given market; first, "is competition feasible?" and then "is competition desirable?" If we limit ourselves to answer these two questions by either *yes* or *no*, we can display the outcome in a matrix as shown below in Figure 3.

[37] It should also be noted that the compressor costs together with other operating costs typically make a minor part of total discounted lifecycle cost as the major cost contribution is related to pipeline investment, which is a sunk cost as the pipeline cannot be recovered or converted.

A market where no individual market player can manipulate the market characterizes the *usual case*.[38] Such a market will in theory limit price growth and any possibility for a producer to collect monopoly rent is ruled out.

There are many industries and markets however, which are not suited for full competition due to many reasons. One major reason in the context of natural gas transportation is that pipeline systems are considerable "sunk costs". Sunk costs are investments characterized by being neither reversible nor convertible into alternative usage or business opportunities for the investor.

		Is competition desirable?	
		Yes	**No**
Is competition feasible?	**Yes**	Usual case	Cream skimming
	No	Barriers to entry	Natural monopoly

Figure 3. Relationship between possibility and desirability of competition

The notion of *natural monopoly* has often been applied to describe the nature of such industries and the argument is that a transportation system, in theory, will exercise market power and collect monopoly profit if left unregulated. The transportation system will seek to maximize its profit at a throughput level where marginal cost equals marginal revenue. At this given level of output, the demand curve will specify the transportation tariff, which may be set at a price above average costs. As there are shippers in the transportation market with a willingness to utilize the system if tariffs were set below the monopoly level but above the marginal cost, welfare loss is occurring.

If competition was to be introduced in a natural monopoly, the transportation system owner may create technical or economic *barriers to entry* for potential competitors. If a competitor attempts to enter the existing market by constructing a new pipeline, the incumbent transportation system owner will have the option of increasing the throughput and reducing the tariff until it eliminates potential willingness to construct a new competing pipeline, provided there is excess capacity in the system.

Lastly, if competition was to be allowed in such markets there is a risk that the supplier of transportation services will limit its offer to those market segments that are most profitable and leave out the more costly and distant

[38] For a further discussion of this issue please see Armstrong et al (1994)

markets, a notion referred to as *cream skimming*. One counteraction often applied by authorities and regulatory bodies is to request the transportation system owners to deliver their services at a regulated tariff to defined geographic regions. This obligation is typically a condition for concession, often referred to as a "public service obligation".

A main question thus remains: how shall we obtain more competition and market liberalization in a market characterized as a natural monopoly? The core notion and answer to the questions is illustrated in Figure 4 below.

As Figure 4 illustrates the notion is to allow for increased competition between the producers (sellers) of the goods or services (e.g. power, natural gas or railway carriage) in the market. But due to the nature of these types of services, being dependent on a network, the network access and cost of usage must be regulated. These two factors form the core element of the EU gas directive and the new Norwegian regulation as detailed in the previous sections of this paper. In the following sections a theoretical approach is presented for how to deal with these issues, as an ideal scenario.

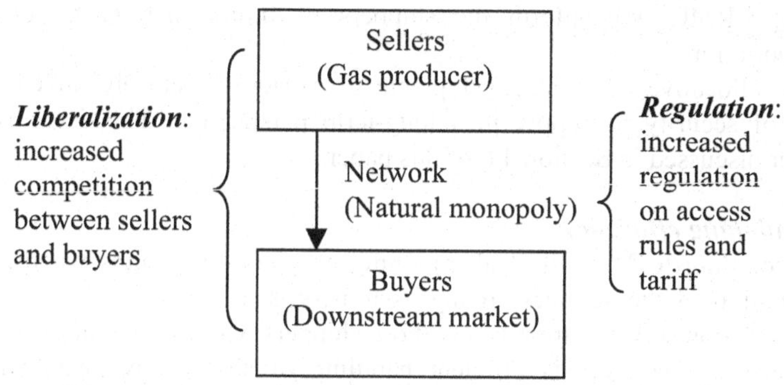

Figure 4. Main elements of the new regulatory regime

4.5 The need for regulating tariff and access rules

4.5.1 Ideal economic efficiency criteria

In this section we describe a few, but essential economic efficiency criteria applicable when designing a tariff regime.

Dynamic efficiency

Dynamic efficiency is a measure of a firm's ability to respond to changing market demands by producing more and better products and finding ways of producing at lower cost, see IEA (1994). Another way of putting it is that innovation and investments in cost reductions are socially efficient.

Allocative efficiency

Allocative efficiency defines the extent to which "the quantity of product and service supplied is efficient"[39]. In a broad perspective this means that all customers who are willing to pay a price equal to or above the marginal cost of production and transportation shall be supplied with gas. When it comes to transportation services only, the aim is to have sufficient capacity to serve all shippers with a willingness to pay a tariff equal to or above marginal transportation costs. Allocative inefficiency exists if this aim is not met or if there is excess capacity due to lack of willingness to pay the given tariff.

One theoretical approach for enhancing allocative efficiency is a tariff regime based on long-run marginal costs (LRMC) where the marginal costs accurately reflect the development of new capacity in a region over a given time frame. If such information is available this may be a feasible approach, as the LRMC will inform the shippers of future incremental costs of transportation. [40]

An allocative efficient development of capacity is closely linked to the issue of security of supply in a long-term perspective. This issue is thus further discussed in Section 4.6 of this paper.

Rationing efficiency

According to Mansell et al. rationing efficiency measures whether "the distribution of the services among customers is efficient". This means that transportation services are given to those shippers who earn the most by using the service. The goal is efficient handling of short daily peak demands efficiently under the assumption of a given system.

The criterion is relevant in a short-run perspective and the criterion may be quite useful for measuring the efficiency of potential auction principles as well as capacity allocation and booking rules in general.

Cost efficiency

Cost efficiency includes the concept of providing the services at the lowest possible cost. The efficiency criterion may also include managerial efficiency. This criterion is relevant in a short-term perspective when it comes to the variable operational costs as well as fixed operational and maintenance costs.

[39] Mansell et al. (1995)

[40] These principles are applied in the UK's "network" code according to Madden (1997) at Financial Times, see page 56. The tariff is based on a LRMC evaluation over future 10 years demands in a given region.

The measure also is applicable in a long-run perspective related to minimizing cost of new capacity.

Efficient product selection

The type of services offered is efficient; i.e., the "menu" of services is differentiated to match the services demanded by shippers. This measure is relevant in a short-term perspective, even though these types of services likely will develop over time.

4.5.2 Optimal theoretical tariff and access rules

The main purpose of the tariff regime – in conceptualized terms – is thus to specify the services provided to shippers such as the booking rules and capacity rights, and to specify the tariff applicable for a specific service. A number of authors, such as Cave and Doyle (1994), Armstrong et al. (1994), Mansell and Church (1995), the IEA Study (1994) reports, and Hope (2000) discuss these issues theoretically and practically.[41]

In the following section we will discuss different tariff options in light of the efficiency criteria.

Marginal-, average cost-, and rate of return – tariff

If we assume some specific ideal conditions, it is well known that [42] for a given capacity, a tariff equal to marginal cost of transportation maximizes social welfare - the sum of consumer's and producer's surplus. Figure 5 offers a standard illustration. We now assume that the pipeline (transportation system) capacity is fixed and equal to K. For simplicity, short run marginal cost is assumed to be flow-independent and equal to a up to K. Marginal capacity expansion cost is equal to b per unit and also indicated. Long run marginal cost is then $a + b$. Any other fixed and sunk costs that induce decreasing both short and long run average cost (natural monopoly) are also disregarded here.

[41] Cave and Doyle (1994), Armstrong et al. (1994), and Mansell and Church (1995). See also the study reports from IEA and Hope (2000) see page 177.

[42] See Von der Fehr (1996) or Bjørkvoll (1996)

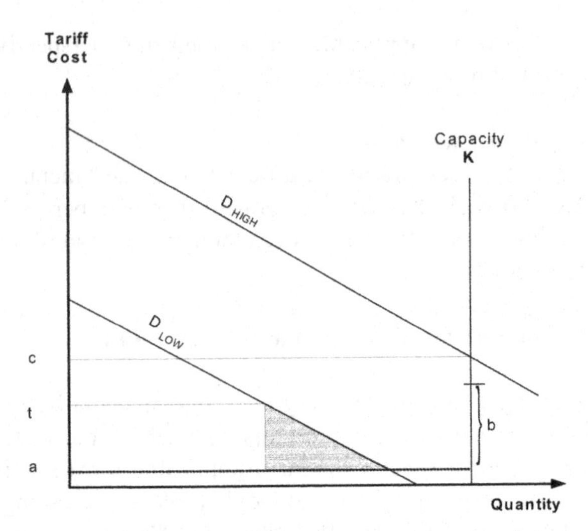

Figure 5. Relationship between demand and cost

If demand happens to be low, as in the summer and indicated by D_{LOW}, the tariff should be equal to the short run marginal cost *a*. If the tariff is higher than *a* – indicated by *t* in figure 5, the result will be a deadweight loss (hatched) as the willingness to pay will be greater than the relevant transportation cost *a*. Tariff equal to *a* will cover short run cost only and will not contribute to capacity and other fixed cost recovery.

If demand happens to be greater than *K* for a tariff equal to *a*, the tariff should be increased to equalize demand and supply at the capacity level *K*. For the demand D_{HIGH} , the optimal tariff is *c*. For the D_{HIGH} demand case, a tariff below *c* will induce excess demand and call for a non-price mechanism to pick out customers - preferably those with the highest willingness to pay.

The clearing tariff equal to *c* includes a shadow price, equal to $c - a$, which is the consumers net willingness to pay for a marginal capacity expansion.

This is the marginal cost pricing scheme and it offers the same result as a competitive spot market, in which atomistic actors offer capacity. The marginal cost scheme or "spot market" maximizes welfare and this scheme is regularly used as a benchmark when appraising other sales and tariff mechanisms. Evidently, the marginal cost-pricing scheme is a congestion pricing scheme and this contribute to capacity cost recovery.

Regarding optimal investment, the capacity should be fixed so that expected willingness to pay is equal to the long run marginal cost (here: $a + b$). In the case of a natural monopoly with overall declining average costs the long term marginal cost is less than average cost and the marginal cost pricing scheme and optimal investment principle will inevitable induce a loss. This will typically be the case if we add a fixed cost to the cost structure indicated in figure 5. The marginal cost pricing may ask for a benevolent investor as the optimal investment rule may expect to bring a loss for the natural monopoly owner.

The marginal cost-pricing scheme seems rather risky to the firm, especially with the cost structure in figure 5 with short run marginal cost well below long run average cost up to the capacity limit. If demand (e.g. willingness to pay) turns out to be slightly less than expected, the marginal cost pricing scheme will induce a loss, and if actual demand turns out to be somewhat below expected demand, the tariff will be equal to short run marginal cost a only.

Indeed, imposing the marginal cost-pricing scheme on a monopoly the company's investment incentives will adversely be affected. As the company will maximize the expected profit, the monopolist will build less capacity than without any regulation. Cutting back capacity makes congestion and high tariffs more likely and unprofitable tariffs equal to short run marginal cost less likely. Under a purely marginal cost-pricing regime, cutting back capacity and creating congestion will be the only way to secure the monopoly rent. Although optimal in the short run, marginal cost pricing is indeed bad in the long run[43]. If the monopoly is allowed to charge more than short run marginal cost when capacity is not limiting, it may improve investment incentives and hence overall efficiency.

Marginal cost pricing has been debated on several grounds other than those pointed out here. Much of the criticism of marginal cost pricing is connected with governmental financing of the firm's deficit and how pricing schemes in different ways attenuate, or create inappropriate incentives for cost reduction.

If we now introduce the notion of a balanced budget, the concept of average cost pricing will serve the purpose. By imposing a balanced budget the firm must demonstrate its viability. However, this is not the problem we face. There is typically no doubt about the company's overall profitability, so the budget constraint should not be introduced to demonstrate profitability, but to constrain tariffs.

The average cost based tariff covers all costs by definition so the budget constraint does not necessarily lead to cost effectiveness. Rather to the contrary, cost incentives are relatively ineffective, as costs very well can be passed on to shippers by increasing the tariffs. In addition, rationing efficiency is a less than ideal solution as the average cost and hence the tariff is inversely related to flow. If the pipeline is not allowed to make a profit,

[43] Cost structure equivalent to the structure in figure 5 and uncertain demand - see Bjørkvoll (1994)

incentives for investment will also be blurred; hence the average cost pricing does not enhance allocative efficiency.

Despite these shortcomings, the average cost scheme, slightly adapted and known as rate of return regulation, is widely used in network industries. In short, the rate of return regulation is an average cost regulation in which the regulator allows for an additional regulated return on investment. It is well known from the economic literature that rate of return regulation has poor incentives regarding cost efficiency[44]. Gold plating is for example a possible measure to increase the rate-base, tariffs and hence profit. We also note that for a given pipeline the average costs will decline as spare capacity is utilized. As cost and tariff are closely linked, the tariff under the rate of return scheme will tend to be high if flow is steadily below capacity and at a minimum if the pipe is congested year round. Obviously, the rationing efficiency is not taken care of under this scheme. .

Finally, it should be noted however, that the rate of return regulation not necessarily exclude congestion pricing, two part tariffs or other more "advanced" tariff schemes in order to enhance rationing efficiency.

Since the 1980's, the incentive and information issues have been the principal theme in regulation economics. Laffont and Tirole (1993) offer an excellent presentation on this matter. Despite the theoretical progress, traditional rate of return schemes are still being implemented today. There are a number of possible explanations for this situation, which we will not go into here. According to Laffont and Tirole, theorists possibly have a long way to go to make their models more realistic and operational[45]

Tariff schemes employed in the gas transportation industry.

Both natural gas as well as gas transportation service have traditionally been traded under take or pay contracts. In the gas transportation industry, deregulation typically brings along a transition from bilateral long term contracting to transparency and non-discriminatory behavior. Transportation suppliers typically offer a menu of often short-term transportation contracts with different degrees of priority and flexibility. In some systems shippers may also bid for capacity rights.

In addition to the often applied firm and interruptible services, a number of other contracts are used. The no-notice services and off peak services are two examples. The no-notice service is a firm service giving the shipper the option to withdraw additional quantities of gas on a no-notice basis. The off

[44] The seminal paper is: Averch, H. & Johnson, L., 1962: "Behavior of the Firm Under Regulatory Constraint", *American Economic Review,* vol 52, No 3, page 1053-1069 (19)
[45] Laffont and Tirole (1993) *Foreword page xvii*

peak service is a kind of firm services that may be interrupted a number of days during peak periods[46].

The usual transportation tariff formula includes first, a capacity charge that gives booking rights for a certain capacity and secondly, a variable part charged per unit of gas actually transported, in order to cover variable costs.

In the same way, offering other services than firm and interruptible transportation may improve on welfare. [47] This is due to characteristics on both the demand and the supply side. In the case of demand, customers have different levels of willingness to pay for not being cut off, whereas on the supply side, curtailment varies from no curtailment to just serving the firm customers. Hence, it is possible for the transporters to offer a menu of priority classes and let customers buy a number, or a basket, of transportation rights with different priorities that match their needs. Such priority pricing may be equivalent to the spot market solution or the marginal cost pricing schemes and can indeed be welfare maximizing [48].

4.5.3 Comparison; theory and the Norwegian approach

In this sub-section we have chosen some few issues, among many, for comparing the theory against the Norwegian approach. One should further note that the access regime and the tariff regulation were implemented less than a year ago and thus experiences are limited.

As was clarified in Section 3, the Norwegian regime provides the possibility of requiring capacity in two markets, the primary and the secondary market. In the primary market long term, medium term and short-term rights are traded. In the primary market capacity rights are traded at tariffs determined and published by Norwegian authorities. Tariffs in the primary market are fixed for each individual and predefined zone of the transportation system.

Unused capacity rights bought in the primary market shall, under given conditions, be released to the secondary market. In the secondary market transportation rights may be traded at any price at the marketplace. The marketplace is facilitated by Gassco or by bilateral agreements.

The primary market tariff consists of a fixed part and a variable part. The fixed part has to be paid for the right to transport one unit of gas. The fixed part is also called the capital element (chapter 3.5). The tariffs are stipulated

[46] (For details see the Columbia Gas Transmission Corp - http://www.columbiagastrans.com/tco_transport.html.)

[47] This is one of the basic ideas within the literature on nonlinear pricing – se for example Willig (1978).

[48] For priority pricing and other equivalent schemes see for example Robert Wilson's book *Nonlinear pricing*

to yield a pre-tax rate of return equal to 7 % p.a. [49]. The variable (operating) part is paid for each unit of gas actually transported. The variable part is equal to estimated average operating unit costs. The variable part is not equal to short run marginal costs [50] but the difference seems limited. In fact, for the "dry gas zone"– the "blue area" in figure 2, where the capital element is approximately 15 øre [51] per Sm3, the average operating unit cost is close to 2 øre per Sm3 which is equal to 13% of the capital element. [52]

The capacity allocation mechanism is straightforward when capacity is not scarce. Capacity is then offered on the "first come, first served" principle. Tariffs are fixed and independent of demand and access is nondiscriminatory.

Although, the regulation implies that the tariff is rigid downwards (fixed), a tariff almost equal to short run marginal cost is possible if one or several shippers first buy transportation rights in the primary market and thereafter find that the booked capacity will not be utilized and release the rights to the secondary market when, at the same time, pipeline capacity is not scarce.

When capacity is scarce, the owners are first allocated capacity according to their requests, limited by:

i) two times their equity interest in the transportation system
ii) their "qualified need"

Hence, the regulation offers owners possibilities to book all available capacity.

If capacity is still scarce after this "first round of allocation" amongst owners, capacity is allocated by the Capacity Allocation Key (CAK). CAK is not intended to allocate capacity according to willingness to pay [53]. However, the possibility of releasing the assigned capacity to the secondary market will have an effect. The price for released capacity will constitute the shipper's opportunity cost and capacity may be reallocated according to willingness to pay.

The CAK is used when capacity is scarce and spot market tariffs would have been higher than the fixed and published tariffs. Simultaneously then,

[49] The pre tax rate of return is expected to be 7% - see regulations December 20th 2002: http://www.dep.no/archive/oedvedlegg/01/02/depar055.pdf

[50] The variable part comprise fixed operating costs which tends to make the variable part greater than short run marginal cost. Besides the short run marginal cost is increasing with throughput (which we here consider to be of minor importance).

[51] October 2003: Approximately 2 cent or 1,8 Eurocent per Sm^3.

[52] The figures are taken from the Gassco web site, see: http://www. Gassco.no

[53] According to the authorities, one reason for not allocating capacity in primary market according to willingness to pay (i.e. by implementing a market mechanism), is that for given tariffs, owners and non-owners will facing different marginal cost. See regulation of December 20th 2002: http://www.dep.no/archive/oedvedlegg/01/02/depar055.pdf

the CAK-mechanism distributes the scarcity payments. The CAK formula is a function of "qualified need" and "Ability to Use" [54] and only indirectly a function of the owner share.

Owners finding the CAK favorable, may in a game theoretic perspective consider it wise not to cash in by releasing capacity to the secondary market, but rather use the rights even when such use is not optimal in the short run.

Apparently, it is not fully clear how the capacity allocation mechanism will work and partly based on some hearing round submissions the authorities have decided, with effect by January 1st 2008, to reconsider the owner's preferential rights (twice their owner share) and the effect of this provision will be especially monitored. [55]

In peak periods, third parties typically will have to acquire transportation rights in the secondary market. As "the price in the secondary market is the market-clearing price and is not subject to control from the Ministry" (ch.3.3.4) in theory, the secondary market will provide rationing efficiency. For third parties access in peak periods, the functioning of the secondary market is crucial and a prerequisite for the secondary market to function is thus that capacity is released.

However, the Capacity Allocation Key (CAK) allocates capacity to a number of participants (owners). Hence, supply in the secondary market will be more "atomistic" than in the usual textbook monopoly situation. As actors may make different judgments, capacity may be released and thus initiate the secondary market. It is however, too early to tell how well this secondary market will function.

So far we have ignored the fact the marginal transportation cost is different for owners and other shippers. If the transportation system has just one owner, it is quite simple as the owner pays the tariff with one hand and receives the revenue with the other. In fact, the owner faces the marginal transportation cost only, while the other shippers have to pay the tariff. This applies as long as capacity is not scarce. If capacity is scarce, the owner has to make a choice, at least on the margin, whether he should transport his own gas or the third parties quantity. In this case, the market-clearing tariff will be the relevant marginal cost for both owner (opportunity cost) and the other shippers.

There are a number of owners; some hold small shares and some hold larger shares. Paying one € for transportation in the primary market, the

[54] The "Ability to use" is an element of the CAK and it thus adds further restrictions to the problem of optimizing network flow. In general, adding restrictions to an optimization problem will reduce the opportunity set and hence the flexibility that the network offers to split contracts and physical flow. If the restrictions are effective they will inevitably also cut off solutions that would have been optimal in the absence of this restrictions.

[55] http://odin.dep.no/archive/oedvedlegg/01/02/kglre054.pdf

owner will receive his ownership share of this payment. When capacity is scarce, the possibility to release capacity to the secondary market does modify this argument. By releasing capacity, the stakeholder (e.g. owner) will receive 100% of the difference between the fixed primary market tariffs and the secondary market tariff. As a result the relative difference between opportunity cost and tariff is reduced when the market clearing tariffs are raised[56]. Although not perfect, the secondary market then will provide rationing efficiency.

Regarding investment incentives of the Norwegian regulations, we have indicated that regulating a monopolist by imposing marginal cost pricing scheme yields poor incentives. Maximizing expected profit, the monopolist will build less capacity than he would if there was no regulation. In this way, unprofitable tariffs equal to short run marginal cost are made less likely.

By the fixed tariffs in the primary market, the Norwegian regulations exclude tariffs equal to short run marginal costs or "free riding" shippers.

As tariffs in the secondary market are market clearing, the owners also collect the full scarcity payments. Hence, the investment incentives should be better than under a marginal cost-pricing regime. The owners rights are pursued, as they are given priority when capacity is scarce, nor are they being inadequately compensated by low tariffs.

Indeed, investment incentives are explicitly of major importance to the Norwegian authorities[57]. To new investors this may be an important signal.

4.5.4 Dynamic efficiency in a liberalized regime

In a liberalized regime, an important consideration is how to finance innovation and technological change in order to secure an efficient diffusion of new technology and in order to increase *dynamic efficiency*. One solution to the problem, adopted in the Norwegian context, is a requirement that shippers shall pay a fixed percentage of the tariff to help cover for research and development costs.

Further, and as a result of the nature of this industry, dynamic efficiency is basically a long-term related question due to several reasons. The sunk costs obviously limits the possibilities for introducing new technology into old systems. If new technology is introduced the old must be phased out. It may thus be more cost effective to operate and utilize the old technology. Further, new investments are throughput driven. No new investments are done if no

[56] The absolute difference between opportunity cost and market clearing tariff is equal to ownership share times primary market tariff.

[57] See regulation of December 20th 2002: http://www.dep.no/archive/oedvedlegg/01/02/depar055.pdf

more gas is sold in the market. Finally, new capacity is developed in large incremental steps. As a consequence of these facts, diffusion of innovation and technological change will largely take place in connection with the development of new capacity, at least in connection with "hardware" components of the systems.

5. CONCLUDING REMARKS

In this paper the authors have focused on the main question; how has the EU gas directive impacted on gas exporters? We have shown that core elements of the Gas Directive are allocated to rules for how to obtain access to essential systems. In a Norwegian context, the main question is thus; how shall Norwegian authorities regulate access to the Norwegian transportation system for natural gas. Such regulation is off course of great importance, as the gas exporters are fully dependent on such transportation services in order to sell their gas in the downstream market.

We have shown that the EU Gas Directive has resulted in an amendment of the Norwegian legislation. The Petroleum Act has been revised and new Regulations are issued to incorporate the requirements of the Gas Directive.

The relatively brief discussion of the new Norwegian regime for gas transport, offered in this paper, has demonstrated a regulated system of third party access (TPA) covering the upstream network. The Ministry has appointed a state owned single purpose company, Gassco AS, as independent system operator for the network. The Ministry has further determined the tariffs and it has retained the right to approve terms and conditions for gas transport.

The Norwegian regulation ensures third parties access to the transportation system by requiring capacity by means of a primary and a secondary market. The authorities publish the tariff for the primary market. In the secondary market, the tariffs are not regulated and they are thus market-cleared. Based on this the authors will indicate that the secondary market, in theory, will provide some degree of rationing efficiency.

Finally, and due to the fact that the Gas Directive in essence regulates access to systems as it's main instrument, no attempt is made by the authors to analyze how gas exporters may benefit in the downstream gas market as a result of the regulatory transition.

6. REFERENCES

Armstrong, M., Cowan, S., and Vickers, J. (1994), *Regulatory Reform, Economic Analysis and British Experience*, The MIT Press, Cambridge, USA.

Bjørkvoll, T (1994) Natural gas pipeline transportation; Costs, tariffs and capacity. (In Norwegian) Doctoral dissertation, 1994:73 ISBN 82-7119-678-2

Dahl, H. J. (2001), Norwegian Natural Gas Transportation Systems. Operations in a Liberalized European Gas Market, Doctoral dissertation, "001:04, ISBN 82-7984-198-9

Dahl, H. J. and Osmundsen, P., (2002), *Cost Structure in Natural Gas Distribution*, Conference Proceedings at the 25th IAEE International Conference, Aberdeen

Laffont, J-J and J. Tirole (1993) *A theory of Incentives in procurement and Regulation.* MIT Press

Mansell, R.L. and Church, J.R. (1995), *Traditional and Incentive Regulation, Application to Natural Gas Pipelines in Canada*, The Van Horne Institute, Toronto, Ontario, Canada.

Willig, Robert D: (1978) Pareto-superior Nonlinear Outlay Schedules, Bell
Journal of Economics Spr. 1978; 9(1): 56-69.

Wilson,-Robert (1993) Nonlinear pricing, Oxford University Press in association with the Electric Power Research Institute, 1993;

Sankar, U. Public sector pricing: theory and applications. IEA Trust for Research and Development, 1992

THE IMPEDIMENTS TO ESTABLISH AN ENERGY-REGIME IN CAUCASIA

Gokhan BACIK and Bulent ARAS
Basic: Lecturer at Department of International Relations, Fatih University, Istanbul.
Aras: Assoc. Prof. and Dean of Graduate School of Social Sciences, Fatih University, Istanbul

Abstract: The aim of this paper is to analyze whether there is a satisfying ground for a functioning energy regime in Caucasia. In this paper, our basic assumption is as follows: Given the problematic nature of the regional politics, it is hardly possible to enhance regional cooperation through bilateral relations. The nature of regional politics, here, limits actors' behaviors.

Key words: Energy regime, regional politics, regional cooperation, bilateral relations

1. INTRODUCTION

Several regional problems have come to the fore in the post Cold War Era. Caucasia has been one of them. Historically speaking, general trends of instability and economic deficiency have shaped the international image of Caucasia for years. Many different and difficult problems do still exist in the region. Put categorically, the problems of the region can be categorized as follows:

- Problems concerning *domestic sovereignty*: The Abkhazian problem in Georgia, several political unrests happened in different Caucasian countries such as Azerbaijan. Here, we use Stephen Krasner's definition of domestic sovereignty. It refers to the organization of public authority within a state and to the level of effective control exercised by those holding authority [1]. In other words, the

J. Hetland and T. Gochitashvili (eds.),
Security of Natural Gas Supply through Transit Countries, 343–354.

Caucasian states do still have different problems in terms of attaining and establishing a full domestic order in their territories. Typically, those countries face important problems during the transition eras. The "presidential crisis" of Azerbaijan is a recent case. Any structural change, including power transfers, might turn to be great crisis in such countries.

- Problems concerning *international legal sovereignty*: This type of sovereignty refers to the mutual recognition of states. This recognition entails the territorial unity of recognized state. From this perspective several problems also exist in the region. The Nagorno Karabakh problem between Armenia and Azerbaijan is one of the most important cases. Along with this well known case several other region-level problems might be analyzed within the same context. Though no official document confirms the case, the problems between Turkey and Armenia have the same nature. These countries' problems are about the very existence of statehood. Though no territorial problem exists between them, there is no diplomatic relationship between Turkey and Armenia. In the same context, the war in Chechnya is another great problem in the region. This problem weakens the domestic sovereignty of Russia. At the same time, the Chechen resistance does not recognize the international legal sovereignty of the Russian Federation. Generally speaking, there is region-wide fear about legal sovereignty. The Caucasian countries are still bound by fears of invasion, foreign intervention.

Having summarized the cited theoretical categories, a more practical discussion might be instructive. The Caucasian region has been under two types of foreign influence since the end of the Cold War. First have been the great powers such as the Russian and the American influence. Second has been the regional medium sized powers' (such as Turkey and Iran) competition over Caucasia. Therefore, the Caucasian region has been shaped by competing foreign policies as well. Both domestic and foreign parameters are acting on the region's politics. Once this multi-level political game is defined in the region, it should be noted that decision-making and political solution mechanism, here, are extremely complex, slow and risky.

A functioning energy regime in the area seems a logical solution to the region-wide problems. Countries facing economic and political problems like Georgia, Azerbaijan and Armenia are not capable of planning and practicing region-level policies, which would enhance regional cooperation. What is worse, the region is under the influence of different and competing policies and states. Each country has a different project on the same issue. For the lack of a functioning region-level regime of any type is a reality, power is the only determining fact in regional configurations. This generates two important problems: First, the instable structure of the region is kept untouched. Second, a region-wide indeterminacy impedes bilateral relations. Therefore any regional project quickly turns to be an area of competition between regional and international actors. The lack of a functioning regional regime yet generates odd cases. Countries with important natural recourses cannot rescue themselves from important economic problems. Thanks to the lack of a functioning regime, an anachronistic mode of diplomacy (secret negotiations, playing one against other, the abuse of political problems for economic interests) still exists in this region.

The problematic nature of the region has just been summarized above. Considering the picture drawn so far, this paper aims to analyze whether an energy regime is feasible in the area. Given the problems alike, is there a satisfying ground for a functioning security regime in the region? Our conclusion is negative, at least, in the short and medium runs and we will explain this position in the following parts of this chapter.

2. INTERNATIONAL REGIMES

A regime is defined as sets of implicit or explicit principles, norms, rules and decision-making procedures around which actors' expectations converge in a given area of international relations [2]. The convergence of expectations means that participants in the international system have similar ideas about what rules will govern their mutual participation: everyone expects to play by the same rules. Several other definitions can be listed as well. According to Nye and Keohane, regimes are "sets of governing arrangements [that include] networks of rules, norms, and procedures that regularize behavior and control its effects" [3]. As seen from the definition, a regime is about accepted norms, rules and procedures. Practically speaking, a regime means almost the end of indeterminacy in a defined region. A regime thus gives way for shared expectations enhanced by formal laws, rules and organizations. A regime aims to end randomized behaviors in a defined area. A regime also aims to make commitments credible. Regimes might be about territorial and non-territorial issues. The World Intellectual Property Organization (WIPO), for instance, is a well-known non-territorial regime. Another well-known worldwide regime is GATT (General Agreement on Trade and Tariffs).

A regime, in general, should have four defining elements [4]:

1. Principles: These principles shape the members' perception of world (or the related region). Through these common principles a

common ground is created among member countries. Principles are beliefs of fact, causation, and rectitude [5].

2. Norms: Members of a regime should have general standards of behavior, and identify the rights and obligation of states.
3. Rules: Rules are designed to reconcile conflicts that may exist.
4. Decision-making procedures identify specific prescriptions for behavior.

A regime depends on a simple logic: There is a high level of interdependency among world states. Therefore it is now difficult to solve problems through bilateral mechanism. There is even no bilateral problem at all. All problems have been re-contextualised in broader contexts. The international community does interfere any problem no matter how bilateral it is. A bilateral solution fails in finding a solution since many other countries might still be contending. A regime aims to solve problems at regime-level rather than endless bilateral talks. A variety of norms, in a regime, aim to guide the behavior of regime members in such a way to produce collective outcomes which are in harmony with the goals and shared convictions that are specified in the regime principles [6].

3. WHY AN ENERGY REGIME?

Regional potentials have been sacrificed because of bilateral long processes. The lack of a region-level functioning regime impedes the success of regional countries. Many projects thus have been slowed down. The Blue Stream Project can be given as an apt case. Ranging from geopolitical concerns to corruption, the Blue Stream Project from the very beginning has faced several confrontations. Why? First, as a traditional motto of diplomacy, foreign policy is shaped by national interest. And national interest is strongly affected by the societal roots of foreign policy. Considering the Blue Stream Project between Turkey and Russia, from the very beginning it has been tied by policymakers to some other important non-technical concerns. This project has meant many other things for Turkey. It was never perceived as a simple technical agreement on how to find new energy sources for Turkey. Second, competition or cooperation on energy policy in the region again has strongly been related to important international concerns. It is obvious that, "political events in the Caucasus, including the Chechen war, Nagorno-Karabakh (Armenian-Azerbaijan) and Georgian-Abkhazian conflicts, in Turkey, will have as much influence on export projects as the reserves of the Caspian region..." [7]. Energy needs to be understood as a keyword associated with security and economy based on the emerging international behavioral model in the region. Within this context, the Blue Stream Project has been criticized in Turkey both for its perceived national and international shortcomings. The lack of a functioning energy regime in the region creates a chaotic atmosphere in which a quick decision making process is impossible. On the other hand it is a fact that there is a new regional division line in the region, which is associated with energy projects. When the United States, Azerbaijan and Georgia support a proposal, both Russia and Iran object, since these countries as having negative political implications perceive any project, which is explicitly backed by the United States [8]. Such conflictual divisions may this time quickly become sources of new regional conflicts. It should be remembered that the Georgian president has stated that it was Russia behind the rebels in Western Georgia who attacked the pipeline construction projects in his country [9]. Since "the region has more national borders per square mile than any other on earth, and inside each small nation are a dozen or so tiny ones struggling to get out", a perfect overall cooperation strategy, which does not exclude any of the region's country, is needed [10]. Following the words of Charles William Maynes [11]:

> *"It is doubtful that this region can develop as all would wish unless there can be more regional cooperation. But another key condition must also be met. Not only must the three countries at the very heart of the region -- Armenia, Azerbaijan, and Georgia -- begin to cooperate with one another in a regional setting, but the three regional powers of significance -- Iran, Russia, and Turkey -- must believe that the arrangements reached exclude none and reward all. Otherwise, Western policy in the area rests on quicksand. Each of the three regional powers is influential enough in the region to play the role of spoiler by subverting any proposed settlement. And the only way the United States and its allies can prevent this is to protect a threatened country by offering a security guarantee."*

The same project has also given way to several conflicts between Turkey and Turkmenistan [12]. As is known, "a Joint Technical Study Agreement for the "Right (Northeastern) Bank of Amu Derya River" region was signed as an extension of the studies on Block-V on September 15, 1998. The studies started in October 1998 in accordance with this agreement and were completed in 1999 with the conclusion that the gas-condensate reserves in the block area were not commercially viable under current economic conditions. Beginning in May 1999, Turkmenistan, fully intent on accelerating activities for the purpose of supplying gas to the Trans-Caspian pipeline, authorized a strategic alliance agreement. According to the terms of this agreement, the Turkmen State companies of Turkmenneft, Turkmengaz and Shell Company to explore and develop seven onshore blocks will form a consortium. TPAO/Turkish Petroleum Corporation, at the urging of Turkmen authorities, expressed its intention to take part in the consortium. Shell Company signed a Memorandum of Understanding for a strategic alliance with the Turkmen government. The follow-up evaluation studies reduced the number of prospective blocks to six. The signing of a "Production Sharing Agreement" (PSA) is expected between Shell Company and Turkmen counterparts in February 2000. TPAO/ Turkish Petroleum Corporation is pursuant to Turkmen government on participation in the consortium." [13] The Trans-Caspian Gas Pipeline Project (TCGP), which aims to bring the Turkmen gas to Turkey, has been showed as the main alternative to the Blue Stream Project. According to the agreement signed between Turkey and Turkmenistan on 29 November 1998, the two sides planned to construct a new pipeline to carry 30 billion cubic meters Turkmen gas to Turkey. There are a number of problems about this project. Of all, the first problem is the status of the Caspian Sea. The discussions between Iran, Turkmenistan and Azerbaijan are still ongoing without any final solution. Recently Turkmenistan will not take part in negotiations of Caspian Sea issues if Iran is not included in the meeting. Second, as a big project the TCGP needs at least three further years to be constructed. Besides it is difficult to find financial support for the project.

Put shortly, as seen in the above Blue Stream case, the lack of a functioning regime here impedes the energy-based cooperation. A great potential, which might be a key step for forward other types of cooperation is being flawed for there is no legal and regional basis for cooperation.

4. THE CAUCASIAN GEOPOLITICS

Despite its huge natural resources, the Caspian region has hosted many regional problems. The developments since the end of the Cold War has shown that a region-level energy regime is needed for stability [14]. But, is there a satisfying ground for the Caspian energy regime? There are both positive and negative facts in the region. Along with material facts, positive and negative cultural facts are also acting here. A potentiality of an energy-regime is highly related to these effects.

The main characteristic of these countries that bind them together is their physical proximity. There are significant differences between the economies of these countries. Main motivation behind the establishment of such an organization was from the regional proximity, which would in turn lead to gains from trade. The participating countries are divergent enough in terms of their production capabilities both in manufacturing and service sectors and factors of production to enjoy such gains as well as the natural resources that are unevenly distributed among them.

Twelve years after the collapse of the Soviet Union, the South Caucasus area remains the most complicated conflict zone of the entire Eurasian region [15]. Though the South Caucasus may be compared to the Balkans in order to understand the severity of conflicts and human suffering there, none of the conflicts in the former give signs of a lasting solution in the foreseeable future. Internal weaknesses and fault lines, in addition to security dilemmas created by external factors, continue to threaten stability in the region.

All three South Caucasian states -- Armenia, Azerbaijan and Georgia -- have separately witnessed a nationalist movement at home and political stability has been held hostage to these state and minority nationalisms. Problems emerged due to a rise in nationalist fervor of the Nagorno-Karabakh Armenians in Azerbaijan; the Ossetians, Ajars and Abkhaz in Georgia; the Azeris in Armenia; the Azeris and Armenians in Georgia; and the Talysh and Lezgins in Azerbaijan. The most problematic among these has been the Nagorno-Karabakh dispute between Azerbaijan and Armenia, which led to Armenians taking 23% of Azeri territory, the death of 35000 people, and the displacement of over 1 million Azeri formerly living in the Karabakh area. Subsequent to this conflict, the Abkhaz people's search for autonomy has led to conflict with the central Georgian authority, already resulting in hundreds of displaced people.

Severe economic results of the ongoing conflicts and the non-democratic, authoritarian governments of these countries have compounded these problems. Indeed the problematic political and economic situations have been used to justify authoritarian leaders and their possible successors in near future. The Armenian population suffers from an economic crisis and also the lacks of the necessary infrastructure and energy resources for industrial development. Georgia seems to be in a better condition, but still has many structural problems hindering an adequately functioning economy. Azerbaijan's situation is somewhat different since this country has rich natural gas and oil resources. However, due to the unending struggle over the final routes of pipelines, which will carry Azeri natural riches to the global market, and mismanagement of current gain from these resources, the Azerbaijani dream of being the "Kuwait of the Caucasus" seems a far-off possibility.

The chance for a change in leadership in all three countries is very low. In Azerbaijan it is almost certain, and indeed can be easily observed from the pictures placed everywhere in Baku, that Heidar Aliyev's son will succeed him. In Georgia, there seems (in 2003) to be no alternative to Eduard Shverdnadze and Western policies toward Tblisi strengthen this vision.

Armenia's position is almost the same and the strong nationalist segments of Armenian society and the Armenian diaspora at least will try their best to keep a nationalist leader in office. Democracy remains elusive and delayed due to the political and economic problems in the region. Thus, there is growing social unrest and discontent in these states and the best signs of these social problems have been assassination and coup attempts (with rumors of foreign support) against leadership in all the South Caucasian states.

The main external players -- that have key stakes in the South Caucasus and are involved in this region -- have been Turkey, Iran and Russia [16]. Among these three, Iranian involvement has been considered the most problematic because of the perceived threat of the spread of fundamentalism in Azerbaijan. However, it has become evident that the main Iranian motives are to find new, in particular economic, partners to counter American attempts to isolate it and to have a share from Caspian riches. The Russian administration has defined this area as its backyard, naming near abroad, and has aimed to control the South Caucasus through Russian military existence in the region, which is vital for economic and military security. Turkish policy toward the region has been motivated by its search for energy resources, desire to decrease Russian military forces to the level dictated by CFE and by its linguistic, cultural and historical ties with the peoples of the region.

Diaspora peoples of Caucasian origin in Turkey have pushed Ankara to pursue an activist policy in the region, sometimes creating problems as in the case of the Abkhaz question in Georgia. As a landmark example in this regard, recent years witnessed an escalating conflict between Georgian and Abkhaz origin diaspora communities in Turkey. Related to this, historical problems and especially the Armenian allegations of Turkish genocide of Armenians in 1915 keeps current relations under the shadow of the past, preventing normalization of relations between Turkey and Armenia. A similar development occurred between Azerbaijan and Iran regarding the Azeri-origin population of northern Iran, with nationalist Azeris' call for a greater Azerbaijan including Iranian Azeris.

It appears that the countries surrounding these three regional states and shifting alliances will play important roles in the future of the region. Armenia feels encircled by Turkic states to the east and west, namely Turkey and Azerbaijan, and seeks to ally itself with Iran and Russia. Azerbaijan and Georgia have turned their face to the West, and consider Turkey a gateway to the West and also an important ally to counter possible Russian hegemony.

In addition to these alliances, a series of geopolitical relationships are emerging in Eurasia, resulting in a security dilemma. As indicated, on one side are Russia and Iran, along with a series of smaller powers, including Greece and Armenia. On the other side are Turkey, Azerbaijan, Georgia, the Ukraine and, as recent developments indicate, Israel. Also on the rise in the region is the influence and engagement of the United States and the European Union, both of which seek to tap into the vast energy reserves of the Caspian region. In addition to all these, NATO, OSCE and multinational oil companies are there to stay in the new geopolitical picture of the South Caucasus.

In this area, the main U.S. goals are to guarantee a share of Caspian riches and provide security for the U.S. firms operating in the region [17]. For this purpose, they have promoted Turkey's position as a long-time ally and have also attempted to curb Russian and Iranian influence in the region. The joint U.S. and Turkish attempt to make the Baku-Ceyhan pipeline route -- which goes through Azerbaijan, Georgia and the Turkish Ceyhan port in the Mediterranean -- the final route is mainly motivated by political and security reasons. However, the decision over the final route depends on the decision of multinational oil companies, which give priority to economic rationale and they continue to delay this final decision.

European involvement has been through special aid programs and NATO's Peace for Partnership program, which includes the three states of the South Caucasus. NATO's involvement deserves further attention since, in particular, Azerbaijan and Georgia aim to be full members in this organization. Both have declared their willingness to host NATO bases in their territories and Georgia even demanded a Bosnia or Dayton-type solution to the Abkhaz problem. If one looks at the new geopolitical picture, in which all security matters, including energy and pipeline security, would be determined, the roles of the European Union, NATO and the U.S. seem confused. These three actors do not have a clear policy and some actions pursued by NATO members may even be understood as NATO's policy. It is not possible to find a unified, common interest agenda for all members of NATO. For example, there is no agreement over Turkey's moves in this area among NATO members. It is also highly questionable that NATO's use of military force would be appropriate -- if NATO reaches a consensus -- even under a worst-case scenario, for example, in case of a terrorist attack on oil and gas pipelines. This confusion may lead to NATO or U.S. involvement in local or interstate conflicts. In addition, this area is increasingly attracting the attention of Japan and China, but for the moment their involvement consists of only small business initiatives.

5. CONCLUSION

The future of political stability in the South Caucasus will depend largely on the Armenian-Azeri conflict, which gives no positive sign of a lasting solution, even after a series of talks held in Paris and the Florida resort of Key West in the first months of 2001. In both countries, there are strong opposition groups, which play watchdog roles to prevent any positive move toward peace. In addition, there is a danger of a similar conflict over the Armenian-populated Javakethia district of Georgia. Another important issue is competition over energy resources and pipeline routes between the two aforementioned constellations of states. Controversially, this competition itself threatens energy and pipeline security, which would be self-destructive in the end for all related parties, but mainly for the South Caucasian states. Energy and pipeline security rests on the national forces of the Caucasian states and the prospect for external interference seems highly restricted at this point. U.S. policy toward Iran may change the entire geopolitical picture if Washington lifts its sanctions, which seems like a far possibility. However, if this happens, the plans of pipeline routes and energy projects are likely to be completely reassessed.

Another development that might change the current picture would be Turkish-Armenian rapprochement, but again this seems to depend on finding a lasting solution to the Karabakh question and overcoming the burden of history between these states. Greece and France show increasing interest in this region, but their involvement may further complicate the intriguing shape of regional politics. Involvement of NATO and the European Union member states will always carry the potential risk of misperception of their policies and their attitude toward Caucasian states. The emerging security environment is thus one in which two blocs of states are in increasing competition with one another and, under such conditions, all players are likely to view events in a zero-sum fashion in which a gain by one side is perceived as a loss by the other. This security structure is not likely to yield a major war in this area, but one may expect small border incidences, state forces' clash with local nationalist groups and coup attempts in following decade. Unfortunately, the Caucasus emerged in such an environment and their future does not depend solely on the choices of the South Caucasian peoples, but also on a number of regional and international actors. These are the main obstacles to establish an international energy regime in this region, even if it is likely to serve for the interest of the all involved countries.

Energy-based map of the region.

Table 1 Caspian Sea Region Oil Production and Exports
(thousand barrels per day)

Country	Production (1990)	Est. Production (2000)	Possible Production (2010)	Net Exports (1990)	Est. Net Exports (2000)	Possible Net Exports (2010)
Azerbaijan	350	200	1,100	-272	0	500
Kazakhstan	251	314.3	1,100	-257	-176.6	350
Iran	0	0	0	0	0	0
Russia	219	30	N/A	N/A	N/A	N/A
Turkmenistan	3,100	1,642	3,900	2,539	1,381	3,300
Total	**3,920**	**2,072**	**6,100**	**2,010**	**1,204.4**	**4,150**

Source: Energy Information Administration (http://www.eia.doe.gov/)

Table 2 Caspian Sea Region Natural Gas Production and Exports
(billion cubic feet per year)

Country	Production (1990)	Est. Production (2001)	Possible Production (2010)	Net Exports (1990)	Est. Net Exports (2001)	Possible Net Exports (2010)
Azerbaijan	259	311.2	1,200	77	175.2	1,000
Kazakhstan	602	811	2,000	109	631	1,700
Iran	0	0	0	0	0	0
Russia	144	11	300	0	7	300
Turkmenistan	125	159	200	69	107	150
Total	**1,130**	**1,292.2**	**3,700**	**255**	**920.2**	**3,150**

Source: Energy Information Administration

REFERENCES AND NOTES

1. Stephan D. Krasner, *Sovereignty Organized Hypocrisy* (Princeton: Princeton University Press, 1999), p. 9.

2. A. Hasenclever, Peter Mayer and Volker Rittberger, *Theories of International Regimes* (Cambridge: Cambridge University Press, 1997), p. 9.
3. Stephen D. Krasner, (eds.), *International Regimes*, (Cambridge, Mass., Cornell University Press, 1993), p. 2.
4. Baylis, p. 303.
5. Hasenclever et al., p. 9
6. Ibid. p. 9.
7. Aleksandr Akimov, "Oil and Gas in the Caspian Sea Region: An Overview of Cooperation and Conflict", http://www.cpss.org/casianw/akim.txt
8. Nancy Lubin, "Pipe Dreams", *Harvard International Review*, (Winter-Spring, 2000), p. 67.
9. "By passing Russia", *The Economist*, 17 April 1999, p. 56.
10. Jennifer Schuessler, "Museum of Nations", *The American Scholar*, Vol. 70, No. 1, (Winter, 2001), p. 145.
11. Charles William Maynes, "A New Strategy for Old Foes and New Friends", *World Policy Journal*, Vol., 17, No. 2, (Summer, 2000), p. 77.
12. Sedat Ergin, *Hurriyet*, 6 November 1999.
13. http://www.tpao.gov.tr/rprte/INT.HTM
14. For a recent detailed study see Bulent Aras and Micheal Croissant, (eds.), *Oil and Geopolitics in Caspian Sea Region* (Westport: Praeger, 1999)
15. For a comparative analysis see, Bulent Aras, "Caspian Region and Middle East Security", *Mediterranean Quarterly*, Vol.13, No.1 (2002), pp.86-109.
16. Bulent Aras, *New Geopolitics of Eurasia and Turkey's Position* (London: Frankcass, 2002), p.21.
17. Bulent Aras, "U.S.-Central Asian Relations", *MERIA Journal*, Vol.1, No.1 (1997).

GAZPROM IN GEORGIA
Some Aspects of Gas Supply Security

Liana JERVALIDZE
The Caspian Area Energy Studies, Partnership for Social Initiatives, Alumna of the Kennan Institute of Advanced Russian Studies

Abstract: This article draws upon experience from the operations of a small Russian-based natural gas supplier Itera, and the Russian gas giant Gazprom basically within the Commonwealth of Independent States (CIS) – and also towards the European Union. The article addresses the energy situation of Georgia in terms of security of gas supply, and the strong dependency of the country on a single gas supplier, Itera. The article also shows how Gazprom is prone to impose Russian policy in the region. On this basis the consequences of admitting Gazprom to become a major supplier of gas to Georgia are assessed. It is further believed that this would not improve the situation of security of supplies, neither would the transparency increase, nor would a competitive and business-friendly arena be established.

Key words: Gazprom, Itera, Caucasus, Georgian energy sector

1. INTRODUCTION

The security of energy supplies constitutes the cornerstone of the security of any nation. The recent history of relationships between the Western countries and countries in the Middle East provides a sound proof thereof. Proceeding from the experience of over-dependence on the energy supplies from the Middle East the United States and the European Union have set a rule of having at least three suppliers as a pre-condition for assuring their energy security. This scheme has proved itself to be a very effective mechanism for several decades. But, this is not the case for countries within the Commonwealth of Independent States (CIS) - in the post Soviet space.

J. Hetland and T. Gochitashvili (eds.),
Security of Natural Gas Supply through Transit Countries, 355–367.

2. THE SITUATION OF THE CAUCASUS REGION

Georgia like many other countries of the CIS such as Armenia, Belarus, Moldova, Ukraine, the Baltic countries and even the energy rich Azerbaijan depend severely on Turkmen gas supplies. Until recently Itera, a Russian-based gas trading company was an exclusive supplier of Turkmen gas to the CIS countries including Georgia. Georgia's over-dependence on this single source will not be altered until 2006 when the Baku-Tbilisi-Erzurum gas pipeline goes on stream. The pipeline will carry Azeri gas to Turkey – with a provision for further extension into Europe. In regard to Azerbaijan this situation may change somewhat earlier than 2006 since Azerbaijan - apart from the Shah Deniz offshore deposit – possesses other gas fields that may be brought on stream earlier than the Baku-Tbilisi-Erzurum pipeline.

Even Armenia, a long-term strategic partner of Russia in Caucasus, seeks to diversify its gas supply sources. With this purpose Armenia has negotiated two draft pipeline projects with Iran, one of which is suggested to come up to the Georgian border. There is also an old gas pipeline from Iran through Azerbaijan to the Georgian border that needs some refurbishment. Thus, Iran can potentially become the number three, if not the number two gas exporter (or re-exporter of Turkmen gas) to the Caucasus provided that Iran manages to attract investments to boost its declining gas production.

So far, until this may happen, Georgia and Armenia will totally depend on a single source of gas. This gas will be traded by a Russian company and supplied through the Russian pipeline system regardless of the gas being Russian, Kazakh or Turkmen gas. Because of the political and economic vulnerability a strong dependence like this is especially dangerous for the newly established Caucasian countries. Russia already exerts a serious leverage over these countries through ethnic conflicts (pertaining to Abkhazia, Nagorni Karabakh and South Ossetia) that have been going on for several years. In addition, owing to the unique position as the only gas supplier to the region, political pressure may be placed on local governments in order to ensure that they make political decisions that are conceived to be friendly to Russia. Therefore, in order for the countries of the region to secure their gas supply the diversification of supply sources is strategically and politically more important than the technical dimension. The security of gas supply has become an important issue for the EU countries - including the eastern part of Europe - as well. The European Union already receives more than 40% of their gas supplies [1] from Russia and its dependency will depend even more on this source if they fail to diversify their import resources [2].

Experts forecast that in about ten years the demand on gas (as a clean energy source) in EU countries will increase by 20%. However, the local gas production within the EU countries is expected to decline. Therefore, a large part of the increasing demand - maybe as much as 60% (according to some estimates) - will probably be met by gas produced by Russia and Central Asian countries. These countries possess some 37,8% of the proved global reserves [3].

According to the Secretary General of the Energy Charter Treaty, Ria Kemper, "one of the main challenges facing EU in the area of energy security will be to ensure the security of its external supplies of gas. It can be achieved primarily by extending the same principles of liberalization and competition that underpin the EU's internal energy market to wider Eurasia space beyond EU. [4]" With a view of achieving this goal, 51 European and Asian countries including Russia and other CIS countries signed the Energy Charter Treaty in 1994. Later on the Energy Charter Treaty member countries have launched negotiations on the Transit Protocol annexed to the Treaty.

According to the missions and objectives of the Energy Charter the charter strives towards the establishment of open, efficient, sustainable and secure energy markets [5].

Despite the fact that Russia signed the Energy Treaty and participated in the negotiations of the Transit Protocol, the Russian gas producer, Gazprom, objected – and still objects. Therefore, the Russian Duma did not ratify the Energy Treaty [6]. Thus, the liberalization of the Russian gas industry and the gas transportation network has been stalled again for better times.

It is likely to believe that the way that the Russian gas industry changes - or does not change - may have a long-term impact on the gas producer in Russia and the Caspian region, and even on the gas consumers everywhere - including the European Union. In the short term this impact represents at threat to South Caucasus, especially Georgia, because Russia is seemingly using its position as a monopolistic gas supplier for imposing its regional energy policy.

3. RUSSIA AS A POWERFUL REGIONAL PLAYER

Russia is an old and powerful regional player having a strategic and political agenda in the Caucasus and Central Asia. The immediate understanding of this agenda is on the one hand to keep these regions within the sphere of its strategic and political influence, and on the other hand, to increase the over-dependence of the European energy market on gas - of Russian, Central Asian and Azerbaijani origin - to be transported exclusively through the Russian pipeline grid.

Therefore, Russia takes as a heavy blow to its vital interests the pipeline projects Baku-Tbilisi-Ceyhan (BTC) and the South Caucasus Pipeline (SCP) that are designed to carry alternative Azeri oil and gas in bypass of Russia through the Caucasus and Turkey to the west. Russia, driven by the desire to preserve its gas market share in Europe and hold under control the Caspian gas trans-boundary energy flows appears to be particularly negative to the South Caucasus Pipeline project. Russia offers to use its yet unloaded Blue Stream pipeline under the Black Sea for carrying Azeri gas to Turkey "instead of building a new and expensive grid" [7].

The United States and the EU countries also have their own agenda in the region. This agenda consists of having an unhindered access to upstream oil and gas reserves in the Caspian region. This includes the development of alternative transportation routes out of the region. It is also within their agenda to promote the building of democratic, market oriented, sustainable civil societies in the region for the purpose of ensuring their own long-term energy supply.

According to the US ambassador Steven Mann, "a fundamental point of east-west energy corridor through the Caucasus and Turkey in bypass of Russia is to support the Caspian countries to achieve a greater level of autonomy and independence through independent energy routes. [8]" Thus, the strategic and political agenda of the west does not always and necessarily coincide with the interests of Russia in the region. Especially the part that provides for the building of democratic, market oriented and sustainable civil societies in the region can be questioned. The reason is that the more the countries of the Caucasus are politically unstable and disintegrated the easier it will be to postpone competitive energy transportation alternatives.

The situation is rather complicated by the fact that Russia itself has been undergoing the same kind of changes as all other post Soviet countries. Indeed there are some significant differences. For example, Russia, in contrast to the Caucasian countries, has a much greater experience of independent statehood, of conducting foreign and international policy, and of the properly functioning state institutions. Russia has also successfully overcome the threat of territorial disintegration that Georgia and Azerbaijan still are facing due to the conflicts in Abhazia, Nagorni Karabakh and South Ossetia. But, the general situation is similar to the situation of the countries of the former Soviet Union characterised by excessive governmental interference with private businesses [9], absence of transparency, deficient legislation and law enforcement practices, and an overspread corruption. These are among the most severe illnesses to which the Russian society is subjected similar to most of the former Soviet Union countries (excepting the Baltic). Under these circumstances – and as long as Russia finds it difficult to admit and stimulate such changes on its domestic market [10] - it is not likely that Russia can give boost to positive changes in its neighbouring partner countries - especially not in a vital sector as the production and transport of gas.

Therefore, the news stating that Gazprom is going to enter the Georgian market in parallel to - or in substitution of - Itera deserves a serious consideration [11]. The Russian gas company, Itera, has been present in the Georgian energy market since 1996. Although being a 100% private company, Itera could seemingly not avoid acting in the interest of Russian regional policy at the detriment of its commercial operations.

4. ITERA, RECORD OF A SINGLE SUPPLIER IN GEORGIA AND FSU

Itera is the only private company that produces significant volumes of gas in Russia (23 billion cubic meters in 2002). Itera has been a major supplier of Turkmen gas to former Soviet countries for several years.

In 2002 Itera delivered 4 billion cubic meters of gas to Azerbaijan, 1.3 billion cubic meters to Armenia, and 1.2 billion cubic meters to Georgia. Apart from being a main supplier of gas, Itera possesses 12 out of 40 gas distribution units, and the chemicals plant, Azot, that is the main gas consumer in Georgia. The Georgian government handed over these assets to Itera despite a very strong objection of the public, in full disregard of fair competition and transparency rules that should normally govern such transfer.

Itera's activities on the CIS market over several years raise many questions whether these activities have been motivated by commercial or political considerations. This statement particularly pertaining to the supply terms, conditions and tariffs that Itera has established with different CIS countries. For example; Itera delivered Turkmen gas to Ukraine at the price of $42, to Belarus at first at $19 then $29, to Armenia at $53, Azerbaijan $52 and to Georgia at $67 including 3% of Sakgas, an affiliate of Itera in Georgia [12]. This scheme points out that Belarus, the closest strategic partner of Russia, is the most favoured, whereas Georgia is the least favoured partner of Itera.

The issue is not whether these prices are high or low, but rather the absence of commercial rationale in them. As Belarus is situated as far from Turkmenistan as Ukraine one should expect that the transport cost is the same. Nevertheless, Belarus used to pay almost half the price that Ukraine had to pay. And despite Georgia is situated much closer to Turkmenistan – and the gas sources - Georgia had to pay the highest price regardless of the lower delivery cost.

The way that Itera acquired assets in Georgia, along with the price and tariffs that it established for its gas supply bring forward some questions that need to be answered before introducing another Russian gas supplier to the Georgian energy market. The crucial question is whether the entry of Gazprom will make the Georgian energy market more secure, transparent, competitive and business-friendly? Any possible answer to this question relates to the recent developments in the Russian gas industry.

5. RECENT DEVELOPMENTS IN RUSSIA'S GAS INDUSTRY

The recent developments in Russia's gas industry point out that liberalization of the Russian gas industry and energy market will not happen soon. A long-awaited restructuring of Russia's gas giant Gazprom has been shelved by president Putin's special decision in December 2002. So far, Gazprom remains a 60% state owned company (the government owns 39% of the shares directly, and 22% through various funds) [13] that has under control 36% of the world's proven gas reserves, [14] 25% of the global gas production, above 40% of the European gas imports (including Eastern Europe), 25% of the Russian state budget revenues, and the whole gas pipeline grid of the Russian Eurasia [15].

Inside Russia Gazprom produces almost 90% of the gas. Independent companies produce the remaining 10%. Gazprom, as the owner of the Russian transportation grid sets terms and conditions for acceding the network. Hence, the independent producers have no choice but selling their gas to Gazprom - at a low price. Because of inaccessibility of export pipelines many Russian oil companies have no other choice but flaring the associated gas.

Two thirds of the gas volumes produced by Gazprom and other small producers has been consumed on internal markets. The gas price on the Russian market constitutes roughly one forth of the world market price and stands at around $ 22 per 1000 cubic meters [16]. Gazprom earns $3 per 1000 cubic meters for gas consumed by industries and only $1 per households domestically [17] and $45 per 1000 cubic meters for gas exported to Europe [18]. Consequently exports represent the main source of Gazprom's cash revenues. This situation incites this giant company to export as much as possible and to seek ways of liberalizing the gas price on the Russian domestic markets.

Experts esteem that Gazprom faces very serious financial challenges since its producing fields are being depleted, and large capital investments are required to bring on stream new gas fields, and the transportation grid needs costly renovation. According to Gazprom's estimates some $2,4 billion need to be invested into the pipeline network by 2006 [19]. Furthermore, Gazprom has to serve an interest on a $12 billion debt on a yearly basis [20]. Obviously these difficulties had been behind the restructuring project rejected by president Putin in December 2002.

Notwithstanding these difficulties Gazprom has been undertaking very costly projects. First of all a 3000 km pipeline estimated at $35-40 billion for the development of the Yamala field including a pipeline project for Europe. Despite experts estimate that the gas from the Yamala fields will find consumers only if the price on the European gas market increases by 100%, or even 200%, works on this project goes on [21]. The same way the Gazprom leadership has been considering options of pipeline construction to China, South Korea and Japan.

Another challenge of Gazprom is the $3.2 billion Blue Stream Pipeline project with a throughput capacity of 16 billion cubic meters that has occurred under heavy pressure. As a result of the collapsing gas demand in Turkey this pipeline operates at a very little capacity - or even stays idle as it did for at least six months in 2003 [22]. The heavy negotiations in June between Gazprom and the Turkish government ended without result and the parties almost decided to go to the court [23].

6. GAZPROM BACK ON FSU MARKETS

Owing to the tremendous gas reserves of Russia and the Caspian region (37.8% of the stated world reserves) the Russian policy is to establish a Euro-Asian gas export organisation - similar to OPEC – to control the gas export of the individual Central Asian producers to the west - basically through the gas pipeline grid under Gazprom control. For this purpose Gazprom held very serious diplomatic and business talks with Ukraine, Belarus and the Central Asian countries in 2002.

As a result of these talks Gasprom made long-term deals with Turkmenistan [24], Kazakhstan and Uzbekistan. According to these deals the Russian domestic market will be supplied by gas coming from these countries at $44-$50 per 1000 cubic meters [25], and the exceeding volumes will be re-exported. These agreements provide for the payment of 50% of the Central Asian gas in hard currency, whereas the remaining 50% will be in kind - covered by Russian goods and services. Also, due to these deals more volumes of Russian gas can be released for exports to the European markets.

But here also serious financial challenges emerged. The pipeline network of Ukraine and Belarus needs serious renovations. The old gas pipeline grid of the Soviet Union in Central Asia should also be totally replaced - like the one connecting Turkmenistan and Kazakhstan, or it should be seriously upgraded - like the one connecting Turkmenistan, Uzbekistan and Kazakhstan [26].

As Gazprom intends to import annually 38 billion cubic meters of gas through these pipelines, Gazprom has made a bid on management and operation of the above-mentioned network. Although, every of the three Central Asian countries has the potential for excess gas production, it is not likely that these pipelines - even if upgraded by Gazprom – will meet all the export and import needs on both sides for the up-coming decade.

So far, by concluding these long-term deals on the purchase of large volumes of gas, and the bidding for the management and operation of Central Asian main gas pipelines, Gazprom blocked the direct access of local producers to the CIS and the European markets. These long-term deals also block an eventual access of foreign trading companies to the Central Asian gas. Further, Gazprom evicted Itera - the second largest Russian gas producing and gas trading company - from almost all the CIS markets, and jeopardized its contract with Turkmenistan on the purchase of Turkmen gas.

At the beginning of 2003 Gazprom seems to have cut Itera's supply of gas to Latvia, Ukraine and Moldova. Gazprom boosts its exports to Lithuania and bids for 34% stake in the Lithuanian utility Lietuvos Dujos [27]. Soon the Baltic countries will become members of EU. Obviously, Gazprom, through entering these markets also seems to target the EU spot market.

In its last efforts to preserve at least some of its market Itera offered 20% of stakes in its producing fields and showed its readiness to share its market with Gazprom everywhere in the CIS [28]. At the beginning of May the Turan Agency reported that Gazprom would take 50% of Itera's gas market in Armenia. Gazprom also seems to exert pressure over Itera in Azerbaijan as well.

Through various deals that Gazprom concluded on gas supply, and management and operation of pipelines, it has emerged as a powerful monopoly in the gas industry and energy market of the CIS. However, through acquiring new assets in these countries, Gazprom is adding up new financial and managerial challenges to the challenges that it already faces at home, inside Russia.

Georgia is the second country after Azerbaijan where Gazprom had not concluded a deal by mid 2003, although negotiations had been underway for some time. Georgia's interest is clear: Should Itera be evicted by Gazprom from the Georgian market as well, the country will need to have an access to uninterrupted gas supplies until alternative Azeri and potentially Iranian gas becomes available in the region.

7. THE ENERGY MARKET OF GEORGIA (AND ARMENIA)

Indeed, the Georgian gas market is too small and barely solvent to justify any serious commercial interest by a company of Gazprom's size and potential to it [29]. As already stated Georgia consumed some 1.2 billion cubic meters of gas delivered by Itera in 2002. Itera used to deliver 1,3 billion cubic meters of gas to Armenia as well through the Georgian pipeline grid. Thus, the current joint gas consumption of these two countries did not exceed 2,5 billion cubic meters per annum by 2003.

In the past Soviet times, when 70% of the population of Georgia and Armenia had access to natural gas, gas supplies came through a highly diversified pipeline system from Azerbaijan, Iran and Russia. By that time neither local governments nor the population had any serious understanding of energy efficiency. Therefore, the consumption rate was very high and reached a peak in 1989 of 6,2 billion cubic meters for Georgia and 5,5 billion cubic meters for Armenia. Experts find it difficult to forecast whether such consumption level will ever be reached in these countries in the foreseeable future.

According to expert estimates, Georgia's gas demand is expected to double by 2007 and reach 2.4 billion cubic meters per annum. By that time the South Caucasus gas Pipeline will be finalized and Georgia will receive 800 million cubic meters of Azeri gas under special terms. Georgia is furthermore free to take more Azeri gas from the South Caucasus Pipeline at commercial conditions.

By 2007 may be the decision on the closure of the Armenian nuclear power station will also be implemented. So far, if the nuclear station is really closed, the demand on gas in Armenia will grow at least by 2,5 billion cubic meters per annum and reach 4 billion cubic meters. If these estimates are correct by 2007 the gas demand in Georgia and Armenia will reach 7 billion cubic meters.

The demand is expected to grow gradually and reach some 8 billion cubic meters per annum. If Georgia and Armenia achieve to diversify their gas supplies, Gazprom may count on a maximum 60-70% of the local gas market. Further, the closeness of these counties to the gas sources in Azerbaijan, Turkmenistan and Iran may contribute to establish a highly competitive regional gas market at prices probably below $70 per 1000 cubic meters [30].

According to the Russian energy strategy by 2020 adopted by the government by August 2002 the domestic gas prices will gradually increase and reach $35-40 per 1000 cubic meters by 2006. If the liberalization of the gas prices goes on beyond 2006 - as is provided for in the said strategy document, the current difference between the Russian domestic gas price of $22-24 (per 1000 cubic meters) and the Turkmen gas traded in Georgia by Itera at $67 will definitely disappear.

To proceed from the above mentioned small volumes of local consumption at a small profit margin would hardly justify the long distance deliveries of gas by Gazprom to Georgia and Armenia. It could therefore be questioned whether Gazprom intends to monopolize the main pipeline grid of Georgia and the regional gas market with a view to have an access to a third market.

8. GAZPROM IN GEORGIA

The Georgian government and Gazprom resumed negotiations on the entry of Gazprom in the Georgian energy market in the beginning of 2003.

Gazprom is not a newcomer in the Caucasus. It already holds under its control the Armenian energy sector according to assets for debt arrangements. In order to have access to its assets in Armenia Gazprom needs a link through Georgia. Therefore, it may be anticipated that Gazprom will bid for the Georgian pipeline grid.

Two options seems to exist for Gazprom in entering the Georgian energy market

1. To become a main supplier of gas to Georgia;
2. To become a main supplier of gas and co-owner or operator of the main pipeline grid.

9. THREATS STEMMING FROM GAZPROM'S POSITION AS A MAIN SUPPLIER OF GAS

Gazprom practices signing long-term contracts on delivery of increased gas volumes on take-or-pay terms like the one with Turkey. If the Georgian government signs a contract with Gazprom on these terms, the country will definitely lose any opportunity to diversify its supply sources. This will contradict the gas supply diversification strategy of the country.

The signing of a long-term supply contract on the delivery of increased volumes of gas with Gazprom does not comply with the Georgian liberalization strategy for the energy market.

10. THREATS STEMMING FROM GAZPROM'S ENTRY INTO THE MANAGEMENT AND OPERATION OF THE MAIN PIPELINE GRID OF GEORGIA

The main pipeline grid of Georgia crosses the country from north to south to the Armenian border. Although this is not an old system it needs substantial refurbishment. According to numerous international estimates the upgrading of this pipeline requires some $200-280 million. Gazprom is the only company that has manifested its interest to invest in this undertaking.

- Provided that it is targeting a third market - potentially the north-eastern part of Turkey - Gazprom wants to invest in a rehabilitation and upgrading of the Georgian main pipeline grid at the current throughout capacity of 9,5 billion cubic meters per annum (18 billion cubic meters per annum designed capacity). Otherwise the upgrading of this pipeline for a small regional market would make no sense.
- Gazprom is operating two pipelines to Turkey, one entering from Bulgaria, whereas the other one is the controversial Blue Stream pipeline. Any third link to be established by Gazprom would be considered as an attempt to monopolize the Turkish energy market at the detriment of the South Caucasus Pipeline that was designed to diversify the Turkish gas supplies with the potential for further extension to the European Union.

11. CONCLUSION

The entry of Gazprom in the Georgian energy market on special terms and long-term strategic partnership as a main supplier – or eventually as owner and operator of the main pipeline grid - will cause serious strategic and political implications. In due course it will lead to in-dept changes in the Georgian foreign policy.

Then, the presence of Gazprom will not mean that the Georgian energy market will become a secure, transparent, competitive and business-friendly arena. This is evident from what is going on in the Russian gas industry and the CIS where Gazprom concluded long-term purchase agreements linked with pipeline management and operation deals. This also becomes evident when considering the way that Gazprom treats the Russian companies in Russia and the CIS countries.

And finally the entry of Gazprom in the Georgian energy market - as a main supplier, operator and manager of the main pipeline - does not comply with the energy market liberalization and gas supply diversification strategy of Georgia. It contradicts Georgia's liabilities under the east-west energy corridor agreements. And, it does not fall under the market liberalization trends that are prone to govern the western energy market.

REFERENCES AND NOTES

---

1. Gas Markets: Liberalisation in Europe, by Ria Kemper, The Energy Charter Treaty, The Utilities Journal, December 2002
2. Security of Gas Supply, Meeting of the Governing Board at Ministerial Level, 28-29 May 2003, IEA
3. British Petroleum Statistical Review of World Energy, Natural Gas Reserves, p. 20, 2001
4. Ria Kemper, Secretary General of the Energy Charter Treaty, Gas Markets: Liberalisation in Europe, The Utilities Journal p24, December 2002
5. Energy Charter Transit Protocol
6. Speech by Aleksei Miller, Chairman of the Board of Gazprom to the European Parliament's "Energy Choices for Europe" conference, Brussels, 5 March 2003
7. Victor Kaluzhni, President Putin's Representative to the Caspian Region has repeatedly made such an offer in his presentation on The Caspian Oil and Gas Show, June 4-5 2003
8. NewsBase, June 2003-Caspian Special Report, page 3
9. Economic Structure, Russia, Economist.com/countries.../profile
10. President Putin by a special decision of December 2002 postponed the decision on reforming Gazprom
11. The Georgian government informed the public about ongoing negotiations on the entry of Gazprom into the Georgian gas market in March 2003. (Liana J.)
12. Teimuraz Gochitashvili, Roman Gotsiridze. Georgia: Blue fuel as a lever of regional policy. CENTRAL ASIA AND THE CAUCASUS Journal of Social and Political Studies. Central Asia and the Caucasus information and Analytical Center, Sweden, 2002, 4
13. Gazprom Faces Life in Liberalised Market, interview with Boris Fiodorov, member of Gazprom's board of directors representing interests of minority shareholders, Neftegaz, 2/2003
14. eia.doe.gov. Country Analysis Briefs, Russia
15. British Petroleum Statistical Review of World Energy 2002

16. Baku, Turan Energy, April 30th. 2003
17. FSU oil & Gas Monitor, p. 9, 29 January 2003,
18. Russia'a Gas Production, Exports Future Hinges on Dramatic Changes Needed at Gazprom, by A. M2.Bahtiari, Oil and GAS journal, March 10, 2003
19. FSU Oil & Gas Monitor, June 2003, p 6
20. Russia'a Gas Production, Exports Future Hinges on Dramatic Changes Needed at Gazprom, by A. M2.Bahtiari, Oil and GAS journal, March 10, 2003
21. Speech by Aleksei Miller, Chairman of the Board of Gazprom to the European Parliament's "Energy Choices for Europe" conference, Brussels, 5 March 2003
22. Energy Compass, EIG, March 2003
23. Gazprom in Court, Caspian Business News, June 30, 2003
24. Russia-Turkmenistan, Energy Intelligence Group, April 21, 2003
25. Turan Energy, Baku, April 30th. 2003
26. Itera May Not Be Only Casualty Of Gazprom's Deals with Kazakhstan and Uzbekistan, By Jennifer Delay, FSU, JAN. 2003
27. Gazprom To Boost Gas Supplies To Lithuania, FSU, January 2003
28. Itera Seeks Strategic Partnership with Gazprom, Nefte Compass, May 8, 2003
29. Gas tariff collection rate is very low among the Georgian population. Gas consumption expenses have been mostly covered by the budget. This is a real source of corruption in Georgia. Itera used to disrupt gas supplies several times even in winter times because of payment disarrays-Liana J.
30. Georgia will receive 800 million cubic meters of gas through South Caucasian Pipeline per annum out of which 500 million at $55 per 1000 cubic meters.

PART V: OTHER ASPECTS OF PIPELINE TRANSPORTATION

FOSSIL FUELS LONG DISTANCE PIPELINE TRANSPORT

Pavel VLASAK [*], Zdenek CHARA [*], Vyacheslav BERMAN [**]
[*] *Institute of Hydrodynamics Academy of Sciences of the Czech Republic*
[**] *Institute of Hydromechanics National Academy of Sciences of Ukraine*

Abstract: The paper deals with pipeline transport of fossil fuels. A general analysis of different ways of pipeline transport of coal, heavy viscous crude oil and natural gas or its products, capsule pipeline transport of coal and solidified crude oil in LNG is introduced. An advantage of common carrier transport system, based on LNG and liquid hydrocarbons used as carrier fluid, is discussed.

Key words: Pipeline transport, fossil fuels, common carrier, LNG, crude oil, coal

1. INTRODUCTION

Fossil fuels can be divided into solid, liquid and gaseous. The first one is mainly represented by coal creating in present about 25 % of global fossil fuels consumption. Coal energy density changes according to its quality; it contains usually un-negligible amount of sulphur, during combustion sulphur oxide, carbon dioxide, nitrogen oxide and other chemical pollutants are generated. Oil becomes the main energy source, which covers nearly 50 % of global energy production. Compared with coal oil has some important advantages. Oil has high energy density, it is easily transported, stored and consumed. Due to more favourable ratio of carbon and hydrogen atoms it produces during combustion less greenhouse gases than coal, less nitrogen oxide, sulphur oxide and other pollutants. Apart from the mentioned advantages, proportion of oil on global energy consumption is slightly

J. Hetland and T. Gochitashvili (eds.),
Security of Natural Gas Supply through Transit Countries, 371–382.

decreasing, since the role of the most advantageous fossil fuel plays natural gas. It becomes a very important chemical raw material; it is very easily used as energetic source for heating as well as for power generation. High energy density, low production of greenhouse gases, high efficiency during electricity production (nearly 60 % compared with to 50 % for oil and 45 % for coal) are reasons of increasing interest about natural gas in present.

Advantages of the use of natural gas, propane-butane gas or other hydrocarbon derivatives instead of coal or oil are given by more favourable effect of these fuels on environment, less transportation cost and higher efficiency during burning and power or heat generation.

Using of different products of natural gas becomes also very interesting for the future. These products can bring even better handling, storing and transport quality than that of natural gas. Liquefied (LNG), compressed (CNG) natural gas or propane-butane gas seems to be the most interesting gas derivative in the near future. Advantage of LNG is in higher energy density, considerable volume reduction, relatively easy storage, and easy transport and storage ability, less risk due to ignitability or flammability. LNG can be easily directly used as fuel for combustion motors or converged into gas. It is non-toxic and cannot explode in open environment. Disadvantage of LNG is a very low temperature necessary for gas liquefaction, - 162 °C for normal pressure condition, or high pressure for higher temperature (gas liquefaction needs pressure about 2.5 MPa for normal temperature). These disadvantages could be avoided by transition of natural gas or other hydrocarbons to propane-butane gas or methanol, which are liquid for normal temperature and reasonable pressure condition.

Diversity of purveyors, transport system and also diversity of sources contribute significantly to security of energy supply and less dependence on monopoly distributor of fossil fuels and primary energy sources can significantly help economical prosperity of regions without own sources, e.g. South Caucasus, Balkans, Central and East Europe. It also helps to avoid different commercial and/or political pressure on the final consumer. From this reason a possibility of using different ways of fossil fuels transport will be discussed.

With annually increasing volumes of various fossil fuels to be handled and/or transported an interest arises in the new economical and operationally safe kinds of transport, including the pipeline one. Preliminary investigation has shown that application of a special kind of pipeline installations can be very useful for solving the problem to supply different countries by fossil fuels from distant finding place [1,2]. The hydraulic, pneumatic and capsules pipeline transport systems may be successfully used to solve this task in the near future. Of course, the state of knowledge in the field of pipeline transport requires applying also experimental research beside a numerical calculation for the design of special transport installations, especially in the case of common carrier systems using liquefied natural gas (LNG) or methanol as carrier liquid.

The pipeline transport in comparison with the conventional one has several expressive advantages, especially in hardly accessible mountains and wastelands regions or in highly populated industrial areas. It can be fully mechanised and automated, it needs very low human activity in the direct contact with transported products, and it can be completely closed with a very small contact with environment, [3,4]. It can transport a large quantity of material and has only negligible demands for space. It should simultaneously ensure high operational efficiency, safety and reduction of total costs of transport.

For successful and economical exploitation and utilisation of isolated sources of fossil fuel, located in remote and/or hardly accessible regions, the problem of operationally safe and economically acceptable way of transport becomes usually one of the decisive parameters. The pipeline transport seems to be one of the promising and powerful systems that can serve for long distance transport of fossil fuels, from dozens to thousands kilometres and can compete with the conventional railway and barges transport.

Because of substantial reserves of coal, situated in distant locations from consumers, the limited sources of oil and natural gas, expected growing price of oil and gas a considerable development and progress in the coal pipelining can be expected especially in the USA, in the former Soviet Union and also in other countries in a near future. One of prospective ways of energy transportation, i.e. gas, liquid and also solid fossil fuels transport, seems to be the transport of solidified oil in LNG and pipelining of coal in methanol.

2. COAL PIPELINING

The world-wide knowledge in design, construction and operation of hydraulic pipeline systems based on experimental investigation and operational experiences as well as on theoretical calculations has shown the efficiency and prospect of the pipeline transport of coal in comparison with other kinds of transport. However, at a level of technical projects a series of weaknesses and/or imperfections in contemporary used systems of the coal pipelining with water used as a carrier liquid have been detected. One of them is a rather difficult, energy consuming and expensive process of dehydration of coal slurry before combustion. Besides, transport of a great quantity of „ballast“ - the carrier water - also requires considerable power inputs and usually also expensive water treatment before reutilization or discharging the water into the natural streams. In addition, water is at present very valuable and hardly available medium.

Because of permanently problematic supply of natural oil and gas, often in consequence with economical and political pressure, the present time brings a new interest to solve the problem of optimisation of pipeline transport of coal. We could pick out three main directions of modern long distance coal pipelining.

The first way is *pipeline transport of concentrated coal-water slurry*. It assumes that utilisation of coal particle distribution preparation and/or using the various additives can substantially increase mass concentration of coal-water slurry till 70-80 %. For example, Black Messa Pipeline realised in USA transports 4.8 million tons of coal per year over the distance 439 km from the colliery in North Arizona to power plant Mohave near the border of Nevada and California with mass ratio of coal to water about 1 : 1, without using any additives, [5]. Recently, in the USA, Italy, China, Russia and other countries several coal pipelines were already realised. Even Oil and gas should be effectively transported as hydrate slurries, [6].

Substantially smaller content of water in slurry could be reached - approximately 75 % of coal and 25 % of water due to using the special additives - fluidising and stabilising agents. The high concentration chemically stabilised coal-water slurry with optimised grain size distribution - called *coal-water-fuel* behaves as homogeneous fluid. It is transportable over long distances and stable. There is no limitation concerning the critical velocity or critical hydraulic gradient. Another advantage is the possibility of long time storage without solid-liquid phase separation and of course no difficulties in pipeline restart. The coal-water-fuel could be directly combusted without de-watering similarly as heavy fuel oil, [7].

The second way is a *capsule pipeline transport*. Hydraulic capsule pipeline and pneumatic capsule pipeline is the transport of freight encapsulated to cylindrical or spherical bodies, so called capsules (with diameter only slightly less than the pipe), conveyed through pipeline by liquid or gas, respectively. Hodgson & Charles [8] and Jensen [9,10] referred to capsule pipelining as the third generation of pipelining.

The third way of coal pipelining intensification supposes using *the carrier liquid different than water*. Oil products (crude oil, residual fuel oil, kerosene, fuel or Diesel oil and various mineral oils) or hydrocarbons (methanol, ethanol, carbonic acid, and other liquid organic compositions) can be used as a carrier liquid.

3. METHANOL AS CARRIER LIQUID

The physical properties of the carrier liquid, especially density and viscosity, can essentially influence both main hydrodynamic parameters of the mixture flow, i.e. critical velocity V_{CR} and the hydraulic gradient I.

As an example, the dependence of hydrodynamic parameters of critical regime for the sand-water mixture and sand-glycol (80 % glycol solution in water) mixture is illustrated in Figure 1. The volumetric concentration C of the mixtures varied from 5 to 30 %, a mean diameter of sand was $d = 0.25$ mm and a pipe diameter was $D = 50.4$ mm. The values of the critical velocity V_{CR} and critical pressure gradient I_{CR} differ substantially according to the used carrier liquid.

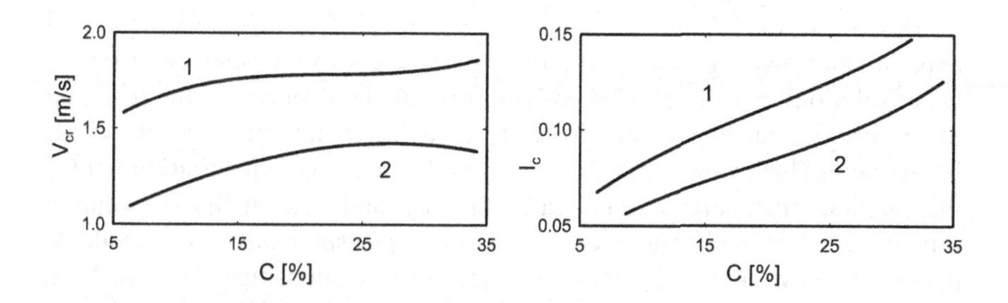

Figure 1. The effect of volume concentration C on critical velocity V_{CR} and critical hydraulic gradient ICR for sand-water mixture (1) and sand-glycol solution mixture (2)

It follows from the foregoing research concerning the most effective carrier liquid for the coal hydrotransport that the majority of authors recommend to use methanol and its water solutions as a carrier liquid instead of water, [11-19]. A choice of the methanol as a carrier liquid for the transport of coal is not accidental. According to [20] and [21] the simple and fairly reliable production of methanol can put it forward in the category of the cheapest products of organic synthesis in the near future.

An annual world production of methanol increased already at present in average about 15%. The majority of methanol is now produced from petroleum or natural gas since it is about 25-50 % cheaper than production from coal. However, according to the economical prediction the difference in prices should be essentially decreased in the near future. Methanol as a carrier liquid for the coal hydrotransport offers the following scheme of the coal pipelining.

In the region of coalmines the installation could be built where part of the yield coal could be processed on methanol and it can be used as a carrier liquid for pipeline transport of the remaining part of coal. One of the following ways of the methanol exploitation could be used at the output terminal of the transport line.

The mixture of coal and methanol (or its water solution) can be directly burned in a boiler of the thermal power plant or preliminary separated. In this case after separation of coal and methanol the first can be used as a fuel and the second as a raw material for various purposes.

The possibility of reutilization of methanol separated from the mixture for multiple uses in the transport process should be also considered.

Using of the methanol as a carrier liquid is also very advantageous due to its low temperature of freezing. It allows to lead the pipeline superficially over a ground surface even in the arctic regions what could play an important role for using the methanol for pipeline transport of coal in the northern and mountainous areas of the USA (Alaska), Canada and former Soviet Union, [18].

Based on the above mentioned, the programme of theoretical and experimental investigation of the main parameters of coal-methanol (or its water solution) mixture pipeline transport should be opened. As the first step of the programme the comparison of power consumption (dependency of hydraulic gradient I on slurry flow velocity V and solid concentration C_s) for the pipeline transport of coal-water mixture and coal-methanol solution mixture was realised. The special laboratory measurements were made to define unknown input data of semi-empirical relationships, i.e. the limit volumetric concentration C_m and the coefficient of mechanical friction of coal in the water or water-methanol solution k_o. The resultant comparison of the hydraulic gradient I of the coal-water and coal-methanol solution mixture flow is presented in Figure 2, where density of coal was ρ_c = 1480 kg/m^3, diameter of the pipe was D = 0.103 mm, the maximal grain size of coal d_{max} was less than 0.25 mm, volumetric concentration - C = 20 %.

The power consumption of the coal-methanol mixture is for about 20 % less than that of coal-water mixture for the same value of transport velocity (except the regime close to the critical velocity V_{CR}). The reason of it is probably lower viscosity of methanol than that of the water. Similar results were also confirmed experimentally by [17].

However, the fundamental economical criterion of any transport system is not the power consumption, but the total transport cost per unit mass and distance. Goedde [18] made the comparison of the total unit transport cost of the coal-methanol mixture pipelining with coal-water mixture pipelining, railway and barges transport. He pointed out that the transport of coal in methanol could be significantly profitable and competitive compared with the rest of the above-mentioned kinds of transport. Therefore, the attention should be paid to the investigation of optimal models of hydrotransport of coal-methanol mixtures.

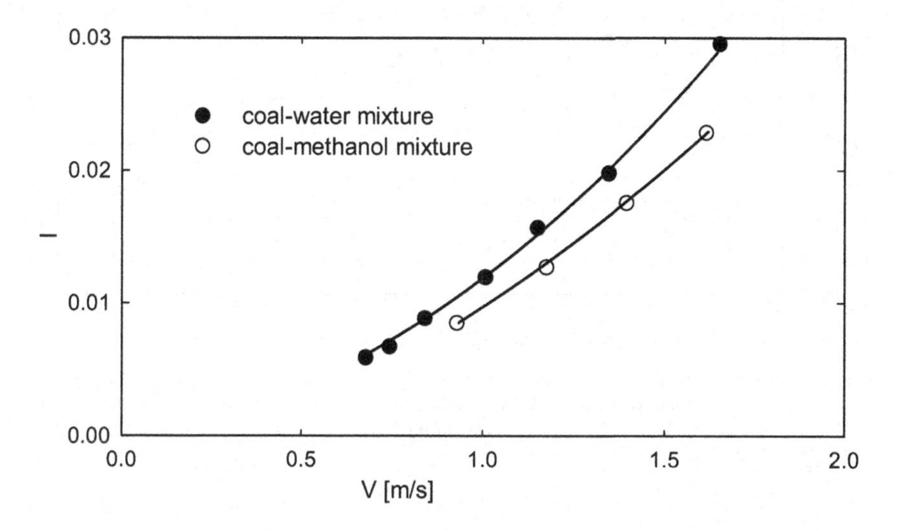

Figure 2. Comparison of hydraulic gradient I of coal-water mixture (1) and coal-methanol mixture (2)

A serious problem results from the fact that methanol is a toxic fluid (extreme permissible concentration of methanol in the air is 5 mg/m^3) and handle it is possible only in specialised laboratories. For the pipeline laboratory research it is useful to find such substitute of methanol, which is absolutely harmless and its physical properties (density and viscosity) strictly correspond to methanol.

Based on our experiments such substitute has been found. The exact specification on the substitute is know-how of the Institute of hydromechanics National Academy of Sciences of the Ukraine at Kiev. The comparison of measured values of dynamic viscosity of the methanol solution substitute and the data available in the literature for real methanol solution are presented in Figure 3 (the value *C* is the weight concentration of methanol in a solution). Since the measurements were realised within the temperature range T = 18-20°C, a fairly good agreement between methanol and its substitute was proved. Of course, to use the substitute is not limited on this temperature range; it should be used also for very low temperatures. The preliminary measurements with the substitute used as the carrier fluid also confirmed results illustrated in Figure 1.

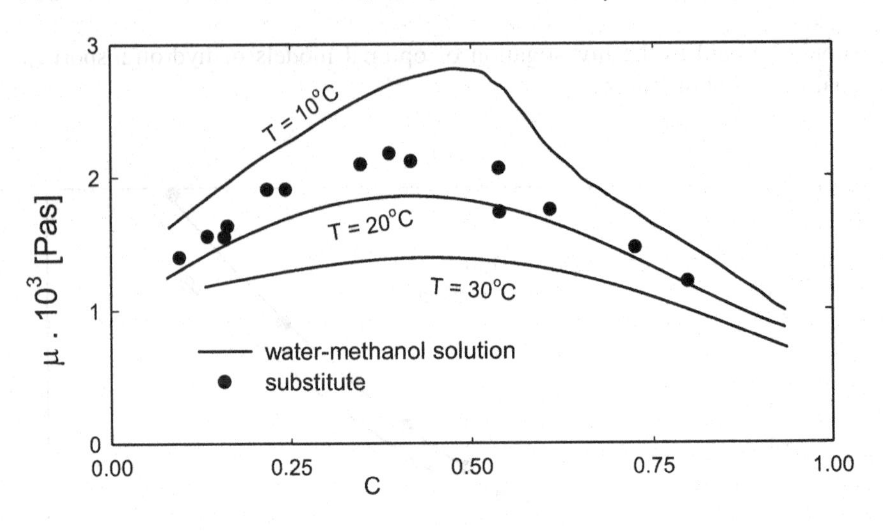

Figure 3. Comparison of dynamic viscosity μ of a real water-methanol solution and its substitute

4. COAL-LOG PIPELINE TRANSPORT

Capsule pipeline transport of coal, so called coal-log pipeline transport seems to be very close to commercial exploitation in the USA, [22,23]. Coal is formed into cylindrical bodies (by using the high pressure without or with addition of some glue), which are conveyed by water in a pipe. Diameter of the body d_c is about 90 % of the transport pipe diameter D, length of the body L_c is about two times of body diameters d_c. The concentration of solids can reach a very high value (up to 80 %) and process of separation of coal and water is very easy and without additional expenses.

For carrier liquid velocity about 3 m/s capsules are conveyed in water practically without contact with the pipe wall. This flow pattern satisfies minimum power consumption and also protects the capsules and the pipe against degradation and wear. In addition, water consumption is three or fourth times less than that for slurry hydrotransport and pollution of water is negligible. Using polymeric or surfactant drag reducing additives could significantly decrease the power consumption what is advantageous especially for long distance pipelining, [24].

Coal log technology makes possible to transport up to twice more solid fuel in the same diameter of pipe than for slurry transport. Except low carrier liquid and energy consumption, separation and de-watering of coal are much easier and cheaper and due to low content of water in coal the combustion process is more efficient.

It follows from economical comparison that coal-log pipeline is cheaper than truck transport for distances longer than 65 km and pipe diameter D = 200 mm. For D = 500 m even for distance over 25 km. Compared with railway, the transport cost is on the level of unit trains. For large quantity of coal it could be even less. Another advantage is given by fact that length of pipeline is usually at least about 30 % lower than that of the railway. From environmental protection point of view, capsule pipeline, similarly as slurry pipeline, is dust free and noiseless. In spite of these advantages the utilisation of coal-log pipeline system could expect only for transport of coal from new mines to power stations, especially in mountains areas without railways and highways or in heavy populated and industrial areas, where railway is overloaded.

5. COMMON CARRIER PIPELINING

The most profitable application of pipeline transport is the use of a system known as a common carrier where both the transported material and carrier fluid can be transported from the same locality and are both exploited at the point of destination. Transport of crude oil and natural gas from deposits in Arctic region is very promising. One possibility can be found in transport of solidified crude oil (SCO) moving as capsules in LNG in common pipeline, where LNG serves as the carrier liquid. The idea of applying capsule pipeline technology to this case is feasible since liquefaction of natural gas is an established technology and very low solubility of crude oil in LNG was proved, [10]. Simultaneously another technical problem could be solved using this technology. Due to high viscosity of the oil considerable part of energy changes to heat and thus conventional oil pipelines can introduce serious environmental problems in the Arctic region, arising due to the relatively warmer pipeline compared with surrounding land. Heat dissipation causes melting of the surrounding permafrost with the danger for pipeline foundation, which could erode and the line can break down.

Another interesting application is transport of highly viscous oil or oil products (e.g. lubricant, asphalt) in rigid or plastic containers transported through pipe by water or low viscosity oil product. A preliminary analysis has shown substantial saving of capital cost and/or operational cost by employing the common carrier transporting system.

The comparison of hydraulic capsule pipeline transport with pipeline transport of oil products and slurry hydrotransport is particularly interesting.

It was demonstrated that power consumption for transport of the same quantity of bulk cargo placed in capsules is less than in case of slurry hydrotransport, especially, for coarse-grained material of density of coal and different types of industrial wastes.

As follows from Figure 4, for coarse coal transport the power consumption of slurry pipelining reaches from 300 % to 100 % higher values than that in case of hydraulic capsule pipeline transport. The energy reduction increases with reduction of the operational velocity.

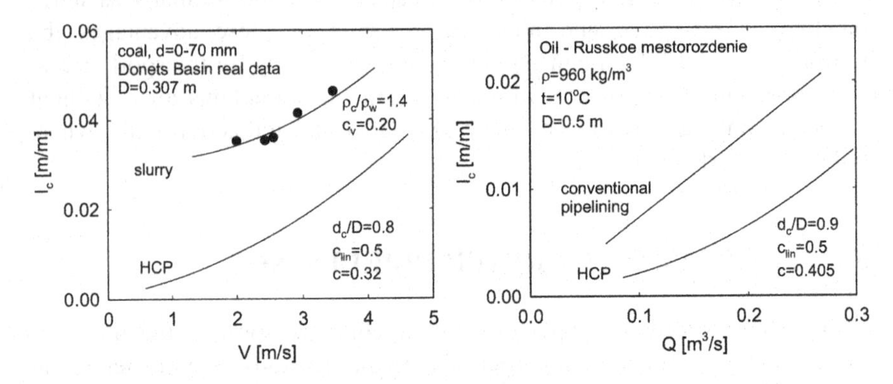

Figure 4. Comparison of power consumption reduction for HCP and conventional transport of coal and viscous liquid

Similar result brings comparison of power consumption reduction for transport of encapsulated viscous liquid (Russian oil) conveyed by water with conventional pipeline transport of the oil. Transport of viscous oil and oil products by means of capsule pipelining may again provide power consumption reduction from 50% to 70%, the reduction increases with operational velocity. Since for low temperature the oil viscosity significantly increases hydraulic capsule pipeline transport of highly viscous oil and oil products for long distances in arctic conditions can be economically attractive. Capsule pipeline transport could be recommended as suitable transport especially for longer distances when power consumption becomes the most important for operational cost.

6. CONCLUSIONS

Fossil fuels have been very important source of energy. It is necessary to pay attention to development of modern, efficient, environmentally friendly and operationally safe way of fossil fuels transport, handling and storage.

The conveying of coal by fluid media through the pipeline has been proposed as a potential method of bulk transport.

Except traditional slurry transport of coal as coal-water slurry or coal-water fuel, pipeline transport of coal in hydrocarbons seems to bring some advantage especially for arctic areas.

Common carrier pipelining is economically efficient transport. Using the different liquid hydrocarbons as carrier for coal pipelining, as well as common pipeline transport of oil and natural gas or encapsulated (solidified) oil in water, methanol or LNG could be prospective way for future exploitation of distant deposits, especially in arctic conditions.

Very promising, especially for long distances, is coal-log pipeline technology, which brings significant savings of energy and water consumption; increasing of combustion efficiency and decreasing of transported material degradation and pipe wear.

There is sufficient knowledge in the hydrodynamics of slurry and capsule transport to design and to construct a pilot plant or even commercial pipeline for transport of coal or other solidified energy sources in a various carrier media. This may include methanol or LNG. However, additional research should be done for safe and economical functioning of each individual pipeline.

ACKNOWLEDGEMENTS

The work was supported by Academy of Sciences of the Czech Republic under the “Programme of Oriented Research & Development” project No. S 2060007 – Pipeline transport of bulk materials.

REFERENCES

1. Vlasak P., Berman V. & Fadeichev V. (1994), Contribution to the capsule pipeline transport of highly viscous oil and oil products, *J. Hydrol. Hydromech.* 42, No.2-3, p.131.
2. Vlasak P., Berman V. & Chara Z. (2003), Pipeline transport of fossil fuels, In: Engineering Mechanics 2003, (J. Naprstek, C. Fischer eds.), ITAM AS CR, p. 374.
3. Vlasak P., Buchtelova M. (1979), Contemporary trends of development of hydraulic transportation in the world. In: Hydraulic transport of materials through pipes. CSVTS, p.154 (in Czech).
4. Zandi J., Gimm K.K. (1976), Transport of solid commodities via freight pipeline, U.S., Dept. of Transportation, National Technical Information Service, Virginia 22161, USA.
5. Cowper N.T., Wasp E.J. (1973), Slurry pipelines – recent development, In: Transportation Symposium 1973, The Australian Institute of Mining and Metallurgy.

6. Anderson V., Gudmundsson J.S. (1999), Transporting oil and gas as hydrate slurries, In: Hydrotransport 14, (J.F. Richardson ed.), BHR Group Conferences Series 36, p.181.
7. Ercolani D. (1986), Production plants and pipeline systems for Snamprogetti´s coal water slurries, recent experience and current projects in Italy and USSR. In: Hydrotransport 10, (A.P. Burns ed.), BHRA, p.19
8. Hodgson G.W., Charles M.E. (1963), The pipeline flow of capsules, Part 1, The concept of capsule pipelining. *Can. J. Chem. Eng.* 41, No.2, p.43.
9. Jensen E.J. (ed.) (1974), TDA-RCA Capsule Pipeline Project. *Alberta Research Information Series*, No. **67**.
10. Jensen E.J. (ed.) (1975), TDA-RCA Capsule Pipeline Project. *Alberta Research Information Series,* No. **63**.
11. Alger G., Simons D. (1968), Fall velocity of irregular shaped particles. *J. Hydr. Div., Proc.of ASCE,* Vol. 94, NHY3, p.721.
12. Snoek P., Gandhi R. & Weston M. (1979), Alternatives are studied for moving coal pipeline. *J. Oil and Gas,* No.8, p.95.
13. Grace W. (1980), Studies Colorado California Slurry Line. *Coal Age*, 85, No.4, p.45.
14. Hydraulischer Rohrleitungstransport von Feststoffen (1980), *Techn. Heute*, 21, 10, p.412.
15. Hydraulischer Rohrleitungstransport von Feststoffen (1980), *Techn. Heute*, 33, 11, p.16.
16. Wilta P. (1980), Slurry pipeline projects along to completion. *J. Chemical Engineering*, 87, No.12, p.75.
17. Aude T., Chapman J. (1981), Coal/methanol slurry lines show promise. *J Oil and Gas*, 79, No.28, p.135.
18. Goedde E. (1981), Potential for slurry pipelines for transportation of minerals in Europe. *J. Pipeline*, 53, No.7, p.14.
19. Lauzon M. A. (1982), A new black gold for pipelines? *Can. Chem. Process*, 66, No.1, p.29.
20. Moiseev I. (1982), The methanol tree. *J. Chemistry and Life*, 18, p. 21. (in Russian)
21. Nekhaev A. (1982), More in detail about methanol. *J. Chemistry and Life*, 18, p. 27. (in Russian)
22. Liu H. (1996), Coal Log Pipeline Design and Economics. In: Hydrotransport 13, (J.F.Richardson ed.), BHR Group Conference Series No. 20, p. 779.
23. Liu H. (2002), Coal log úpipeline technology – latest developments. In: Hydrotransport 15, (N. Heywood ed.), BHR Group, Vol. I, p.263.
24. Vlasak P. (1995), The Toms effect in capsule-liquid flows, In: 8th Int. Freight Pipeline Society Symposium", p. 93.

LOGICAL BASIS, LOGISTICS AND ARCHITECTURE OF MODULAR SYSTEMS

Technology for Natural Gas Supply Chains Design and Engineering

Michael KERVALISHVILI, Maya CHKONIA, George GEGELIA, Paata KERVALISHVILI
Georgian Technical University, Georgia

Abstract: Distributed systems and networks of differentiated services are fair in the way that different types of traffic can be associated with different network services, and so with different quality levels.

In the systematic development of distributed systems it is necessary to use the basic system models using the interface, distribution and state transition approaches. Each of these fundamental parameters is very helpful and plays an important role in the systems development process. For large systems, the development is carried out through several levels of abstraction. And the same time for creation of such kind of development processes of the modular systems it is obvious to estimate and select the refinement steps, which give the possibility to build the effective multilevel and multidimensional integrated distribution.

The transition way from abstraction to concrete distributed systems logistics and architecture includes - and is based on - Supervisory Control and Data Acquisition System (SCADA) technologies. A system that is used to manage networks for energy carriers must allow two-way communication between the suppliers and the consumers. One of the most advantaged new directions of SCADA solutions is the remote control of energy supply and oil and gas distribution networks. This technology is mainly based on the substation automation and substation monitoring systems.

Key words: Systems, control, supervision, SCADA

J. Hetland and T. Gochitashvili (eds.),
Security of Natural Gas Supply through Transit Countries, 383–392.

1. INTRODUCTION

Information and communication systems strategy – designated ICSS - for oil and gas distribution pipelines is based on a system architecture that includes four components:

- process control system,
- safety system,
- console operator interface and
- portable operator interface.

Relevant control modes are: pipeline mode with executive, automatic and manual regimes. At the same time the system possesses the technology of control location changeover, active and local control possibilities.

Pipeline integrity monitoring systems are more suitable for the gas supply networks. It generally includes areas such as: routing, material selection, river crossings, pipe coatings, cathodic protection, microscada monitoring. The main operations of the integrated monitoring systems are based on the surveillance monitoring, including patrol along the pipeline route and supervision of all activities carried out close to - or on the pipeline - and liaison works with local owners, tenants, and authorities. It is also important to monitor the working parameters of the cathodic protection systems, the interference checks and mitigation. For gas and oil pipelines the SCADA system is very oriented for precise checking of the flow rate of the energy carriers and related pressure levels.

Modular systems based SCADA architecture provides:

- operational interface to support operation of complete gas pipeline network;
- measurement and control of product delivered to consumers;
- continuity of supply to consumers;
- optimization of the operating efficiency of the gas pipeline network;
- maximization of the safety to personnel and equipment;
- simplicity of operation, easy of maintenance, reliability.

For a scientific discipline such a modular systems development - which is the mathematical basis of farther building of logistics and architecture of distributed networks design and engineering - it is a very important to define the components for a relevant mathematical model. It is well known that the most convenient way to manipulate and compose components is the usage of components with following characteristics of the component [1-3]:

- It is interactive;

- It is connected to its environment exclusively by its interface formed by named and typed channels;
- It receives input messages from its environment on its input channels and generates output messages to its environment on its output channels;
- It is nondeterministic;
- Its interaction with its environment takes place in a global time frame as the basic condition.

Among the main basic conditions for the investigations:

- every actions happened in discrete time;
- the interaction of the components within the network is performing by parallel composition operator;
- development of components is going through several and different steps of refinement, mainly – glass box and interaction refinement.

2. MODELS AND COMPUTATIONAL EXPERIMENTS

It is well known that the glass box refinement works with the relation of property refinement and the special terms that are representing the refining components $X:S_1$, $Y:S_2$ / $X':T_1$, Y': T_2 $\{T_1:= S_1, T_2:=S_2\}$. Figure 1 shows a state transition diagram for the normal state transition function with data defining state t_1 and t_2.

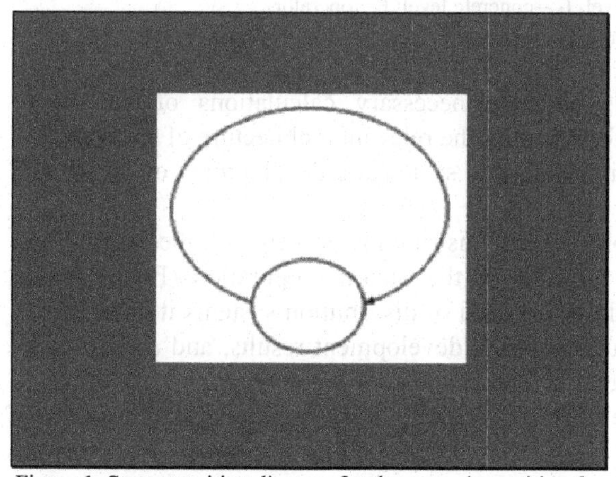

Figure 1. State transition diagram for the normal transition function

The next step of the logical base is dedicated to the interaction refinement, which includes several parts that demonstrate abstractions, representations number of simulations, etc. The interaction refinement allows refinement of the components for the input and output channels. The principals of the interaction refinement are visualized by the scheme in Figure 2 for the so-called U-simulation.

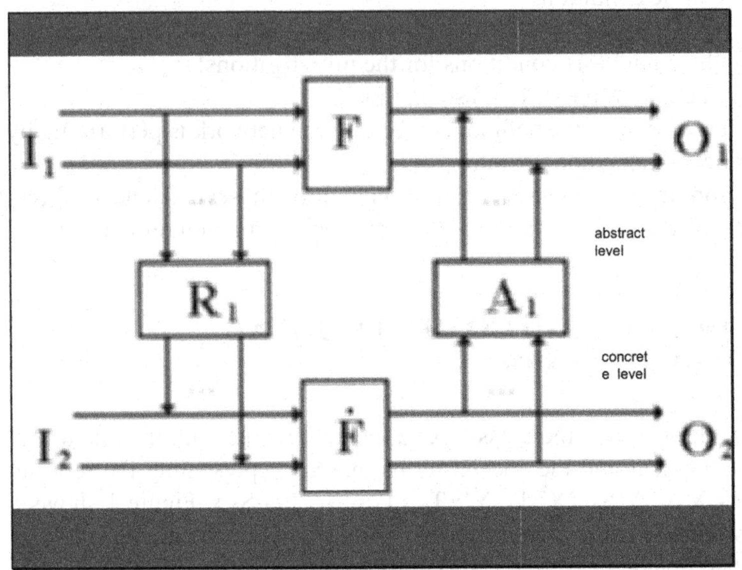

Figure 2. Visualization of the Interaction Refinement (Simulation). (I_1 – abstract level, I_2 – concrete level; F – operator, R_1 and A_1 refinement modules)

Mathematical models and the necessary calculations of the above mentioned logical steps followed by the relevant architecture of software and real systems are very suitable for the systematic development of distributed networks.

The practical example of real distribution systems is the natural gas supply network scheme offered by the Statoil corporation (Figure 3). In order to effectively build up this kind of distribution systems it is necessary to make use of the modular systems development results, and also to make use of the base of virtual reality simulation methodologies in order to prepare the architecture that responds to all requirements and needs of distributed systems.

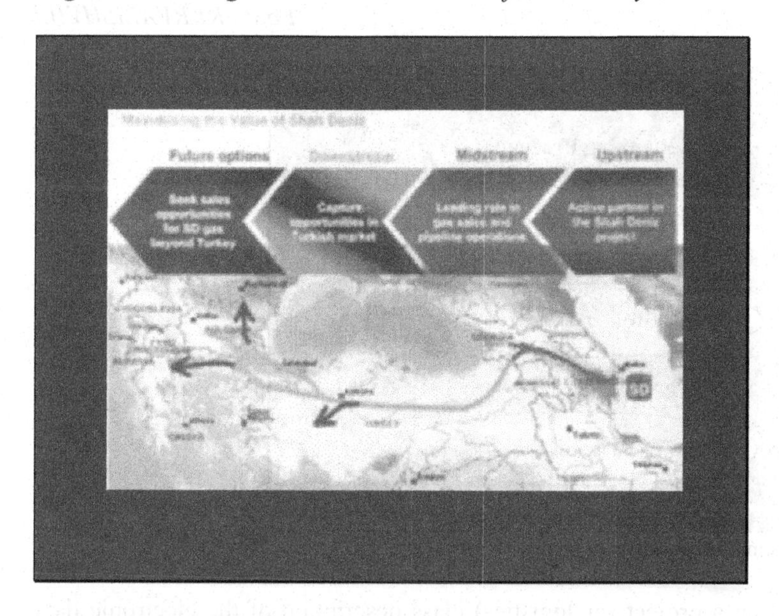

Figure 3. Logical System of the Gas Value Chain
(Example for Shah Deniz Gas distribution system)

The next stage - from theory to practice in distributed system development - is to include logistics planning and material flow simulation (liquid or gas). Based on virtual reality simulation systems the integrated simulation system will combine the virtual reality and the material distribution network simulation.

For the general level of planning, the material flow and optimisation are two significant tasks. Both of them have many different underlying technical and environmental requirements and constrains. Feasibility studies and evaluation of scenarios are common strategies for reaching the demanding solution. For the planning level (Figure 4), as well as a more detailed level, the simulation tools give important input for the evaluation of possible solutions. Despite the close linkage in reality between the material-flow and the layout [4,5], the lack of data integration between of those systems leads to a variety of independent solutions. This means redundancy and overhead of data-acquisition and data management.

The integration of more sophisticate simulation systems like virtual reality systems is also possible. Electronic data management systems, that provide an integration platform for both the layout and the material-flow based data models, give an important contribution to the whole planning process.

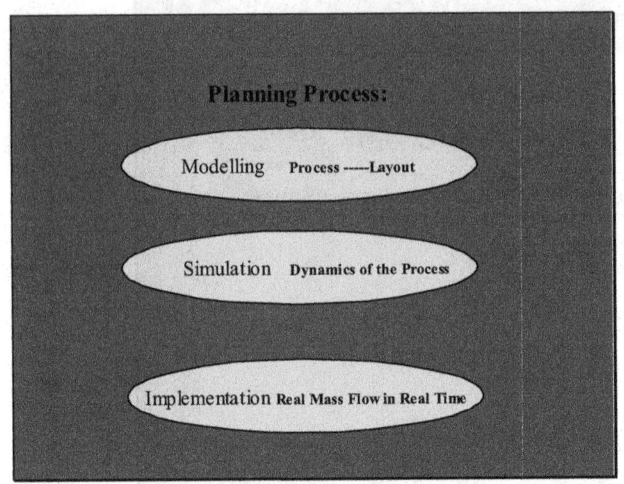

Figure 4. Simplest scheme of planning process

Based on a project (or logistics) class description of the electronic data management system, and the library of technical components it is possible to simulate any system in connection with the planning tasks. Figure 5 shows the integration scheme of different simulation systems for the planning aspects.

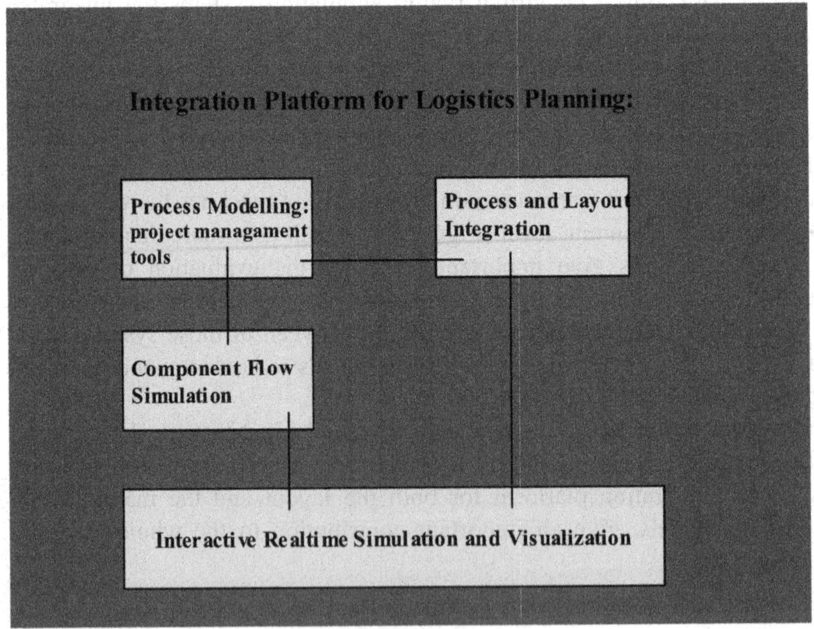

Figure 5. Integrating platform for logistic planning

From the understanding of virtual reality as a virtual place of work - where the user can carry out all steps of development - interactive planning seems feasible within this environment. Prerequisite to this scenario is a real time simulation environment for the simulation of technological systems, particularly for distributed networks. One part of simulation model is based on a vertical flow of information, whereas another part of the model is based on the material flow (Figure 6).

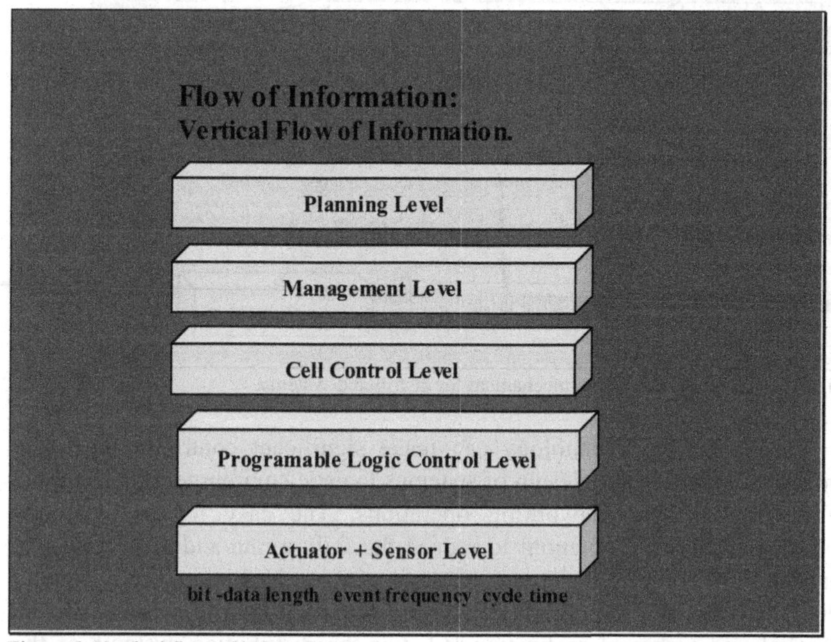

Figure 6. Vertical flow of information

Logistic environments are very complex with different design aspects like architecture, mechanics and the logic. For this task most convenient is to establish an integration tool for the final collection, simulation and real-time presentation in a virtual environment.

Communication with the simulation tool should be done by bit or word oriented shared memory areas. Complex technological systems for real operations consist of several physical components, controllers, sensors and actuators, which define their behaviour. Information runs between these parts via links with bar systems. Different types of sensors, actuators and other micro machines are available on the market - corresponding to real sensors like light bars, distance accelerators and other instruments. The positioning can be done graphically or textually (Figure 7).

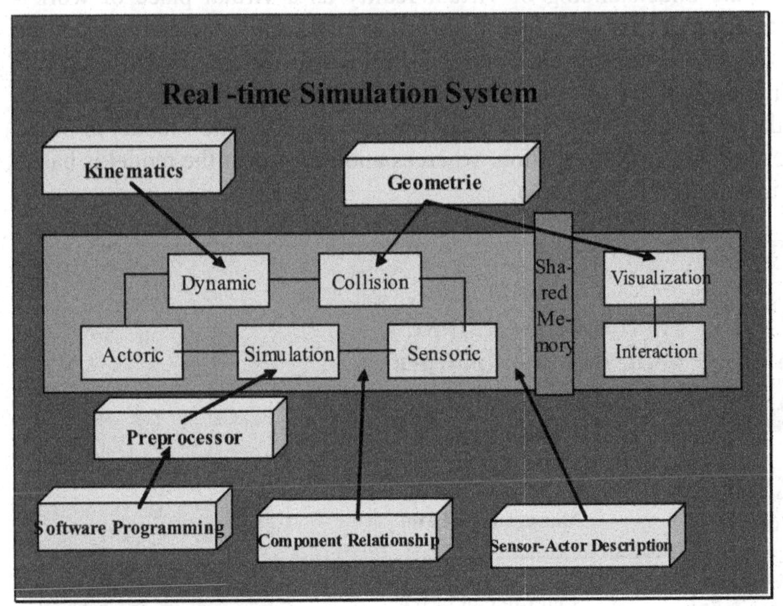

Figure 7. Real time simulation mechanism for distributed systems.

Virtual Reality Technology can make significant contributions to the process of planning and design of spacious logistic environment for the tasks of distributing and networking operations. The easy to use graphical environment has the potential to reduce the failure rate and accelerate the realization of complex and spacious environments. Next steps are the use of sensor and actuator simulation in combination with control systems for the physical and technical realization [6]. For the distribution systems of the energy carriers' - particularly for the gas pipelines - the better way is to use the large volume gas metering methodology and relevant instruments. Among them there are orifice plate systems designed in accordance with the novel standards (ISO 5167 and higher) with an inaccuracy not exceeding +/- of 0,1% of measured flow at a 95% confidence level (e.g. single chamber orifice fittings, dual gas chromatographs for gas composition and others). At the same time for the modular based control systems it is quite necessary to obtain significantly more precise measuring data, which should be recorded by means of recently developed multifunctional sensors and transducers such as coriolis metering microsensors and their microsystems.

3. CONCLUSIONS

Using the elaborated theoretical and computing approaches a most effective pipeline management system has been conceptually elaborated with control modes at local level, station level and pipeline level. Local control is manually implemented from the device itself. The automatic or manual control of an installation is affected from its local micro-SCADA system. The supervisory pipeline control is effected from the active pipeline control centre.

The subsequent step of the modular systems approach as applied to concrete industrial engineering is the creation of solutions for integrating networking gateways designated the Common Semantic Models and the Generalized Macro Models.

For the networking systems of natural gas supply chains, the modular systems provide the designers and the process planners with a different (and independent) toolkit without restricting the application area. This is conceived as a new high-level tool for the representation of engineering knowledge and support that can be useful for the reduction of time and cost of the development and implementation of integrated distribution systems.

The described workbench is oriented to rapid application development (RAD) which nowadays is a most promising part of the information technology.

ACKNOWLEDGMENTS.

A deep appreciation for the most helpful support, comments and suggestions to: Professor Nils Martensson - Chalmers University of Technology of Goeteborg, Sweden; Professor Hamlet Meladze - Tbilisi State University (Georgia), Professor Rolf-Dieter Schraft –Fraunhofer IPA (Germany), and to many other fellow colleagues.

REFERENCES

1. M. Broy. A Logical Basis for Modular Systems Engineering. In book: Calculational System Design. IOS Press. 1999, pp.101- 123.
2. P. Mendes, H. Schulzrinne, E. Monteiro. Session-Aware Popularity-Based Resource Allocation for Assured Differentiated Services. IEEE Communications magazine, September 2002, pp.104-111.

3. A. Carzaniga, D. Rosenblum, A. Wolf. Design and Evaluation of a Wide-Area Event Notification Service. ACM Trans. Comp. Syst., Vol. 19, N 3, 2001, pp. 332-383.
4. T. Fluig, K. Grefen. Virtual Environment for Integrative Factory and Logistics Planning with Virtual Reality. In book: Chainging the Ways We Work. IOS Press, 1998, pp. 610-619.
5. D. Neuber. Spezifikation und Implementierung von Datenstrukturen zur echtzeitorientierten Simulation von Speicherprogrammierbaren Steuerungen, Frounhofer IPA, Februer 1995, Stuttgart, Deutschland.
6. P. Kervalishvili. Sensors, Actuators and their Systems for Pipeline Networks. Georgian Engineering News. Vol.3. 2003, pp.14-23.

SOME ASPECTS OF RELIABILITY OF GAS MAINS

Leon MAKHARADZE, Teimuraz GOCHITASHVILI, Malkhaz KUTSIA
Makharadze and Gochitashvili: Professors at the Mining Institute of Academy of Sciences; Kutsia is affiliated to the Sukhumi State University, GEORGIA

Abstract: The article presents the general overview of Georgian gas mains, which are characterized by a complex profile and other technological difficulties. The calculation method used to assess the reliability of pipelines is adopted for gas mains in parallel and consecutive connections of the compressors and other elements of the system with the pipeline segments. The offered method of analysis can be used for the gas mains that are planned or under construction, and whose rehabilitations and development are subjected.

Key words: Pipeline, Compressor, Reliability, Failure

1. GENERAL OVERVIEW AND RISK FACTORS

Natural gas is one of the most significant primary energy resources in the world. The consumption of natural gas has gradually increased all over the world approximately at 25 % over the last 10 years. According to forecasts the consumption is expected to grow faster in the nearest future [1, 2]. This requires further development of networks for gas transportation. The diversification of pipelines from the regions producing large amounts of natural gas (e.g. Russia and the Caspian region) to the basic consumer regions and countries (e.g. Turkey and Europe) is one of the key issues of the European energy policy – and will be so for the next decades. As the pipelines are prone to promote a guaranteed access of the producer countries to the international market, they will also foster economic development and increase the incomes of the transit countries. Moreover the transport of natural gas through the territories of neighbouring countries can often

J. Hetland and T. Gochitashvili (eds.),
Security of Natural Gas Supply through Transit Countries, 393–403.

significantly mitigate the problems of political character that sometimes are conceived to be more valuable factors than the economic development.

In order to connect remote energy resources new supply chains are needed. Quite often this implies the constructions of new main pipelines through transit countries, with increased reliability. The reliability of main pipelines is connected to various risks that may cause failure, or promote the termination of system functioning. The fundamentals among them are the following main risk elements:

1. Political risk which may imply termination of the construction of new pipelines, or the termination of the operation of existing main pipelines because of political instability in the region.
2. Commercial risk that is connected to changing market conditions (for example the price of natural gas for various reasons).
3. Supply risk factor that might be connected to the depletion of existing gas reservoirs, or with wrong forecast for resources of new findings.
4. Technical or technological risks as the supply of natural gas can be stopped as a consequence of low reliability of a pipeline, related infrastructure (armature, compressors, pumps etc.) or the entire system.

Influence of the political instability on the reliability of the functioning of main pipelines in countries with transitive economy is especially typical. For example, the supply of natural gas to the various states of the former Soviet Union was stopped several times in the period 1992-2000. Similar cases occurred in the Middle East where according to the 1991-1995 data each eighth main pipeline was not functioning appropriately.

To increase the reliability of main pipelines due to political risks, it is recommended to support the:

- Diversification of energy supply including several transit means and routes from various sources of supply. Hence, the Caspian gas constitutes a rational option for Europe to diversify its energy supply - in addition to the traditional sources of supply;
- Establishment of strong and stable political and mutual relation between producing countries, transit countries and consumer countries can also become a strong basis for a regional political safety. This is one of the basic predetermining factors for a safe and reliable transportation of natural gas through transit countries. It is necessary to note that reliability of transit pipelines implies the several risk factors that provide political stability of the region, including the permanent readiness of a powerful, coordinated action of military sources along the entire pipeline, irrespective of the belonging of the territories;

- Development and enforcement of appropriate actions of the international transit legislation and regulations. These actions should become the authoritative leveller of the maximum restriction for the use of the transit means for political reasons, and for the highest reliability of the operating pipelines.

For this purpose several international consents are already developed, the most progressive of which is considered the Energy Charter Treaty (ECT). In the framework of the ECT the signatory parties have been agreed on many important issues to support the transit pipelines reliable functioning.

The commercial risk depends on the tariff policy, transit expenses and cost of energy resources at final consumer. Incomes that can receive the producer and the transit countries, in many ways depend on a delivery distance and the transit tariff.

The cost of the natural gas transit can vary within a wide range; Diversification of supply, increase of capacities of existing transit pipelines and the creation of new routes to support the reduction of political and economic risks. But implementation of these activities requires large-scale investments and increases indirect expenses for new transit pipelines. Hence, the natural gas transportation through new main pipelines will become more expensive than the transportation through traditional routes. Furthermore, the delivery of natural gas from new sources to the international market will sharpen a competition between producers and, thus, promote a price reduction at the end-users. Therefore the incomes received by transit countries using new pipelines, especially during the transition period appear to become lower than initially expected. First of all this requires a maximal decrease of transportation expenses, especially for a transition period, and also the creation of equal rights and conditions for foreign and local investors.

2. GEORGIAN GAS MAINS AND TECHNOLOGICAL RISK FACTORS

The natural gas supply chain of Georgia was connected to the gas mains of several supplier countries (Azerbaijan, Russia, Iran, and Turkmenistan). After the break up of the Soviet Empire these connections were also broken that resulted in a substantially reduced supply that, eventually, also affected the demand and consumption of natural gas. According to forecasts for the nearest future, the increase of demand for natural gas will appear. Besides Georgia gains the status as transit country as the main South Caucasus Pipeline System will be constructed on its territory [2].

Since 1958 several gas mains for a total of 1940 km were constructed in Georgia. Only 10-20 % of the total planned throughput of Georgian gas mains is used nowadays (2003). The main specifications of Georgian gas mains are presented in Table 1.

Complexity of the relief of Georgia causes difficulties of gas mains construction. The most complex profile has the North-South Caucasus pipeline system. This system is crossing of the Caucasus Mountains with certain parts located on the Caucasian ridge. The highest segment of the profile is located at 2400 m altitude. Such conditions cause special requirements for its operation. Besides a significant difference between altitudes, the temperature variation must also be taken into consideration. The similar complex profile will have the South Caucasian Pipeline system as well with maximum altitude point above 2000 m.

Table 1. Georgian Gas Mains

Name of Pipeline System	Diameter, mm	Length, km	Construction Period
North-South Caucasus	1200	135	1988-1994
Kazakh-Saguramo	1000	112	1980
Karadag-Tbilisi	800-700-500	110	1959-1968
Vladikavkaz-Tbilisi	700	266	1963-1966
Saguramo-Kutaisi	700-500	370	1967-1975
Kutaisi- Sukhumi	700-500	338	1986-1989
Rustavi-Telavi-Jinvali	500-300-200	370	1969-1975
Red Bridge-Tsalka-Alastani	500-300	180	1978-1990
Gori-Khashuri-Bakuriani	500-300	59	1972-1989
Total	300-1200	1940	1959-1994

Georgian gas mains are complex hydro-mechanical systems and their operation reliability depends on many factors. Complexity of gas mains caused by profile and long distances of transportation requires the creation of multistage systems with consecutive connections of compressor stations in pipeline segments, several take off points for various consumers and other armature for control and regulation of gas flow.

Technological complexities of natural gas transportation over a long distance requires additional factors to be taken into account, from which the basic is the degree of its compression along the pipeline route causing a corresponding variability of density of the transported media. Therefore the reliability of gas mains in many ways depends on the functioning of the compressor stations. Even the most extreme situation arises when compressor stations are consecutively included in the operation of gas pipelines. The analysis of the reliability of such pipelines is given below.

3. RELIABILITY OF PIPELINE SYSTEMS

The technological reliability of gas mains is meant as a condition when: in the given concrete operating conditions, for the certain time intervals or the full period of operation, the system should develop the planned parameters.

Failure of the various components of gas mains can cause changes of the parameters of pipeline system drastically. Therefore it is necessary to define the probability of their state that demands the definition of the characteristics of the reliability of composite elements. Following the character, the quantity and nature of failure of the specified elements the gas main may turn into various states. Transition from one state to another occurs under the influence of sequential failures and the restorations of these elements.

Let us consider the plan of a gas main system in which some working compressor stations are included, and which have the reserved and repairing aggregates as well. The plan foresees interchange ability reserve and repairing units with working ones, when the latter stops because of failure, repair or maintenance [3]. For the first stage of calculation it is supposed that:

- The compressor together with the engine represents one single element (unit).
- Other elements that are included in a pipeline system are considered absolutely reliable.

Through α and β the intensities of failures for the first and consecutively included compressors are addressed, but through γ and μ the intensities of their restoration are designated respectively.

In Figure 1 the labelled state of the graph show of the compressor stations of a gas main system corresponding to the different conditions of the compressor stations. Conditions differ from each other by character and quantity of failed units. The number of a condition specifies the compressor failure: $K = 3i+j$, $i = 0, 1, 2$; $j = 0, 1, 2$. Here i designates the quantity of failed consecutively included compressor stations, and j - quantity of failed basic compressor stations. For example state 0 means that all compressors are efficient, but states 1 and 3 mean conditions when failed is one the basic or one the consecutively included compressor, accordingly. States 2, 5, 6 and 7 represent failures of the compressor station when damaged are: two basic compressors; two basic and one consecutively included compressor; two consecutively included compressors; two consecutively included compressors and one basic compressor. Arrows show opportunities of transition from one to other possible state [4].

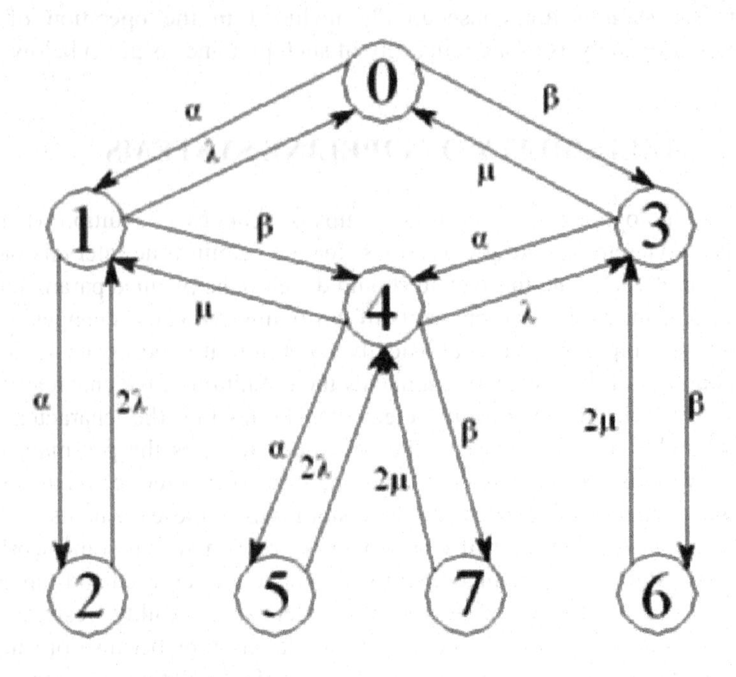

Figure 1. Labelled state of graph of the compressor station

Through K_k (t) the probability of compressor station at the state K for the moment t (k= 0, 1, 2, 3, 4, 5, 6, 7) is designated. In that case the corresponding system of the differential equations looks like this:

$$\left.\begin{aligned}
&P_0^1(t)=-(\alpha+\beta)P_O(t)+\lambda P_1(t)+\mu P_3(t);\\
&P_1^1(t)=-(\lambda+\alpha+\beta)P_1(t)+\alpha P_O(t)+2\lambda P_2(t)+\mu P_4(t);\\
&P^1{}_2(t)=-2\lambda P_2(t)+\alpha P_1(t);\\
&P^1{}_3(t)=-(\mu+\alpha+\beta)P_3(t)+\beta P_O(t)+\lambda P_4(t)+2\mu P_6(t);\\
&P^1{}_4(t)=-(\alpha+\beta+\lambda+\mu)P_4(t)+\beta P_1(t)+\alpha P_3(t)+2\lambda P_5(t)+2\mu P_7(t);\\
&P^1{}_5(t)=-2\lambda P_5(t)+\alpha P_4(t);\\
&P^1{}_6(t)=-2\mu P_6(t)+\beta P_3(t);\\
&P^1{}_7(t)=-2\mu P_7+\beta P_4(t);\\
&\sum_{K=0}^{7} P_k(t)=1.
\end{aligned}\right\} \quad (1)$$

The numbers 0, 1, 2 and 3 of central parts of graph correspond to the working condition of compressor stations; therefore to find the probabilities of appropriate conditions is important. In this case concerning the stationary probabilities, the system of the algebraic equations is received from the system of the differential equations (1):

$$\left.\begin{aligned}&-(\alpha+\beta)P_0+\lambda P_1+\mu P_3=0\\&\alpha P_0-(\beta+\lambda)P_1+\mu P_4=0;\\&\beta P_0-(\alpha+\mu)P_3+\lambda P_4=0;\\&P_0+\frac{\alpha+2\lambda}{2\lambda}P_1+\frac{\beta+2\mu}{2\mu}P_3+\frac{\alpha\mu+\beta\lambda+2\lambda\mu}{2\mu\lambda}P_4=1.\end{aligned}\right\}\qquad(2)$$

Solution of this system gives:

$$P_0=\frac{\lambda\mu}{\gamma};\ P_1=\frac{\alpha\mu}{\gamma};\ P_3=\frac{\lambda\beta}{\gamma};\ P_4=\frac{\alpha\beta}{\gamma},\qquad(3)$$

$$\text{where}\quad \gamma=\frac{1}{2}\left[\frac{(\mu+\beta)(\lambda^3-\alpha^3)}{\lambda(\lambda-\alpha)}+\frac{(\lambda+\alpha)(\mu^3-\beta^3)}{\mu(\mu-\beta)}\right]\qquad(4)$$

But the factor of availability of the compressor station will read:

$$K_1=P_0+P_1+P_2+P_3=\frac{(\alpha+\lambda)(\beta+\mu)}{\gamma}.\qquad(5)$$

If the real picture is taken into account - the circumstance, that except for the compressor stations all other elements of the gas main system are considered absolutely reliable, the calculated value of this factor should be increased by a correction factor. Then, the availability of compressor station will be defined by the formula [4]:

$$K^1{}_1=K_1.K,\qquad(6)$$

where K=0,9999 - 0,999999 is the appropriate correction factor.

As a rule the natural gas delivered from the production wells contains various solid contaminants and condensates, which can sharply and negatively influence the reliability of the system. It is necessary to remove these impurities from gas before feeding the gas into the transportation system. For this purpose the gas processing plant is equipped with filtering devices.

By the theoretical analysis the filtering units are considered as independent elements. Proceeding from this, the availability factor of the compressor is determined as a derivative of availability factors of these two units. As in the previous case, at the first stage of calculations the same assumptions are accepted according to which all other elements included in the system are deemed absolutely reliable.

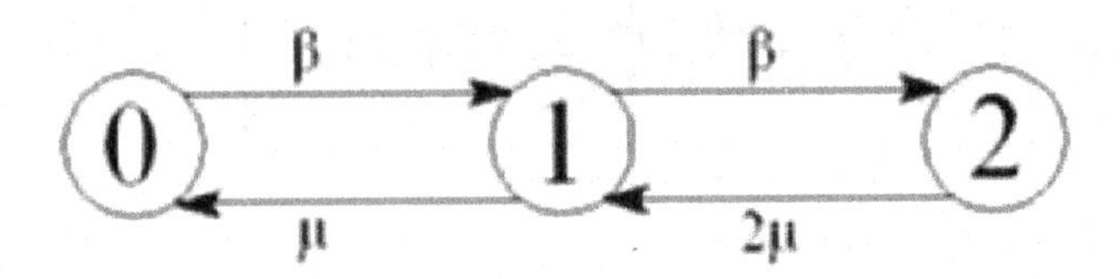

Figure 2. Labelled state of graph of gas pipeline system including gas processing unit

Figure 2 shows the labelled state of graph, considering the interchange ability of failing filtering units. In an examined case it is considered, that the filtering unit consists as a minimum of two filters - one working, while the other is a stand-by reserve. Numbers 0 and 1 on the figure designate states of serviceability of the filtering units: both filters are in service; and one filter is in service but another – failed. Number 2 means failure of the filtering unit, i.e. such condition when both filters are failing. β and μ designate intensity of the failure and the restoration respectively.

The system of the differential equations for the shown state of graph looks like this:

$$\left.\begin{array}{l} P^1{}_0(t)=-\beta P_0(t)+\mu P_1(t); \\ P^1{}_1(t)=-(\beta+\mu)P_1(t)+\beta P_0(t)+2\mu P_2(t); \\ P^1{}_2(t)=-2\mu P_2(t)+\beta P_1(t); \\ P_0(t)+P_1(t)+P_2(t)=1. \end{array}\right\} \qquad (7)$$

The following values are drawn from the system of these equations for stationary probabilities P_0, P_1, and P_2:

$$\left.\begin{aligned} P_0 &= \left(1+\frac{\beta}{\mu}+\frac{\beta^2}{2\mu^2}\right)^{-1}; \\ P_1 &= \frac{\beta}{\mu}P_0; \\ P_2 &= \frac{\beta^2}{2\mu^2}P_0. \end{aligned}\right\} \quad (8)$$

The formula for the calculation of factor availability in this case looks like this:

$$K_2=P_0+P_1=\left(1+\frac{\beta}{\mu}\right)\left(1+\frac{\beta}{\mu}+\frac{\beta^2}{2\mu^2}\right)^{-1}. \quad (9)$$

Assuming that some elements of the system are absolutely reliable, which differs from the reality, it is derived:

$$K'_2 = K_2 . K, \quad (10)$$

where K_2 constitutes the factor of availability of the filtering unit; K=0,9999 - 0,999999 - correction factor.

In practice gas main pipeline systems have as in parallel functioning, as in consecutive functioning compressors. In such cases, by calculation of the factor of availability of the compressor station, it is necessary to pay attention to the simultaneous work of at least two compressors. It should be made assumptions as well, that in such cases when only one compressor is in service that causes obvious reduction of technological parameters (pressure and output), and however the system as a whole is in the state of serviceability. Labelled state of the graph for such cases is shown in Figure 3

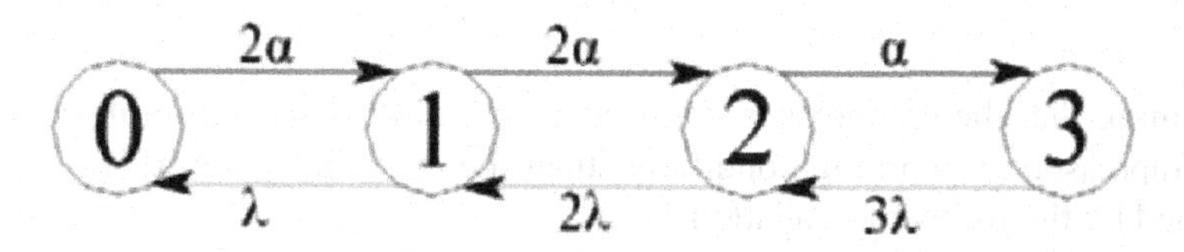

Figure 3. Labelled state of graph for complicated gas pipeline system

The figures in the circles designate quantity of failed (and under restoration) compressors, α and λ - intensity of failure and restoration of compressors respectively. For the examined case the system of the differential equations looks like this:

$$\left.\begin{aligned}
&P_0^1(t)=-2\alpha P_0(t)+\lambda P_1(t);\\
&P^1_1(t)=-(2\alpha+\lambda)P_1(t)+2\alpha P_0(t)+2\lambda P_2(t);\\
&P^1_2(t)=-(\alpha+2\lambda)P_2(t)+2\alpha P_1(t)+3\lambda P_3(t);\\
&P^1_3(t)=-3\lambda P_3(t)+\alpha P_2(t);\\
&P_0(t)+P_1(t)+P_2(t)+P_3(t)=1
\end{aligned}\right\} \qquad (11)$$

Compressor stations in a state of serviceability P_0, P_1, and P_2 are drawn from this system of the equations:

$$\left.\begin{aligned}
&P_0= 3\lambda^3\left(3\lambda^3+6\alpha\lambda^2+6\alpha^2\lambda+2\alpha^3\right)^{-1};\\
&P_1=\frac{2\alpha}{\lambda}P_0;\\
&P_2=\frac{2\alpha^2}{\lambda^2}P_0;\\
&P_3=\frac{2\alpha^3}{3\lambda^3}P_0.
\end{aligned}\right\} \qquad (12)$$

Assuming that the compressor station is in the state of serviceability and both compressors in working conditions, then the availability factor K_3 will be defined by the following equation:

$$K_{3.1}=P_0+P_1= 3\lambda^2(\lambda+2\alpha)\left(3\lambda^3+6\alpha\lambda^2+6\alpha^2\lambda+2\alpha^3\right)^{-1}; \qquad (13)$$

If it is assumed that the compressor station is in the state of serviceability when even only one compressor is in the state of operability, the factor of availability should be defined by the formula:

$$K_{3.2}=P_0+P_1+P_2= 3\lambda\left(\lambda^2+2\alpha\lambda+2\alpha^2\right)\left(3\lambda^3+6\alpha\lambda^2+6\alpha^2\lambda+2\alpha^3\right)^{-1}; \quad (14)$$

If we take into account the cases analogous to the previous ones, that other elements and units of the gas main system are not absolutely reliable in practice, the factors of availability for both cases should be defined by the formulas:

$$K^1_{3.1}=K_{3.1}.K, \quad K^1_{3.2}=K_{3.2}.K. \quad (15)$$

The Above-stated analysis can be used for the gas mains which are planned and under construction, and for those whose rehabilitation and development are subjected.

REFERENCES:

1. Gochitashvili T., Shenoy B. Regional Fuel and Energy Resources and Possible Routes for the Caspian Oil Export. Georgian Strategies Research and Development Centre, Tbilisi, Bulletin №13, 1998. 46 pp.
2. Gochitashvili T. Georgian main Gas pipelines and prospects of natural gas transmission. Tbilisi, 1999. 72 pp.
3. Makharadze L., Gochitashvili T., Kutsia M., Reliability of Oil pipeline. “Georgian Oil and Gas”, №1. Tbilisi, 2000. pp. 131-134.
4. Borokhovich A., Makharadze L., Kutsia M., Gochitashvili T. Reliability of pressure hydro transport systems. University of Krasnoyarsk, 1990. 190 p.

PART VI: RECOMMENDATIONS AND CONCLUSIONS

CO-DIRECTORS SUMMARY OF CONCLUSIONS AND POLICY RECOMMENDATIONS

Jens HETLAND and Teimuraz GOCHITASHVILI
Hetland: Senior Scientist at SINTEF Energy Research AS, Trondheim, Norway, and Professor (II). Gochitashvil: Professor at the Georgian Technical University, Tbilisi, Georgia

1. ENERGY SECURITY STRATEGIC ISSUES

On a short-to-medium term basis the main priorities of an energy security strategy of transit countries are:

- Diversification of imported fuel supply
- Enhanced use of local energy sources (including renewables)
- Improved self-sufficiency by rehabilitation/renovation of existing electricity and gas supply chains
- Strategic reserves planning combining water, liquid fuels and gas
- Restoring commercial viability and creditworthiness of energy companies
- Co-ordination and accountability of political actions and initiatives

The strategy must include appropriate policy changes to improve the investment climate and the market structure, and should address risks for investors that prohibit the implementation of measures aimed at enhanced energy security. There is an urgent need to develop local capabilities pertaining to energy management and risk assessment that are necessary for developing and pursuing policies conductive to energy security. This should be exercised without interference of any kind.

J. Hetland and T. Gochitashvili (eds.),
Security of Natural Gas Supply through Transit Countries, 407–410.

It is believed that the technical risks in terms of energy security at present are outweighed by political and commercial risks, which constitute the main challenge to the security of the region. The immediate recommendations are:

- Efforts should be directed towards initiatives that reduce the political and commercial risks
- Enhanced capabilities in energy management and risk analysis should be applied to techno-economic aspects
- Measures to create competition in the energy sector should be identified and implemented
- Clear and transparent regulation of the energy sector including infrastructure for natural gas should prevail
- A regulatory system consistent with the EU natural gas industry legislation should be pursued
- The provisions of the Energy Charter Treaty and associated documents should be practically implemented
- Gas sales should be separated from the natural gas production through independent companies
- Efforts should be made that may lead to greater competition, transparency, respect of contractual provisions including reduction - and eventually elimination - of non-payments and illegal take-off

Such measures would improve the opportunities for investors in the natural gas sector, and restructure the gas market and also promote the direct private participation. However, in-depth analytical skills, management and administrative capacity will be needed in order to operate the companies, and to resolve problems such as non-payment and administrative interference in commercial activities.

2. RISK MITIGATION STRATEGIES IN TRANSIT GAS PIPELINES

Actions that are likely to significantly reduce investor risks - and thus improve the security of energy supply - should include:

- Multilateral, trilateral and bilateral treaties or protocols that regulate the gas transit
- Multilateral development bank participation with partial risk insurance
- Trustee management of transit fees with drawing rights for claims by the importing nation
- Technical cross-border access guarantees (inspection, repairs)

- Treaty with mandatory international arbitration clauses, and no right to block the transit during dispute resolution (for example ECT, WTO)

3. CONSIDERATION OF TECHNOLOGY

Further actions are recommended towards technology and strategic energy storage, however, within the framework of technical and economic viability. This means that a system analysis would be needed in order to reflect the entire energy supply situation of the transit country. Options are LNG, UGS, LPG, or other fuels for strategic reserves, both in the (non-producer) transit country itself and in cooperation with other nations in the region.

An appropriate networking system, and contractual, operational and financial arrangements will be required in order to optimise the entire energy supply system. Some of these arrangements, while feasible and potentially lucrative, are deemed too risky under the current conditions.

The entire energy sector operates in an environment that lacks modern system monitoring and management tools, such as automated data acquisition, processing and storage systems, analytical tools and decision-support systems. This leads to the absence of transparency and inability to make informed decisions. However, the issue of procuring, installing, commissioning and operating such systems and technologies is dependent on first resolving the primary risks inherent to the energy sector. Unless there are clear signals that such investment will be efficient, there will be little or no interest in investing in security-enhancing technologies.

4. PARTICULAR SHORT-TERM SCENARIOS AND REGIONAL ASPECTS

For Georgia in particular:

- Owing to large cash deficiency in the energy sector a critical situation is predicted for the following winter (2003/2004) to cover the supply of electricity and natural gas to the industry and households. The main risk factors are non-payment and continuing practices of administrative interference in commercial activities
- Without Governmental money - or other sources of financing - a considerable lack of gas and generated electricity will occur, which will result in power cuts and rationing

- It is not expected that significant improvement in collection rates for wholesale deliveries of electricity and natural gas can be achieved during a reasonable forecast period unless a significant change in collection rates and practices of handling cash flows takes place
- Large imports of electricity will be necessary if the repairing of Georgian power plants is not addressed urgently. The lack of capital to maintain the power sector constitutes a major risk factor, and it is deemed quite possible that incidents will occur that may collapse the entire system.

In Armenia:

- The natural gas supply is not diversified
- The gas storage capacity is insufficient
- The domestic industrial reserves of fuel are not developed
- There is only one operating gas duct from Russia via Georgia that exhibits technical, economic and political risks.

5. GEOLOGICAL ASPECTS AND UGS

Assessment studies show that several geological formations could be used for UGS, of which one location (Ninotsminda) has already been positively identified as termed feasible. While the needs for storage capacity should continue to be analysed and clearly defined, the priority is (now) to provide data required to make a business case for the prospective sites. However, until further business cases are being developed companies with relevant expertise in UGS should be technically and financially involved.

The recommendations are to concentrate the initial planning as extension of ongoing projects (EC Tacis and U.S. TDI), by carrying out a simplified reservoir modelling of the selected formation and, most importantly, by making sure that the project structure, and the commercial and managerial aspects are defined in a way that could be of interest to potential investors.

For mountainous areas, accurate information in appropriate format is key. In order to provide relevant geological information, modern tools like GIS and 3D modelling are required.

Tbilisi, 26 May 2003

LIST OF ARW PARTICIPANTS

- including guests and the secretariat

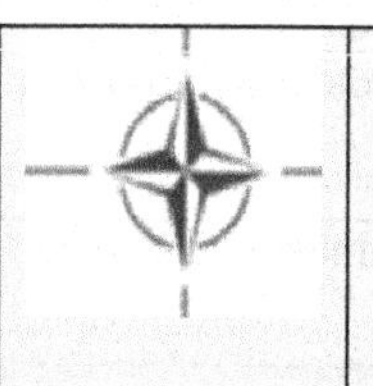

NATO ARW
SECURITY OF NATURAL GAS SUPPLY THROUGH TRANS IT COUNTRIES
22-24 MAY 2003
TBILISI

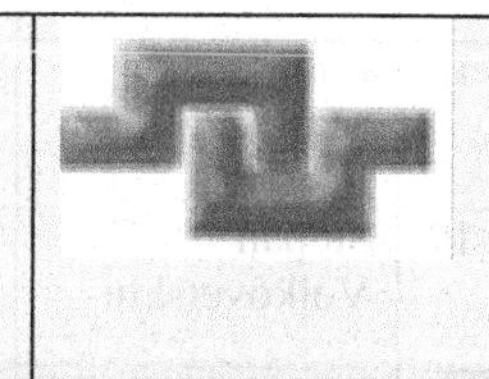

N	*Name*	*Organization*	*Contact information*	*E-mail*
1	Pavel Vlasak	Institute of Hydrodynamics	Czech Republic	vlasak@ih.cas.cz
2	Boyko Nitzov	Energy Charter Secretariat	Brussels	Boyko.Nitzov@encharter.org
3	Mads Christensen	RAMBOLL	Denmark	mads.christensen@gmx.net
4	Fabien Farvet	French Gas Association	France	ffavret@sofregaz.fr
5	Andras Gilicz	Er-Petro	Hungary	mail@er-petro.hu
6	Peter Csontos	Er-Petro	Hungary	mail@er-petro.hu
7	Gyorgy Lorincz	Er-Petro	Hungary	mail@er-petro.hu
8	Mauro Piccolo	Eurekos Ltd	Italy	mpiccolo@eurekos.it
9	Jens Hetland	SINTEF Energy Research	Norway	Jens.Hetland@sintef.no
10	Hans Jorgen Dahl	Norwegian University of Science and Technology	Norway	hjd@gassco.no
11	Jerzy Sobota	Agricultural University Wrotslaw	Poland	sobota@ar.wroc.pl
12	Gokhan Basik	Fatih University	Turkey	gbacik@fatih.edu.tr

J. Hetland and T. Gochitashvili (eds.),
Security of Natural Gas Supply through Transit Countries, 411–413.

		Istanbul		
13	John W.Coker	BP/Statoil	United Kingdom	cokerjw@bp.com
14	Ronald Nathan	Washington Strategic Advisors	USA	rgnripley@sprintmail.com
15	Dean White	PA Government Services Inc	USA	dean.white@paconsulting.com
16	Suren Shatvoryan	Energy Strategy Center	Armenia	piuesc@arminco.com
17	Ganifa Abdullaev	TUSI Ltd	Azerbaijan	tusildtd@intrans.az or ganifa.abdullayev@tusiltd.com
18	Bakhid Mursaliev	International EcoEnergy Academy	Azerbaijan	tusildtd@intrans.az
19	Stepan Volkovetskiu	Oil and Gas University Ukraine	Ukraine	ukraina@access.sanet.ge
20	Youri Rykov	Research Center of the Foundation for East-West cooperation	Russia	engo@orc.ru
21	Teimuraz Gochitashvili	Georgian Technical University	Georgia	t.gochitashvili@geoengineering.ge
22	Omar Kutsnashvili	Georgian Technical University	Georgia	contact@geoengineering.ge
23	Paata Kervalishvili	Georgian Technical University	Georgia	kerval@iberiapac.ge
24	David Rogava	Georgian Technical University	Georgia	opet@eecgeo.org
25	George Vashakmadze	Parliament of Georgia	Georgia	giocdir@gioc.ge
26	Liana Jervalidze	PSI Georgia	Georgia	liana@psigeorgia.org or lika@psigeorgia.org
27	Levon Makharadze	Georgian Technical University	Georgia	opet@eecgeo.org

ARW Guests				
28	David Mirtskhulava	Minister for Fuel and Energy	Georgia	
29	Gogodze Levan	GGIC	Georgia	
30	David Tvalabeishvili	GIOC	Georgia	
31	Archil Magalashvili	GIOC	Georgia	
32	George Mjavanadze	National Security Council	Georgia	
33	Michael Sidamonidze	EU TACIS Project	Georgia	
34	Nana Khundaze	Georgian Technical University	Georgia	
35	Giorgi Natsvlishvili	Georgia for NATO	Georgia	georiafornato@mail. ge
36	Zurab Sikharulidze	IDC Ltd	Georgia	
37	Iveri Kutsnashvili	Ltd Geoengineering	Georgia	
38	David Oniani		Georgia	
39	George Abulashvili	Energy Efficiency Center	Georgia	g_abul@eecgeo.org
ARW Secretariat				
40	Manana Dadiani	Energy Efficiency Center	Georgia	m_dadi@eecgeo.org
41	Liana Garibashvili	Energy Efficiency Center	Georgia	l_gari@eegeo.org
42	Keti Mirianashvili	Energy Efficiency Center	Georgia	eecgeo@eecgeo.org
43	Marina Tabidze	Energy Efficiency Center	Georgia	eecgeo@eecgeo.org

UNITS AND ABBREVIATIONS

1. TECHNICAL AND PHYSICAL UNITS

Although the technical and physical units used in the book are prone to correspond to the SI system, units that are conceived as common terms in the oil and gas industry will - to some extent - appear in various chapters. Units that deviate from the SI Units are defined below:

barg – over-pressure recorded in bars
bcm – billion cubic meter (i.e. 1000 million cubic meters)
boe – barrel of oil equivalent
btoe – billion tons of oil equivalent
kcm – kilo cubic meter
ktoe – kilo tons of oil equivalent
Mcm – million cubic meters
MNm^3/d – million of normal cubic meters per day (recorded at 0 °C)
MSm^3/d – million of standard cubic meters per day (recorded at 15 °C)
Mtoe – million tons of oil equivalent
MTPA – million tons per annum
m_N^3 – normal cubic meters (also given as Nm^3)
Tcm – tera cubic metre (i.e. 10^{12} cubic meters)
toe – ton of oil equivalent
TPA – tons per annum

2. ABBREVIATIONS

The most abbreviations of the book is explained below:

ACG – Azery-Chirag-Guneshli reservoir
BS – Blue Stream pipeline
BTC – Baku-Tbilisi-Jeihan Main Export Pipeline
BUSD – Billion US Dollars (i.e. 1000 million US Dollars)
CCGT – Combined Cycle Gas Turbine technology
CIS – Commonwealth Independent States
CNG – Compressed Natural Gas
CP – Cathodic Protection
EOP – Early Oil Production
FB – Fluidised Bed coal burning technology
FSU – Former Soviet Union
GPP – Gas Processing Plant
GTL – Gas To Liquid
HGA – Host Government Agreement
IEA – International Energy Agency
ITP – Iran-Turkey pipeline
LNG – Liquid Natural Gas
MAOP – Maximum Allowable Operating Pressure
MFCD – Mixed Fluid Cascade Process
MTPA – Million Tons Per Annum
NCS – Norwegian Continental Shelf
NG – Natural Gas
NSC – North-South Caucasus pipeline
PA – Petroleum Act
PR – Petroleum Regulation
PSA – Production Sharing Agreement
SCO – Solidified Crude Oil
SCP – South Caucasian Pipeline system
SOCAR – State Oil Corporation of Republic of Azerbaijan
SPA – Sale & Purchase Agreement
TCP – Trans-Caspian Pipeline
TkIT – Turkmenistan-Iran-Turkey pipeline
TPA – Third Party Access
TR – Tariff Regulation
UGS – Underground Gas Storage facility

INDEX

Topical structure of the book

CURRENT STATUS OF PRODUCER COUNTRIES

MARKET FOR NATURAL GAS

TRANSIT PIPELINES FOR NATURAL GAS

ENERGY SECURITY ISSUES

GAS LIQUEFACTION TECHNOLOGIES

STRATEGIC RESERVES

GEOLOGICAL SURVEY

ECONOMICS AND LEGISLATIVE ASPECTS OF GAS TRANSPORTATION

GEOPOLITICAL IMPLICATIONS

OTHER ASPECTS OF PIPELINE TRANSPORTATION

RECOMMENDATIONS